Product/Process Fingerprint in Micro Manufacturing

Product/Process Fingerprint in Micro Manufacturing

Special Issue Editor

Guido Tosello

MDPI • Basel • Beijing • Wuhan • Barcelona • Belgrade

MDPI

Special Issue Editor
Guido Tosello
Technical University of Denmark
Denmark

Editorial Office
MDPI
St. Alban-Anlage 66
4052 Basel, Switzerland

This is a reprint of articles from the Special Issue published online in the open access journal *Micromachines* (ISSN 2072-666X) from 2018 to 2019 (available at: https://www.mdpi.com/journal/micromachines/special_issues/fingerprint_micro_manufacturing)

For citation purposes, cite each article independently as indicated on the article page online and as indicated below:

LastName, A.A.; LastName, B.B.; LastName, C.C. Article Title. *Journal Name* **Year**, *Article Number*, Page Range.

ISBN 978-3-03921-034-3 (Pbk)
ISBN 978-3-03921-035-0 (PDF)

Cover image courtesy of Dario Loaldi, M.Sc., Technical University of Denmark, Department of Mechanical Engineering.

Contents

About the Special Issue Editor

Guido Tosello, PhD, is Associate Professor at the Technical University of Denmark, Department of Mechanical Engineering, Section of Manufacturing Engineering. Tosello's principal research interests are the analysis, characterization, monitoring, control, optimization, and simulation of precision molding processes at micro/nanoscales of thermoplastic materials. Technologies supporting precision/micro/nano molding processes are of research interest: advanced process chain for micro/nano tools manufacturing, precision and micro additive manufacturing, and dimensional and surface micro/nano metrology. Guido Tosello is the recipient of the "Technical University of Denmark Best PhD Research Work 2008 Prize" for his PhD thesis "Precision Moulding of Polymer Micro Components"; of the 2012 Alan Glanvill Award by The Institute of Materials, Minerals, and Mining (IOM3) (UK), given as recognition for research of merit in the field of polymeric materials; of the Young Research Award 2014 from the Polymer Processing Society (USA) in recognition of scientific achievements and research excellence in polymer processing within 6 years from PhD graduation; and of the Outstanding Reviewer Award 2016 of the Institute of Physics (UK) for his contribution to the Journal of Microengineering and Micromechanics. Guido Tosello is currently the Project Coordinator of the Horizon 2020 European Marie Skłodowska-Curie Innovative Training Networks MICROMAN "Process Fingerprint for Zero-defect Net-shape MICROMANufacturing" (2015–2019) (http://www.microman.mek.dtu.dk/) and DIGIMAN4.0 "DIGItal MANufacturing Technologies for Zero-defect Industry 4.0 Production" (2019-2022) (http://www.digiman4-0.mek.dtu.dk/).

micromachines

MDPI

Editorial

Product/Process Fingerprint in Micro Manufacturing

Guido Tosello

Department of Mechanical Engineering, Technical University of Denmark, Produktionstorvet, Building 427A, 2800 Kgs. Lyngby, Denmark; guto@mek.dtu.dk; Tel.: +45-45-25-4893

Received: 16 May 2019; Accepted: 21 May 2019; Published: 22 May 2019

The continuous trend towards miniaturization and multi-functionality embedded in products and processes calls for an ever-increasing research and innovation effort in the development of micro components and related micro manufacturing technologies. Highly miniaturized systems manufactured by a wide variety of materials find applications in key technological fields such as healthcare devices, micro implants, mobility and communications sensors, optical elements, and micro electromechanical systems.

High-precision and high-accuracy micro component manufacturing can be achieved through post-process (i.e., off-line) and in-process (i.e., on-line) metrology of both process input and output parameters, as well as the geometrical and functional features of the produced micro components. It is of critical importance to reduce metrology and optimization efforts, since process and product quality control can take a significant portion of total production time and cost in micro manufacturing.

To solve this fundamental challenge, research efforts are undertaken in both the industrial and scientific communities to define, investigate, implement and validate the so-called "Product/Process Manufacturing Fingerprint" concept.

The "Product Manufacturing Fingerprint" refers to that unique dimensional and/or functional outcome (e.g., surface topography, form error, etc.) on the produced component that, if kept under control and within specifications, ensures that the entire micro component complies with its specifications.

The "Process Manufacturing Fingerprint" is a specific process parameter or feature to be monitored and controlled in order to maintain the production of products complying with their specifications. Effective process monitoring will control the presence of a specific "Process Manufacturing Fingerprint" based on relevant process variables that are measured by sensors. This allows for real-time process control aiming at zero-defect micro manufacturing of the "Product Micro Fingerprint", and as a consequence of the whole component. By integrating both Product and Process Manufacturing Fingerprint concepts, metrology and optimization efforts are highly reduced and the micro product quality increased, with an obvious improvement of the production yield.

Accordingly, this Special Issue seeks to present research papers focusing on innovative developments and applications in precision micro manufacturing process monitoring and control, as well as microproduct quality assurance and characterization. The focus will be on micro manufacturing process chains and their Product/Process Fingerprint, towards full process optimization and zero-defect micro manufacturing.

The Special Issue consists of 16 original research papers, which cover both fundamental process technology developments, as well as their application, combined with the definition and the validation of Product and Process Manufacturing Fingerprints.

The papers included in the Special Issue address research in two main areas of the Micro Manufacturing Fingerprint:

(1) Definition and application of Product Fingerprints for new product development and functional characterization.

a. In the field of micro optics, Wang et al. [1] presented a study on the development of single composite diffractive optical element for color images generation, and Cao et al. [2] worked on the design and fabrication of an artificial compound eye for multi-spectral imaging.

b. In the field of micro mechanical systems and sensors, Díaz Pérez et al. [3] presented the development of a one-dimensional control system for a linear motor of a two-dimensional nanopositioning stage, and Choi et al. [4] showed how the vibration acceleration can be used as Product Fingerprint on the development of a miniaturized mobile haptic actuator.

(2) Definition and application of the Process Fingerprints for process development, monitoring and control.

a. In the field of injection molding of micro structured components, Giannekas et al. [5] and Loaldi et al. [6] presented studies concerning quality assurance and process control for the manufacturing of polymer microfluidic systems and Fresnel lenses, respectively. With regards to micro injection molding, Baruffi et al. [7] and Luca et al. [8] presented studies related to the correlation between Process-related Fingerprints and the achieved Product Fingerprints (flash marks and flow length, respectively).

b. As far as machining processes are concerned, Process Fingerprints were established and validated for electrical discharge machining (EDM) by Świercz et al. [9], for micro EDM drilling by Bellotti et al. [10], for plasma electrolytic polishing by Danilov et al. [11], for jet electrochemical machining by Yahyavi Zanjani et al. [12], for nanosecond pulsed laser ablation by Cai et al. [13], and for micro grinding by Fook et al. [14].

c. Process Fingerprints were developed for forming and additive processes by Cannella et al. [15] and Guo et al. [16], respectively. The former presented process development and monitoring of electro sinter forging for the manufacturing of miniaturized titanium discs and rings. The latter monitored and characterized electrohydrodynamic jet (e-jet) printing for the establishment of an optimized process window.

We wish to thank all authors who submitted their papers to the Special Issue "Product/Process Fingerprint in Micro Manufacturing". We would also like to acknowledge all the reviewers whose careful and timely reviews ensured the quality of this Special Issue.

A special thanks is due to all the colleagues of the European project MICROMAN ("Process Fingerprint for Zero-defect Net-shape MICROMANufacturing", http://www.microman.mek.dtu.dk/, 2015–2019). MICROMAN is a Marie Skłodowska-Curie Action Innovative Training Network supported by Horizon 2020, the European Union Framework Programme for Research and Innovation (Project ID: 674801). The support and funding from the European Commission is highly appreciated.

Conflicts of Interest: The authors declare no conflict of interest.

References

1. Wang, J.; Liu, L.; Cao, A.; Pang, H.; Xu, C.; Mu, Q.; Chen, J.; Shi, L.; Deng, Q. Generation of Color Images by Utilizing a Single Composite Diffractive Optical Element. *Micromachines* **2018**, *9*, 508. [CrossRef] [PubMed]

2. Cao, A.; Pang, H.; Zhang, M.; Shi, L.; Deng, Q.; Hu, S. Design and Fabrication of an Artificial Compound Eye for Multi-Spectral Imaging. *Micromachines* **2019**, *10*, 208. [CrossRef] [PubMed]

3. Pérez, L.C.D.; Gracia, M.T.; García, J.A.A.; Fabra, J.A.Y.; Pérez, L.D.; García, J.A.; Fabra, J.Y. One-Dimensional Control System for a Linear Motor of a Two-Dimensional Nanopositioning Stage Using Commercial Control Hardware. *Micromachines* **2018**, *9*, 421. [CrossRef] [PubMed]

4. Choi; Kang; Jeon; Lee; Choi, B.; Choi, H.; Kang, Y.-S.; Jeon, Y.; Lee, M.G. 2-Step Drop Impact Analysis of a Miniature Mobile Haptic Actuator Considering High Strain Rate and Damping Effects. *Micromachines* **2019**, *10*, 272. [CrossRef] [PubMed]

5. Giannekas, N.; Kristiansen, P.M.; Zhang, Y.; Tosello, G. Investigation of Product and Process Fingerprints for Fast Quality Assurance in Injection Molding of Micro-Structured Components. *Micromachines* **2018**, *9*, 661. [CrossRef] [PubMed]

6. Loaldi, D.; Quagliotti, D.; Calaon, M.; Parenti, P.; Annoni, M.; Tosello, G. Manufacturing Signatures of Injection Molding and Injection Compression Molding for Micro-Structured Polymer Fresnel Lens Production. *Micromachines* **2018**, *9*, 653. [CrossRef] [PubMed]

7. Baruffi, F.; Calaon, M.; Tosello, G. Micro-Injection Moulding In-Line Quality Assurance Based on Product and Process Fingerprints. *Micromachines* **2018**, *9*, 293. [CrossRef] [PubMed]

8. Luca, A.; Riemer, O. Analysis of the downscaling effect and definition of the process fingerprints in micro injection moulding of spiral geometries. *Micromachines* **2019**, *10*, 335.

9. Świercz, R.; Oniszczuk-Świercz, D.; Chmielewski, T. Multi-Response Optimization of Electrical Discharge Machining Using the Desirability Function. *Micromachines* **2019**, *10*, 72. [CrossRef] [PubMed]

10. Bellotti, M.; Qian, J.; Reynaerts, D. Process Fingerprint in Micro-EDM Drilling. *Micromachines* **2019**, *10*, 240. [CrossRef] [PubMed]

11. Danilov, I.; Hackert-Oschätzchen, M.; Zinecker, M.; Meichsner, G.; Edelmann, J.; Schubert, A. Process Understanding of Plasma Electrolytic Polishing through Multiphysics Simulation and Inline Metrology. *Micromachines* **2019**, *10*, 214. [CrossRef] [PubMed]

12. Zanjani, M.Y.; Hackert-Oschätzchen, M.; Martin, A.; Meichsner, G.; Edelmann, J.; Schubert, A. Process Control in Jet Electrochemical Machining of Stainless Steel through Inline Metrology of Current Density. *Micromachines* **2019**, *10*, 261. [CrossRef] [PubMed]

13. Cai, Y.; Luo, X.; Liu, Z.; Qin, Y.; Chang, W.; Sun, Y. Product and Process Fingerprint for Nanosecond Pulsed Laser Ablated Superhydrophobic Surface. *Micromachines* **2019**, *10*, 177. [CrossRef] [PubMed]

14. Fook, P.; Berger, D.; Riemer, O.; Karpuschewski, B. Structuring of Bioceramics by Micro-Grinding for Dental Implant Applications. *Micromachines* **2019**, *10*, 312. [CrossRef] [PubMed]

15. Cannella, E.; Nielsen, C.V.; Bay, N. On the Process and Product Fingerprints for Electro Sinter Forging (ESF). *Micromachines* **2019**, *10*, 218. [CrossRef] [PubMed]

16. Guo, L.; Duan, Y.; Deng, W.; Guan, Y.; Huang, Y.; Yin, Z. Charged Satellite Drop Avoidance in Electrohydrodynamic Dripping. *Micromachines* **2019**, *10*, 172. [CrossRef] [PubMed]

micromachines

MDPI

Article

Micro-Injection Moulding In-Line Quality Assurance Based on Product and Process Fingerprints

Federico Baruffi *, Matteo Calaon and Guido Tosello

Department of Mechanical Engineering, Technical University of Denmark, Produktionstorvet Building 427A, 2800 Kgs. Lyngby, Denmark; mcal@mek.dtu.dk (M.C.); guto@mek.dtu.dk (G.T.)
* Correspondence: febaru@mek.dtu.dk; Tel.: +45-45-25-4822

Received: 15 May 2018; Accepted: 4 June 2018; Published: 11 June 2018

Abstract: Micro-injection moulding (μIM) is a replication-based process enabling the cost-effective production of complex and net-shaped miniaturized plastic components. The micro-scaled size of such parts poses great challenges in assessing their dimensional quality and often leads to time-consuming and unprofitable off-line measurement procedures. In this work, the authors proposed a novel method to verify the quality of a three-dimensional micro moulded component (nominal volume equal to 0.07 mm^3) based on the combination of optical micro metrology and injection moulding process monitoring. The most significant dimensional features of the micro part were measured using a focus variation microscope. Their dependency on the variation of μIM process parameters was studied with a Design of Experiments (DoE) statistical approach. A correlation study allowed the identification of the product fingerprint, i.e., the dimensional characteristic that was most linked to the overall part quality and critical for product functionality. Injection pressure and velocity curves were recorded during each moulding cycle to identify the process fingerprint, i.e., the most sensitive and quality-related process indicator. The results of the study showed that the dimensional quality of the micro component could be effectively controlled in-line by combining the two fingerprints, thus opening the door for future μIM in-line process optimization and quality assessment.

Keywords: micro-injection moulding; quality assurance; process monitoring; micro metrology

1. Introduction

Microsystems are among the main drivers of the technological evolution introduced by the information age. Consequently, the demand for small components whose dimensions are in the micrometric and nanometric scale has largely increased in numerous engineering fields over the recent decades [1]. In this context, micro components made of thermoplastic polymers became more and more widespread due to the reduced weight, high chemical resistance, low production cost and ease of fabrication, even in complex shapes. Most of these products are nowadays produced by micro-injection moulding (μIM). This process was ideated as miniaturized version of the conventional injection moulding process (IM), with the aim of combining its high productivity with micro manufacturing capabilities [2]. If, on one hand, the two technologies share the same process cycle phases (plasticization, injection, packing, cooling and ejection), on the other, they have substantial differences [3]. Firstly, dedicated μIM machines having separate elements for plasticization and injection have to be used if tolerances in the micrometre range are the production target [4]. Since the positive outcome of any replication process strictly depends on the dimensional accuracy of the master, new micro tooling technologies were developed to manufacture moulds with features having micrometric dimensions [5]. Another discrepancy between IM and μIM relates to the filling of the cavity, which becomes much more challenging in the micro-scale. In fact, as the injected melt volume is extremely small, the surface-to-volume ratio increases, and thus a very fast solidification occurs, hindering the complete filling of the cavity [6]. Therefore, in order to favour the replication capability of the process, high levels of injection speed, melt temperature and mould temperature are typically required [7–9]. Since the levels of these parameters are superiorly limited by the occurrence of polymer degradation, the process window of µIM becomes narrower than that of IM, making process optimization a fundamental step for manufacturing products that comply with design specifications. However, µIM optimization is made difficult by the fact that, when the geometrical characteristics of the components are the response variables, time-consuming experimental investigations based on off-line, high-accuracy dimensional measurements are necessary to tune the process. In fact, numerous features have to be assessed at the same time by means of state-of-the-art measurement systems. Typically, optical instruments are preferable for this task [10–12] because of their contactless nature and sub-micrometric resolution. A possible solution to this issue is the identification of a single measurable characteristic of the part that is strongly statistically correlated to the other measurands and therefore to the overall product quality. By assessing only this dimensional outcome, which can be referred to as the "product fingerprint", the conformance to the specifications of all functional tolerances could be ensured. The product fingerprint must be also sensitive to the variation of process settings in order to work as an optimization tool for µIM; a change of process parameters has to be reflected in a variation of the fingerprint value if an effective control over the process has to be performed.

Another typical quality assurance issue of µIM and other moulding processes is that their extremely high throughput rates do not allow for measurement of all the produced parts with three-dimensional instruments [13], and therefore the production is verified by measuring a few random components extracted from the manufactured batch. This approach, which is favoured by the industry for its cost efficiency, is unsuitable for micro productions, where micrometric tolerances require an extremely high process repeatability and therefore a total quality assurance approach. A solution to this problem is the use of µIM process monitoring. The application of process monitoring to IM has been often reported in literature. Most studies [14–16] agree upon the centrality of cavity pressure as the process variable that best summarizes the evolution of the moulding cycle and determines the final part shrinkage. If, on one hand, the usage of a pressure sensor inside the cavity is generally without risks for a conventional moulded part, on the other, the size of micro plastic components is comparable or even larger than that of typical transducers, therefore impeding their use without drastically changing the cavity shape and the part design [17]. A possible solution to this problem is monitoring the pressure provided by the injection screw or plunger (the so-called hydraulic pressure in IM). In fact, this quantity is always stored in the machine data for each moulding cycle and can be extracted for analysis without the need to use any further sensor. The main drawback of this approach is the substantial difference between cavity and hydraulic pressures due to pressure losses generated within the nozzle and feed system and to the high compressibility of polymer melts. Therefore, the pressure measured at the plunger might not be representative of how the polymer melt is behaving inside the cavity.

In the field of µIM process monitoring, Griffiths et al. examined the effect of the main µIM process parameters on cavity pressure [18], demoulding forces [19], melt temperature [20], and air evacuation from the cavity [21]. They concluded, in all cases, that the selection of different process settings had a relevant impact on the recorded conditions, demonstrating that µIM can be successfully controlled by monitoring suitable process variables. However, the authors did not investigate the correlation between those variables and the dimensional quality of the produced micro components. If a strong correlation between the part dimensions and a monitored process indicator, called the "process fingerprint", is established, the quality assurance can be carried out in-line by only controlling its value. The process fingerprint must be also influenced by the variation of the main µIM process parameters in order to be used as an optimization tool. By finally correlating the product and process fingerprints, an experimental approach to carry out an in-line control of the main part features can be established and implemented, reducing the quality control time and simultaneously enhancing its robustness.

The present paper introduces a study aimed at the identification of the product and process fingerprints for the μIM process of a three-dimensional micro component for medical applications. The most important geometrical part features were selected and the effects of the variation of process parameters were studied by applying a statistical Design of Experiments (DoE) approach. A correlation study was then performed to identify the feature being mostly correlated with the others, i.e., the product fingerprint. In-line process monitoring was also applied during the moulding experiments; injection pressure and injection velocity curves were analysed with respect to process variations and the best process fingerprint candidate was identified. Finally, the correlation between the process and product fingerprints was investigated to establish an effective in-line quality assessment procedure for the micro part.

2. Materials and Methods

2.1. Case Study

The produced part was a polyoxymethylene (POM) micro part used in medical applications. Figure 1 shows its main dimensions and three-dimensional shape. The two structures on the internal surface and the 2° tape of the outer conical surface were designed to facilitate the ejection of the part from the mould. Since its nominal volume equalled 0.07 mm^3 (equivalent to a nominal mass of 0.1 mg) and the dimensional tolerances of the diameters were specified as ±10 μm, the part belonged to the category of micro moulded products according to the standard definitions [22]. Table 1 reports some examples of the volumes of micro moulded components reported in literature. The extremely small size of the part made the use of any in-cavity sensor inapplicable.

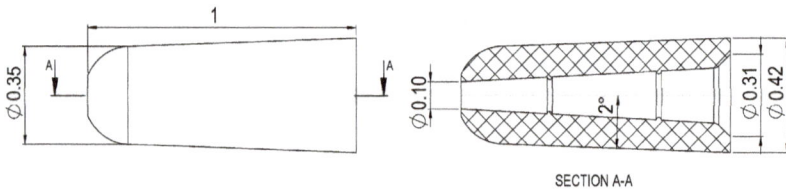

Figure 1. Geometry and nominal dimensions in mm of the micro component.

Table 1. Examples of micro-injection moulding (μIM) components and their volumes.

Micro component	Approximate Volume/mm^3	Reference
Micro filter	1.0	[11]
Micro ring	2.5	[23]
Part for weld line investigation	6.5	[24]
Part with micro pillars	12.0	[17]
Micro gear	14.0	[25]
Toggle for hearing aids	22.0	[26]
Dog-bone tensile bar	28.0	[27]
Thin-walled part	31.5	[28]
Square part for shrinkage evaluation	35.0	[29]
Cylindrical support with micro pillars	110.0	[30]
Disco with micro features and nano features	113.5	[31]

2.2. Mould Design

The micro components were moulded using a replaceable insert made of tool steel mounted in a three-plate mould (see Figure 2a). Such a mould configuration had the main advantage of enabling the automatic separation of the component from the feed system. This feature becomes particularly useful when small parts are produced since it avoids a manual gate separation that inevitably introduces variability in the production process. The structured hole was created by replicating a micro pin protruding from the movable plate of the mould. The insert cavity and the pin were both machined using micro electro-discharge-machining (μEDM). The feed system consisted of a cylindrical sprue with a 5 mm diameter, a conical runner, and a ring gate that had a nominal thickness of 25 μm and was axially symmetric with respect to the part. The ejection was carried out by means of a vacuum gripper mounted on a robot arm, thus ensuring a fully automated μIM process. Figure 3 illustrates the mould design and the location of the part within the mould frame.

Preliminary experiments highlighted the need of a venting channel to achieve complete filling of the cavity. The cause of this was identified as the presence of entrapped air in the cavity, and therefore a circular 5 μm deep venting channel was machined by μEDM on the back of the insert (see Figures 2b and 3c) to obtain consistent part filling. If, on one hand, such modification allowed the issue related to the unfilling to be solved, on the other, it generated a flash defect around the largest outer diameter of the component since the polymer melt was allowed to flow inside the venting channel.

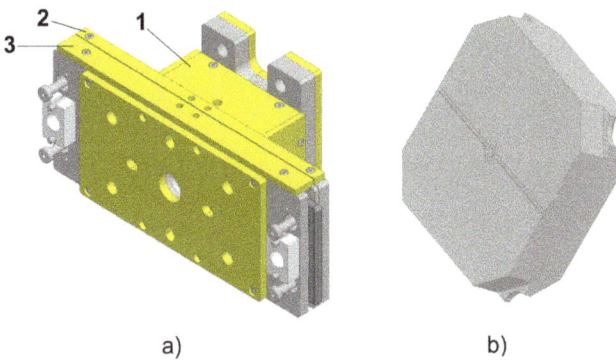

a)　　　　　　　　　　　　　　　b)

Figure 2. (**a**) 3D model of the mould: 1, 2, and 3 indicate ejection, middle, and injection plates respectively. (**b**) Replaceable insert with venting channel machined on the opposite side of the injection point.

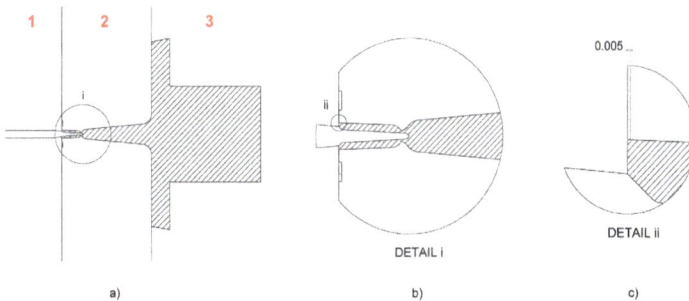

a)　　　　　　　　　　　　　　b)　　　　　　　　　　　　c)

Figure 3. Mould design. (**a**) Mould cross-section showing the part and feed system location (hatched) with respect to the three mould plates (numbered in red). (**b**) Detail of the moulded component and ring gate. (**c**) Detail of the venting channel machined on the insert at the end of the flow path (nominal thickness in mm).

2.3. Experimental Details

Micro moulding experiments were carried out using a state-of-the-art Wittmann Battenfeld MicroPower 15 μIM machine (Kottingbrunn, Austria, maximum injection velocity: 750 mm/s, maximum clamping force: 150 kN). This machine has a plasticization screw and a separate injection plunger; the first has a diameter of 14 mm and the function of melting and homogenizing the polymer, while the second has a diameter of 5 mm and drives the melt inside the cavity with the desired speed and pressure. The material used was an unfilled POM (Hostaform® C 27021, Celanese, Irving, TX, USA). This grade was selected for its peculiar properties, namely its low friction coefficient, good mechanical properties, and extremely low melt viscosity. Table 2 reports the main characteristics of the material and Figure 4 shows its viscosity and *pvT* data.

Table 2. Main properties of the polyoxymethylene (POM) grade.

Property	Unit	Value	Test Method
Density	kg/m^3	1410	ISO 1183
Melt volume rate (*T* of 190 °C, load of 2.16 kg)	cm^3/10min	24	ISO 1133
Melting temperature	°C	166	ISO 11357-1, -2, -3

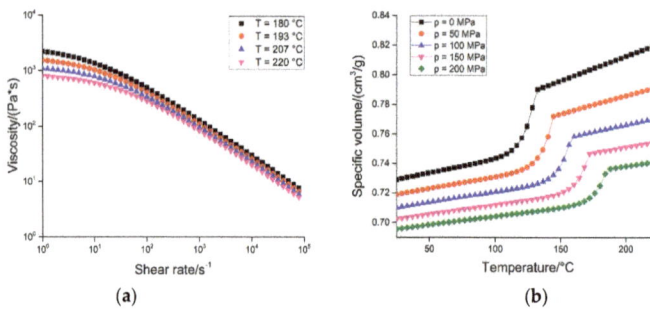

Figure 4. Viscosity plot at different temperatures (**a**) and *pvT* data at different pressures (**b**) for the polyoxymethylene (POM) grade.

In order to identify the best product and process fingerprints, an experimental campaign was carried out by varying the main μIM process parameters. Such a study was necessary since, as already anticipated, both the product and process fingerprints must be sensitive enough to process settings variations in order to function as optimization tools. Moreover, the data gathered through the experimental campaign were also used to determine the indicator that had the highest level of correlation to part quality, that is, the other characteristic needed to carry out an effective in-line quality control based on process monitoring. Figure 5 shows the flowchart representing the general procedure used for determining the best product and process fingerprints.

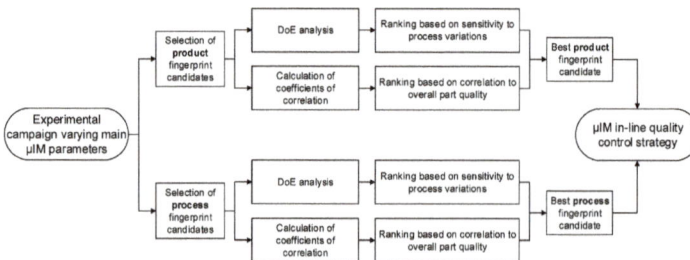

Figure 5. Flowchart representing the identification method for product and process fingerprints. The procedure starts with the selection of product and process fingerprint candidates. Design of Experiments (DoE) and correlation analyses allow for the determination of the best candidates and for the definition of the micro-injection moulding (μIM) in-line control strategy.

A Design of Experiments (DoE) approach was adopted. Four process parameters, namely holding pressure, injection speed, mould temperature and melt temperature were varied according to a two-level full factorial design. Such variables were selected since they are widely reported in literature to be the ones having the largest influence on the replication capability as well as on the level of shrinkage of moulded products [32–36]. The designed experimental plan allowed for the evaluation of the effects of the single process parameters and of their interactions with the maximum resolution. The levels of the investigated variables were set according to preliminary moulding experiments and the material manufacturer's recommendations. Table 3 shows the details of the DoE experimental plan. It can be observed that the holding pressure values were set at a relatively low value for µIM; this was adopted in order to avoid an excessive flash formation that would have disabled the part functionality. For each combination of process parameters, the first ten produced parts were discarded and the following five were kept for evaluation, thus making five DoE replicates available for the following analysis.

Table 3. Design of Experiments (DoE) process settings.

Process Parameter	Symbol	Unit	Low Level	High Level
Holding pressure	p_{hold}	bar	250	500
Injection speed	v_{inj}	mm/s	150	350
Mould temperature	T_{mould}	°C	100	110
Melt temperature	T_{melt}	°C	200	220

2.4. Measurement Strategy and Uncertainty Evaluation

Five dimensional features of the part were selected as product fingerprint candidates and assessed for each of the 80 produced parts. Three of them were functional geometries of the component. These are shown in Figure 6a and are the outer top diameter (ODt), outer bottom diameter (ODb), and inner bottom diameter (IDb). The other two were related to the main defects affecting the part quality: the flash and the gate mark. The flash, as already anticipated, was generated by the presence of the venting channel. The gate mark was caused by the detachment of the feed system from the part by means of the displacement of the middle plate and appeared as an unwanted prolongation in correspondence with the gate area (see Figure 6b). Both defects had to be minimized in order to ensure the part functionality, and therefore their size was a straightforward optimization response and an ideal product fingerprint candidate. The flash and the gate mark were quantified by means of dimensional indicators: the area of the flash, A_{flash}, was used to characterize the flash size, while the length of the gate mark, L_{mark}, was chosen as indicator of the gate mark size (see Figure 6b). Therefore, A_{flash} and L_{mark} both increase when the two defects become larger.

The five product fingerprint candidates were measured on each moulded part with a 3D state-of-the-art focus variation microscope (InfiniteFocus, Alicona Imaging GmbH, Raaba, Austria) using a 20× magnification objective lens. Table 4 reports the main instrument characteristics in the used acquisition mode.

Table 4. Focus variation microscope characteristics.

Instrument Characteristic	Value
Objective magnification	20×
Numerical aperture	0.40
Working distance/mm	13.0
Field of view/µm	714 × 542
Digital lateral resolution/µm	0.44
Declared vertical resolution/nm	0.14

a)

b)

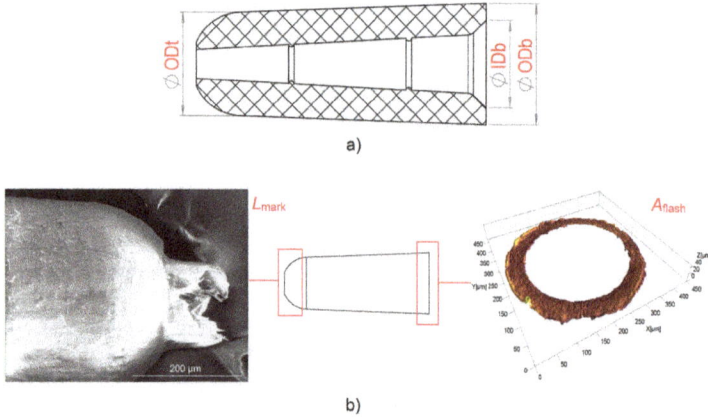

Figure 6. Product fingerprint candidates. (**a**) The three diameters outer top diameter (ODt), inner bottom diameter (IDb) and outer bottom diameter (ODb). (**b**) The two part defects and respective size indicators: the area of the flash (A_{flash}) and the length of the gate mark (L_{mark}).

In order to capture the five measurands, two acquisitions were carried out for each moulded sample. The top side of the part, i.e., its left side in Figure 1, was acquired and then the measurands, ODt and L_{mark} were extrapolated. The bottom of the part, i.e., its right side in Figure 1, was instead acquired to measure ODb, IDb and A_{flash}. The five measurands were then extracted from the 3D optical reconstructions using a dedicated image processing software (MountainsMap®, Digital Surf, Besançon, France). In detail, the processing was carried out as follows:

- ODt: this diameter was extrapolated from a top acquisition by fitting a circle to the desired points (see Figure 7b).
- L_{mark}: this quantity was measured as the vertical distance between the highest point belonging to the gate mark and the flat surface from which it protruded (see Figure 7c). This surface was identified in the same way for all the moulded parts by considering a constant height with respect to the plane on which the circle with the ODt diameter lay.
- IDb: this diameter was extrapolated from a bottom acquisition by fitting a circle to the desired diameter points (see Figure 8b). The 3D acquisition was initially processed by applying a levelling and a threshold along the Z-axis. In particular, the threshold allowed to accurately identify only the points belonging to the flash surface, thus eliminating the undesired influence of points acquired inside the hole.
- ODb: this diameter was extrapolated in the same way as IDb. As illustrated in Figure 8b, the outer perimeter of the flash was not as circular as the ones identifying ODt and IDb, making the ODb measurement less reliable than the other ones. This was most probably caused by an imperfect positioning of the central pin inside the mould cavity that created an unbalance in the polymer melt flow inside the micro channel.
- A_{flash}: the flash area was measured starting from a bottom acquisition by counting the pixels of the flash surface. After the number of pixels, N_p, was determined, a simple equation was used:

$$A_{flash} = N_p \times A_p \tag{1}$$

where A_p is the area of one pixel, equal to 0.442 µm^2. Even though both ODb and A_{flash} are indicators of the size of the flash affecting the bottom of the part, the second one is more representative since it is not affected by any circularity error and therefore is more sensitive to any variation of the defect size.

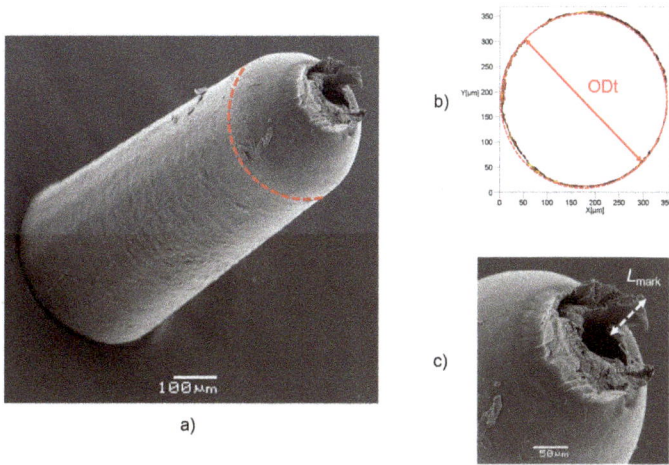

Figure 7. (**a**) SEM image of a moulded part; outer top diameter (ODt) is indicated in red. (**b**) Measurement of ODt. (**c**) Definition of L_{mark}.

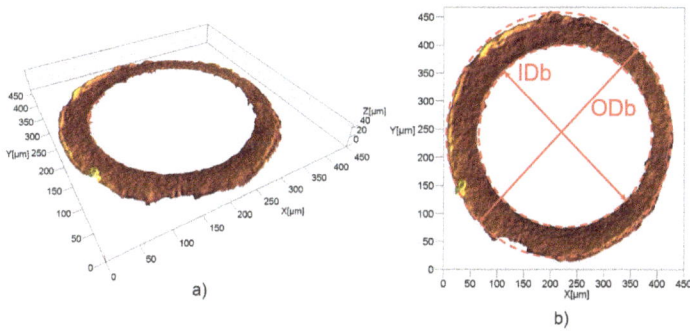

Figure 8. (**a**) 3D acquisition of the bottom of a moulded part. (**b**) Measurement of inner bottom diameter (IDb) and outer bottom diameter (ODb).

For all moulded parts, each acquisition was repeated three times, and the average of each extracted measurand was taken as output in order to minimize the influence of the instrument repeatability.

The mould geometries corresponding to the three diameters were also measured to provide a reference for calculating the replication capability of the µIM process. In fact, when evaluating replication technologies, especially in the micro-scale and nano-scale where machining accuracy becomes a very challenging task, the knowledge of the master dimensions is of paramount importance to assess the process capabilities [23]. The mould diameters corresponding to ODb and IDb were measured with the aforementioned focus variation microscope by acquiring the hole on the insert and the pin. As for the mould feature correspondent to ODt, no direct measurement procedure was applicable due to the inaccessibility of such a feature for any optical or contact instrument. A fast replication media was therefore used to replicate the internal geometry of the insert. In particular, a brown polyvinylsiloxane (PVS) replication media (AccuTrans®, Coltene, Cuyahoga Falls, OH, USA) was casted inside the cavity and measured after solidification. Such a method has been demonstrated to be successfully applicable to indirect measurement tasks where single-digit micrometric accuracy is required [37]. By knowing the mould dimensions, the replication capability Δ_D was defined as follows:

$$\Delta_D = D_{part} - D_{mould} \tag{2}$$

where D_{part} and D_{mould} represent a generic diameter, D, measured on the part and the mould respectively. Δ_D is an indicator of the level of accuracy of the µIM process, and therefore the three variables Δ_{ODb}, Δ_{IDb} and Δ_{ODt} were considered as responses for the experimental analysis in place of ODb, IDb and ODt.

The quality of the dimensional measurements was verified by calculating the measurement uncertainty, U. Such parameter is used to characterize the dispersion of the values that could be reasonably attributed to the measurand [38] and assumes more relevance when the quality of micro products are evaluated. In fact, the uncertainty-to-tolerance ratio, U/T, becomes much larger when higher precision levels, typical of the micro-scale, are demanded [10]. Moreover, the measurement uncertainty could, if large enough, partially or totally hide the effects of the experimental variables on the measured output [24]. In this study, the uncertainty related to the five measurands was calculated following ISO 15530-3 [39], which is based on the use of a calibrated artefact sharing similar characteristics with the actual measurand. Two different artefacts were used in the investigation: a calibrated circle of nominal diameter equal to 250 µm for the uncertainty of ODb, IDb and ODt and a calibrated step height of 1 mm for the uncertainty of L_{mark}. Four uncertainty contributions were considered for the calculation of U: u_{cal}, equal to the one stated in the calibration certificate of the artefact; u_{res}, due to the resolution of the measurement instrument and calculated by considering a rectangular distribution; u_w, related to the material and manufacturing variations of the actual measurand; and u_p, introduced by the measurement procedure and calculated as a standard deviation of 20 repeated measurements on the calibrated artefact. In particular, u_w was calculated as:

$$u_w = \frac{\max(\mathbf{M}) - \min(\mathbf{M})}{2\sqrt{3}} \tag{3}$$

where $\mathbf{M}$ is the vector containing the three measurement repetitions of a generic one of the five measurands. The expanded uncertainty, U, was finally obtained by combining the contributions:

$$U = k \times (u^2{}_{cal} + u^2{}_{res} + u^2{}_w + u^2{}_p)^{1/2} \tag{4}$$

where the coverage factor k was equal to 2 in order to achieve an approximate confidence interval of 95%. As for A_{flash}, the contributions u_{cal} and u_p were not considered since no calibrated artefact for area measurement was available. Table 5 reports the uncertainty budget for the five measurands involved in the study. It can be seen that for the three diameters, the uncertainty-to-tolerance ratio, U/T, ranged between 11% and 12%; such values are satisfying with respect to the upper recommended limit of 20% [26], especially considering the very narrow tolerance range imposed by the design specifications.

Table 5. Mean values of uncertainty contributions and expanded uncertainty for the five measurands.

Uncertainty Contribution	ODt/µm	IDb/µm	ODb/µm	L_{mark}/µm	A_{flash}/µm^2
u_{cal}	0.50	0.50	0.50	0.45	/
u_{res}	0.13	0.13	0.13	0.04	0.19
u_w	0.19	0.20	0.16	0.73	1.5×10^2
u_p	0.12	0.12	0.12	0.37	/
U ($k = 2$)	1.1	1.1	1.1	1.9	3.0×10^2

2.5. Process Monitoring

During each moulding cycle, two process variables were recorded in-line: the pressure, p, and the velocity, v, of the injection plunger. The two were derived from machine data, and therefore no external sensor was used. Such a type of data is easy to access and available to any machine user, making this analysis particularly interesting for industrial productions, when usually no external sensor is mounted within the moulding machine. In particular, the injection pressure was acquired by means of a strain gauge transducer (Sensorplatte microline, X Sensors AG, Diessenhofen, Switzerland) embedded in the machine and mounted on the back of the injection plunger. The speed of the injection plunger was recorded via the speed of the motor driving the plunger through the control unit of the machine. For both the monitored curves, the frequency of acquisition was set at 167 Hz (equivalent to a sampling interval of 6 ms), corresponding to the maximum value allowed by the machine computer. Pressure and velocity were acquired synchronously by the machine, and therefore the recorded data needed no alignment with respect to the time-scale.

From the first analysis of the acquired profiles, it was clear that the moulding parameter that mostly affected the shape of the p and v curves was the holding pressure. In fact, when the high level of p_{hold} was selected, a monotone increase of injection pressure was always observed (see Figure 9a). The increase culminated in a peak caused by the high pressure drop due to the extremely small dimensions of the gate and the mould cavity. After the switch-over point, which was defined on the machine by assigning a threshold value of injection pressure, p readapted to the negative linear profile of the holding pressure that was set through the machine interface. This profile was chosen since it guaranteed a smooth displacement of the injection plunger. Conversely, when the low value of p_{hold} was selected, the injection pressure always showed a point of discontinuity in its profile (see Figure 9b). This behaviour was caused by the plunger suddenly decelerating and then accelerating again. This was because, when moulding with p_{hold} equal to 250 bar, the cavity was not filled at the defined switch-over point, thus causing a pressure decrease. After this, the plunger accelerated again during the holding phase, completed the filling, as witnessed by the presence of the pressure peak, and readapted to the imposed negative holding pressure profile. The different level of p_{hold} influenced the velocity profiles as well. When moulding with a holding pressure of 500 bar, the plunger accelerated until v reached the value of the selected v_{inj} and then stopped in correspondence with the switch-over point (see Figure 10a). On the other hand, when moulding with p_{hold} equal to 250 bar, v suddenly decreased at some point, and then increased again until the final stopping (see Figure 10b). This discrepancy was due, as for the case of p, to the decrease in pressure and the consequent piston deceleration that occurred when the cavity was not filled at the defined switch-over point.

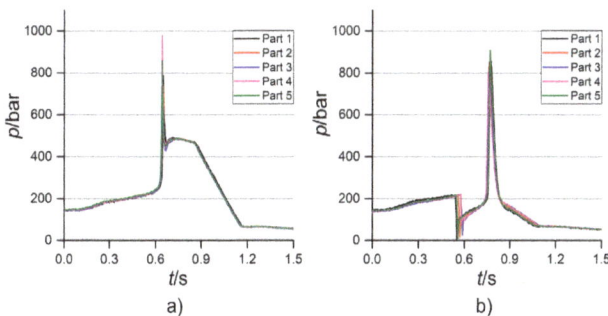

Figure 9. Injection pressure profiles recorded when moulding at: (**a**) high holding pressure and (**b**) low holding pressure. v_{inj} was equal to 150 mm/s in both cases.

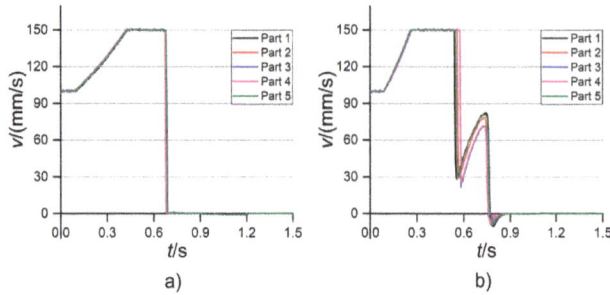

Figure 10. Injection velocity profiles recorded when moulding at: (**a**) high holding pressure and (**b**) low holding pressure. v_{inj} was equal to 150 mm/s in both cases.

Once p and v were acquired, their dependence on µIM parameters was studied by identifying some indicators that acted as process fingerprint candidates. In this case, variables that well characterized the shape and main features of both the pressure and velocity curves were selected. In particular, the process fingerprint candidates derived from the injection pressure curves were the following:

- p_{max}: this quantity was defined as the maximum value for each recorded p curve. Being the pressure peak due to the small size of the channels, this is an indicator related to the filling behaviour of the cavity.
- p_{mean}: this value was calculated as the average pressure in the time interval between the start and the end of the moulding cycle (points A and C in Figure 11a). This quantity provides average information on the pressure acting during one moulding cycle.
- I_p: this is the integral of the pressure in the peak region. The peak region was identified as the time interval spanning from the abrupt increase of p, correspondent to the start of the filling, and the point where the injection pressure adapted to the holding pressure profile given in input to the machine. Therefore, this quantity is related to the amount of energy provided by the injection plunger during the filling phase and is expected to be very process-dependent since the filling of the cavity is usually highly influenced by variations of the µIM process parameters. In particular, I_p was calculated by applying the trapezoidal rule:

$$I_p = dt \times \sum_{t=t_A}^{t_B - dt} \frac{p(t) + p(t + dt)}{2} \tag{5}$$

where t is the time, t_A and t_B are the times correspondent to points A and B respectively (see Figure 11a), and dt is the sampling interval of 6 ms correspondent to the sampling frequency of 167 Hz.

- $I_p/\Delta t$: this quantity is equal to the integral mean of p in the peak region, i.e., I_p divided by the integration range $\Delta t = t_B - t_A$ (see Figure 11a). Such a variable differs from I_p since it is not influenced by the range of integration, i.e., the duration of the filling phase. $I_p/\Delta t$ is thus an indicator of the average p acting during the pressure peak.

From the injection velocity curves, two indicators were extracted:

- v_{mean}: the mean velocity calculated as the average of the v values in the time interval between the start of the acceleration of the plunger and its stop at the switch-over point (point D to point F in Figure 11b), i.e., when v starts decreasing towards a null velocity. This quantity is therefore related to the average velocity that characterized the filling phase of the moulding cycle.

- v_{slope}: the slope of the velocity curve between the start and the end of the acceleration (point D to point E in Figure 11b). This value is equal to the constant acceleration assumed by the plunger to reach the selected v_{inj} value.

The maximum value of v was not taken into account in the investigation since it was equal to the selected v_{inj} and therefore only dependent on that µIM parameter.

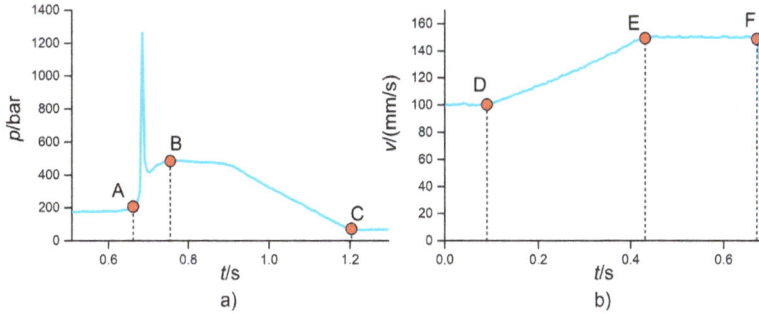

Figure 11. Pressure (**a**) and velocity (**b**) curves with points indicating the time intervals used for calculating the process fingerprints candidates. The position of the points was determined by tracking changes of the slope using the first-order derivative value.

3. Results and Discussion

3.1. Product Fingerprint Analysis

Figure 12 reports the results of the analysis for the five product fingerprint candidates. The main effects plot and the Pareto chart of the effects are shown: these two graphs allow for the determination of which process parameter had a significant influence on the response in a robust statistical way. In both the plots, interval bars are shown. In the main effects plots, they represent the expanded measurement uncertainty (see Table 5) that must be taken into account when evaluating the effects of process variation on a measurand. In fact, the variability of the process could be entirely covered by the measurement uncertainty, especially for micro manufacturing processes where the induced dimensional variations are typically in the micrometre range [23]. In the Pareto charts, the interval bars represent the standard deviation of the effects obtained by running five separate analyses for the five DoE replicates. The evaluation of such a variability is important since it quantifies the repeatability of the conclusions drawn from the Pareto chart; a low standard deviation of an effect leads to the conclusion that the significance of that effect is robust with respect to process repeatability. In particular, an effect whose interval bar overlaps with the significance limit cannot be considered as significant.

Figure 12a shows the results for Δ_{ODt}. For this output, the most significant process parameter was T_{melt}; its increase led to a decrease in replication fidelity by 2.5 µm. Considering the measurement uncertainty, the other three µIM process parameters did not have a relevant influence on the measured output. As for second-order interactions, only the one between T_{melt} and v_{inj} had a relevant impact on Δ_{ODt}, given that its standardized effect was larger than the significance limit and its standard deviation bar was not overlapping with it.

Figure 12b reports the results for the replication of IDb. On average, the replicated IDb was 14.0 µm smaller than the master. This shrinkage level was very similar to that of ODt. The mould temperature was the only significant process parameter. In particular, increasing T_{mould} from 100 °C to 110 °C led to a decrease in replication level. Such a behaviour, which may seem opposite to the usual enhancement of replication obtained when increasing mould temperature, was already observed in the literature when internal features, i.e., pins, were replicated [23]. The Pareto chart of the effects revealed that only the second-order interaction between T_{mould} and T_{melt} was significant.

Δ_{ODb} was more sensitive to process variations than the previous measurands (see Figure 12c). In particular, variations in p_{hold}, v_{inj}, and T_{mould} led to an increase in the output that was bigger than the measurement uncertainty. In this case, the results assumed positive values since ODb is the diameter of the circle that identifies the perimeter of the flash formed at the end of the flow (see Figure 8); the replicated diameter was larger than the correspondent one of the micro cavity. Increasing any of the investigated process parameters had a positive effect on Δ_{ODb} and consequently on the flash size. The reason for this is to be found in polymer rheology. When using the high mould temperature, the viscosity of the polymer melt inside the cavity was reduced, thus providing a better replication of the cavity features and, in this case, a bigger flash. The same effect was obtained when increasing the injection speed; the viscosity of the melt was reduced thanks to the shear thinning behaviour of thermoplastic polymers such as POM, thus achieving a longer flow path inside the venting channel that resulted in a larger flash and consequently a larger ODb. Finally, increasing the holding pressure allowed more material to enter the cavity, thus increasing the flash size. These findings were also confirmed by the Pareto chart, which also showed that the second-order interaction between T_{mould} and p_{hold} had a significant impact on the measured response. Similar effects of the µIM process parameters on the flash size were also reported in [40].

The area of the flash, A_{flash}, showed the same trends as Δ_{ODb} (see Figure 12d). This was expected since the two dimensional outputs both have a direct relationship with the flash size. However, the measurement uncertainty was less influent in this case. Moreover, according to the Pareto chart, p_{hold} can be considered more significant than in the case Δ_{ODb} since the standard deviation of its standardized effect does not overlap with the significance limit. Differently from the previous case, the second-order interaction between the mould and the melt temperature was also influent. Therefore, it can be concluded that A_{flash} was more sensitive to process variations than Δ_{ODb}, thus representing a more suitable product fingerprint candidate based on the flash size.

Figure 12e shows the results for the gate mark length, L_{mark}. The size of this defect was on average equal to 82 µm, which is relevant if compared to the overall component dimensions (see Figure 1). All four investigated process parameters had a significant influence on L_{mark} according to the main effects plot and Pareto chart. In particular, an increase in the variables led to a decrease in the defect size, demonstrating that such a product fingerprint was very sensitive to µIM settings variations. The holding pressure showed the largest impact, since the use of 500 bar provided parts having on average a 30 µm shorter gate mark than those moulded with p_{hold} equal to 250 bar. In order to further inspect the effects of µIM on this defect, SEM images of parts manufactured with different DoE combinations were taken and compared (see Figure 13). It can be observed that when moulding with the low levels of the four parameters, the gate mark was very evident, almost occluding the upper hole of the component and thus disabling its functionality. A zone of deformation was clearly visible around the defect, meaning that the mechanism that generated the defect was a ductile breakage caused by the detachment of the gate from the part. Setting the µIM parameters at a high level allowed a great reduction of the gate mark size (see Figure 13b,c) and of the area of deformation. Finally, selecting the high levels of all the four process variables generated parts almost free of the defect. The great influence of the process conditions on the gate mark size and appearance may have been caused by a change in the crystallinity of the polymer. In particular, moulding with high levels of the process parameters may have impeded the formation of crystals because of the more drastic cooling rate, thus decreasing the mechanical properties and in turn facilitating the brittle detachment of the gate from the part. It is worth noting that p_{hold}, v_{inj}, T_{mould} and T_{melt} had an opposite effect on the size of A_{flash} and L_{mark}, i.e., the two defects affecting the part quality. This means that a simultaneous minimization of the two defects was impossible inside the investigated experimental range.

The analysis of the effects of the µIM parameters on the five product fingerprint candidates allowed some conclusions to be made. In particular, the best candidates with respect to the characteristic of process sensitivity were A_{flash} and L_{mark}; both were greatly influenced by all the four investigated parameters and can therefore serve to tune the process with the aim of optimizing the quality of the produced parts.

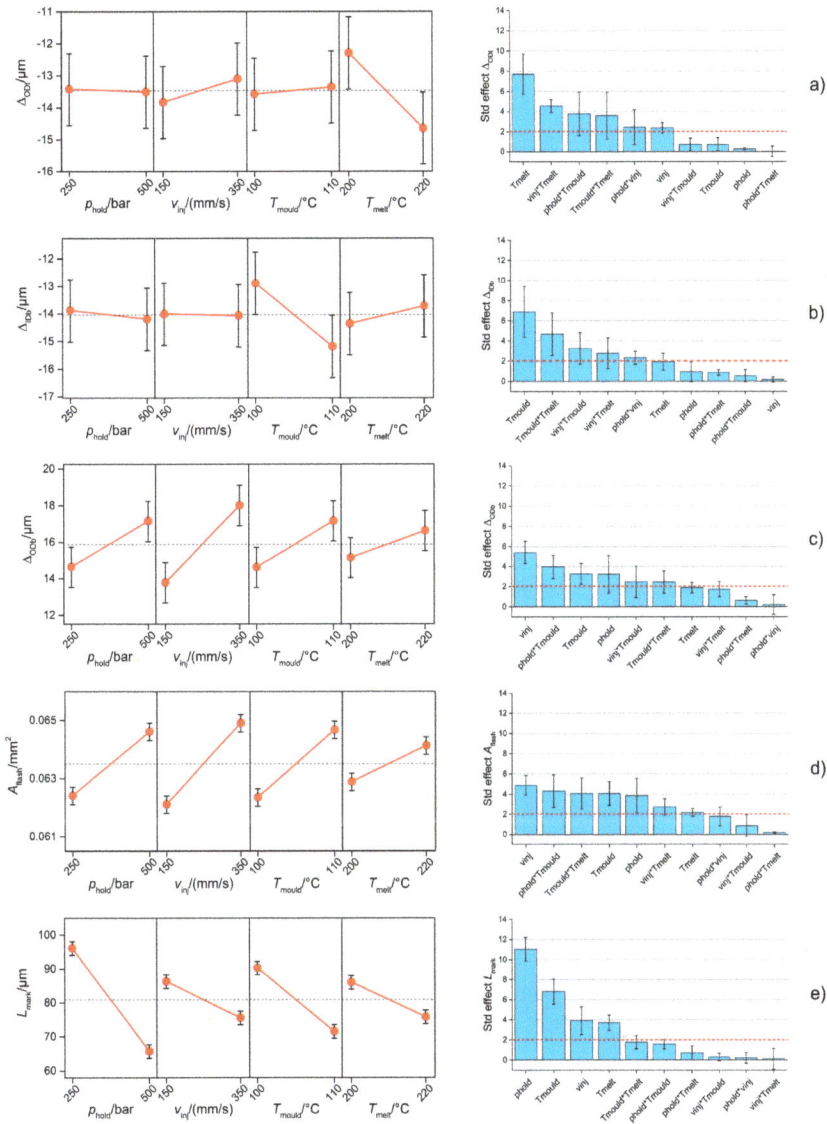

Figure 12. Influence of the µIM process on the five product fingerprint candidates: (**a**) Δ_{ODt}, (**b**) Δ_{IDb}, (**c**) Δ_{ODb}, (**d**) A_{flash} and (**e**) L_{mark}. Main effects plot (left column) and Pareto chart of standardized effects (right column) are reported. The interval bars represent the expanded measurement uncertainties, U, in the main effects plots and the standard deviations of five Pareto analyses for the five DoE replicates in the Pareto charts. The red dashed line in the Pareto chart is the significance level at 95% of confidence level.

Along with the sensitivity to process variation, the other characteristic that an effective product fingerprint must have is the correlation to the overall part quality. This is necessary to guarantee an efficient quality control based only on the measurement of a single fingerprint. In order to identify the best fingerprint candidate with respect to this requirement, a correlation analysis was carried out. In particular, the coefficient of correlation, ϱ, was calculated for each couple of measurands. ϱ was calculated as follows:

$$\varrho(\mathbf{x}, \mathbf{y}) = \frac{\sum(x - \bar{x})(y - \bar{y})}{\sqrt{\sum(x - \bar{x})^2 \sum(y - \bar{y})^2}} \tag{6}$$

where $\mathbf{x}$ and $\mathbf{y}$ are generic vectors containing two datasets and $\bar{x}$ and $\bar{y}$ are their respective mean values. Such a coefficient can vary between -1 and $+1$, where the first describes a perfect negative correlation and the second a perfect positive correlation. A ϱ equal to 0 indicates that no correlation exists between the two examined sets of values. In this study, ϱ was calculated by considering all the 80 values derived from the DoE campaign for each of the five measurands.

Figure 14 shows the correlation coefficients calculated for the ten couples of product fingerprint candidates. It can be seen that the greatest correlation ($\varrho = 0.9$) was the one between Δ_{ODb} and A_{flash}. The two measurands are in fact strictly related to the flash size, as explained before. Therefore, an increase in one determined an increase in the other and vice versa, meaning that controlling only one of them allowed for the monitoring of both the geometrical outputs. A second group of coefficients of correlation ranging from -0.5 to -0.4 can be identified. A_{flash} and L_{mark} shared in fact a ϱ equal to -0.5, demonstrating that a significant amount of correlation existed between these two product fingerprint candidates. This was also mirrored in the main effects plots (see Figure 12) where it was clear how the four investigated process parameters had an opposite effect on the two defect sizes. Significant negative correlations were also observed between Δ_{ODb} and L_{mark} and between Δ_{IDb} and A_{flash}. The second one was due to the geometry of the flash; a decrease in IDb resulted in an increase in the flash area according to its definition (see Figure 8). The other coefficients of correlation were all lower than 0.3 in absolute value and thus negligible if compared to the others.

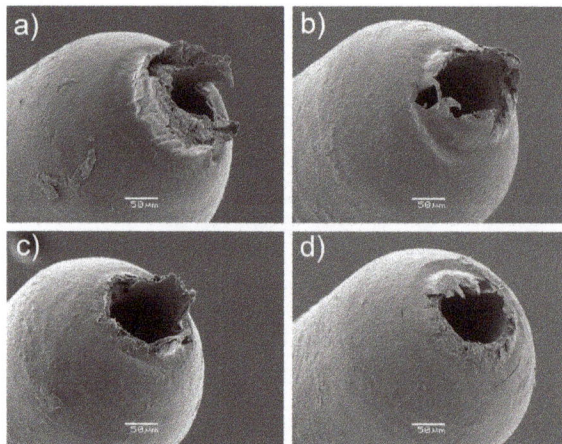

Figure 13. Gate mark appearance for different combinations of process parameters: (**a**) was moulded with low levels of T_{melt}, p_{hold}, T_{mould}, and v_{inj}; (**b**) was moulded with low levels of p_{hold}, T_{mould}, and v_{inj}; (**c**) was moulded with low levels of p_{hold}; and (**d**) was moulded with high levels of the four process parameters.

From this analysis, it can be concluded that the product fingerprint candidate that was the most correlated to the overall part quality was A_{flash} since its values were the ones that showed high levels of ϱ in combination with three other measurands, namely Δ_{ODb}, L_{mark}, and Δ_{IDb}. Therefore, by controlling only this dimensional characteristic of the moulded part, the best control over the overall part quality can be carried out. Considering also the requirement of sensitivity to process settings variations, the best product fingerprint for the specific part under analysis was the flash area, followed by the length of the gate mark. Such quality indicators must be related to in-line, monitored process variables in order to function in a fast and comprehensive μIM assurance strategy.

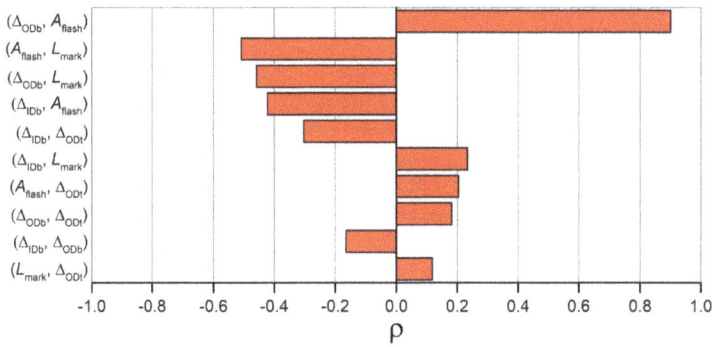

Figure 14. Values of the coefficients of correlation calculated between each couple of measured datasets. Values were sorted largest to smallest in absolute value.

3.2. Process Fingerprint Analysis

A similar type of analysis was carried out to identify the best process fingerprint candidate among the six indicators derived from the monitored pressure and velocity curves with respect to sensitivity to process variations. In this case, the main effects plots were represented with interval bars equal to the standard errors calculated among the values related to the particular combination of process parameters.

Figure 15a shows the results for p_{max}. This indicator depended mainly on the selected p_{hold} and v_{inj} values. Particularly, an increase in both the holding pressure and the injection speed resulted in an increase of the maximum injection pressure. The effect of v_{inj} can be explained by considering that flowing a fluid with a higher speed requires a higher pressure. The effect of p_{hold} is, on the other hand, due to the used modality of switch-over; as already anticipated, the machine was set to switch from the filling phase to the holding one when a certain pressure was reached. Therefore, when selecting a higher p_{hold}, the injection pressure was allowed to rise more before switching to the holding profile. The second-order interaction between p_{hold} and v_{inj} was also significant.

p_{mean} was predominantly influenced by the holding pressure (see Figure 15b). This was caused by the fact that p_{mean} was calculated as the average p among the entire moulding cycle (see Figure 11a). The averaging operation, in fact, minimized the relevance of the peak phase, making the holding phase preponderant, and thus p_{hold} became the most significant term. All the other effects, second-order interactions included, were negligible with respect to the holding pressure significance.

Figure 15c illustrates the effect of μIM process parameters on I_p. This fingerprint candidate was mainly influenced by holding pressure. The second most significant effect was that of v_{inj}. The reason for this is similar to the one explained when commenting on p_{max} dependence on the process variations. In fact, the integral of p in the peak region was strongly influenced by the p peak. However, in this case, there was a larger sensitivity with respect to other process parameters such as T_{mould}, which had a larger impact on the results. However, considering the interval bars of both the main effects plot and the Pareto chart, the mould temperature cannot be considered as significant for I_p.

$I_p/\Delta t$ showed a process dependence very similar to I_p (see Figure 15d). However, the operation of normalizing the pressure integral over the integration range considerably diminished the relevance of

v_{inj} while enhancing that of T_{mould}, which was significant according to the Pareto chart even though the interval bars of the main effects plot slightly overlapped. This may have been caused by the fact that changing the injection velocity setting had an impact on the duration of the peak region and therefore on the filling time. In particular, increasing the mould temperature resulted in an increase in $I_p/\Delta t$. This is somehow unexpected since increasing T_{mould} usually decreases the polymer viscosity and therefore the pressure needed to drive the melt through the cavity channels. As for the previous case, second-order interactions of v_{inj} had a significant effect on this output.

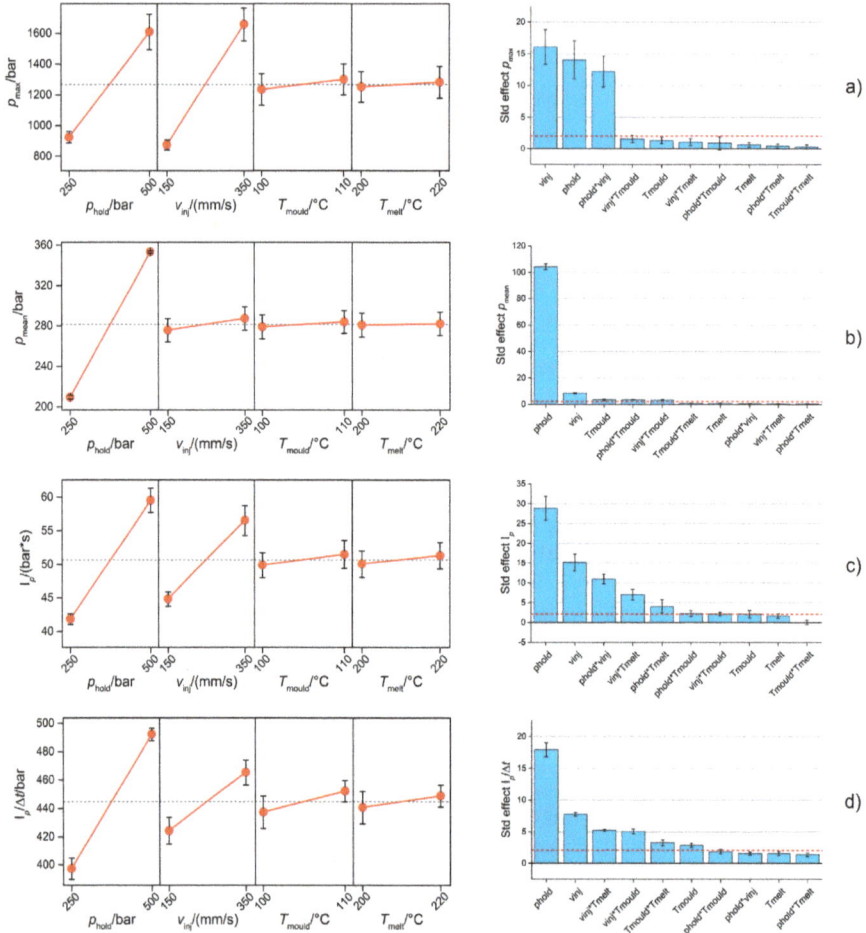

Figure 15. Influence of µIM process on the four process fingerprint candidates derived from monitored injection pressure curves: (**a**) p_{max}, (**b**) p_{mean}, (**c**) I_p and (**d**) $I_p/\Delta t$. Main effects plot (left column) and Pareto chart of standardized effects (right column) are reported. The interval bars represent the standard errors in the main effects plots and the standard deviations of five Pareto analyses for the five DoE replicates in the Pareto charts. The red dashed line in the Pareto chart is the significance level at 95% of confidence level.

Figure 16a shows the results for v_{mean}, whose value was influenced mostly by v_{inj}, p_{hold} and their interaction. Increasing the injection speed had a positive effect on v_{mean}, since a higher velocity plateau was recorded. Increasing the holding pressure had the same effect; this was caused by the fact that when moulding at the high level of p_{hold}, no deceleration of the plunger, and therefore no decrease in speed during the filling phase, was observed (see Figure 10b). The significance of the interaction was caused by the fact that when using high p_{hold}, an increase in injection speed led to a larger increase in v_{mean} than when using a low p_{hold}. In fact, the deceleration behaviour was similar when using both high and low v_{inj}.

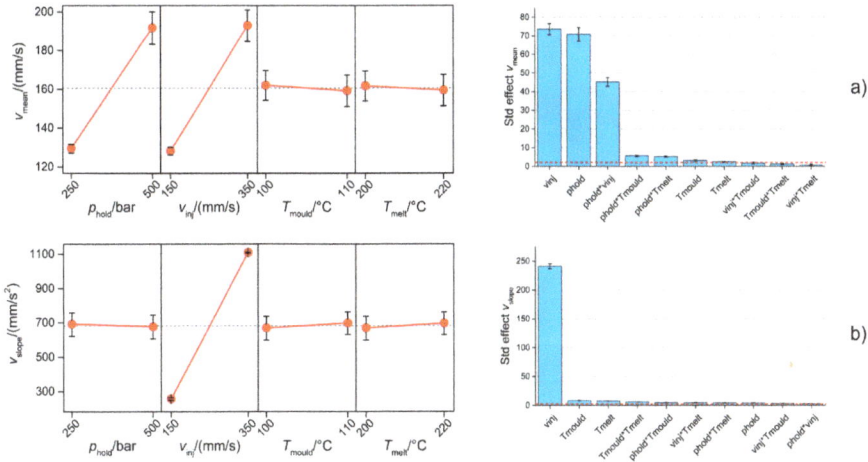

Figure 16. Influence of μIM process on the two process fingerprint candidates derived from monitored injection velocity curves: (**a**) v_{mean} and (**b**) v_{slope}. Main effects plot (left column) and Pareto chart of standardized effects (right column) are reported. The interval bars represent the standard errors in the main effects plots and the standard deviations of five Pareto analyses for the five DoE replicates in the Pareto charts. The red dashed line in the Pareto chart is the significance level at 95% of confidence level.

v_{slope} depended mainly on the value of the injection speed (see Figure 16b). In particular, moulding at a high v_{inj} determined a substantial increase in the plunger acceleration from 260 mm/s^2 to 1100 mm/s^2. This means that when a higher v_{inj} was selected, the machine motor provided the plunger with a higher acceleration in order to reach it.

From this analysis, it can be concluded that holding pressure and injection speed were the most influencing parameters for the six process fingerprint candidates derived from monitored injection pressure and velocity curves. On the other hand, mould and melt temperature variations were in most cases not significant in determining the level of the responses. This was because pressure and velocity were both measured at the injection plunger location. Quantities measured inside the mould with external sensors proved in fact to be more sensitive to T_{melt} and T_{mould} variations [18]. Only the integral mean of the pressure, $I_p/\Delta t$, showed a relevant dependence on T_{mould}, making this indicator the best process fingerprint candidate among those extracted from the p and v curves with respect to the sensitivity to μIM process variations. In fact, by in-line monitoring $I_p/\Delta t$, variations of holding pressure, injection speed and mould temperature can be observed and quantified.

In order to investigate the relation between $I_p/\Delta t$ and the part quality, which is a fundamental characteristic of a process fingerprint to allow an effective in-line quality monitoring, the same approach used for the product fingerprint correlation analysis was adopted. In particular, the coefficient of correlation, ϱ, was calculated using Equation (6) among $I_p/\Delta t$ and the five dimensional measurands.

Figure 17 shows the results. The correlation coefficient values appear as divided in two distinct subgroups: one made of ϱ values calculated for L_{mark}, Δ_{ODb} and A_{flash}, which were all larger than 0.7 in absolute value, and the other made of ϱ values calculated for Δ_{IDb} and Δ_{ODt}. This means that

$I_p/\Delta t$ was highly correlated with the size of the gate mark and the flash, i.e., the defects affecting part quality, and with the dimension of ODb, which was in turn highly related to L_{mark} and A_{flash} (see Figure 14). On the other hand, the other two measurands showed no relevant link with the integral mean of p, being the correlation coefficients values equal to -0.25 and 0.21 for Δ_{IDb} and Δ_{ODt} respectively. These findings proved that $I_p/\Delta t$, besides being sensitive to variations in µIM process parameters, was also correlated to the two best product fingerprint candidates, namely the flash area and the length of the gate mark. Therefore, it can act as the bridge between process monitoring and part quality, representing the link needed to perform a faster and more comprehensive assurance of the investigated moulded component.

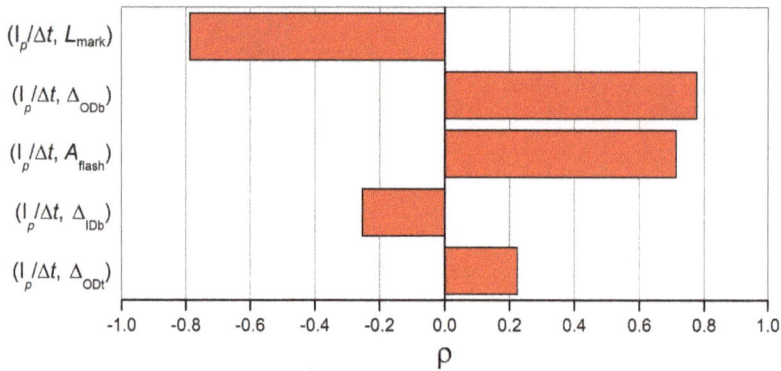

Figure 17. Values of the coefficients of correlation calculated between $I_p/\Delta t$ and the five dimensional measurands. Values were sorted largest to smallest in absolute value.

Figure 18 shows the plots of gate mark and flash sizes against $I_p/\Delta t$. It can be observed that there indeed existed a relation among the data. In particular, there is an almost linear correlation between L_{mark} and the integral mean. The data also show a limited dispersion around the depicted linear trend, demonstrating that the relation between the two variables was robust inside the investigated experimental range. Since the trend is negative, a higher $I_p/\Delta t$ determined a smaller gate mark on the moulded parts. This finding agrees with the opposite slopes of the main effects plots for the two variables (see Figures 12e and 15d). By controlling the process fingerprint, an accurate control on the size of this defect can be performed: $I_p/\Delta t$ must be kept at values around 500 bar if the size of the gate mark has to be minimized. Conversely, $I_p/\Delta t$ equal to 300 bar generated a defect having double the size. In regards to the size of the flash, indicated by the A_{flash} value, a positive relation was observed. Therefore, lower $I_p/\Delta t$ values were needed to minimize the size of this defect; the lower the pressure applied during filling, the smaller the flash on the component. As already anticipated, the minimization of both defects was not possible within the investigated process parameters ranges. In this case, the data were more dispersed than for L_{mark}, resulting in a less precise and accurate prediction. However, the general data trend is still evident, as shown by the high coefficient of correlation of 0.71 (see Figure 17). As A_{flash} correlated to most of the other dimensional measurands (see Figure 14), the optimization of the part quality can be effectively carried out by monitoring $I_p/\Delta t$ for every moulding cycle, thus assessing the quality of all moulded parts.

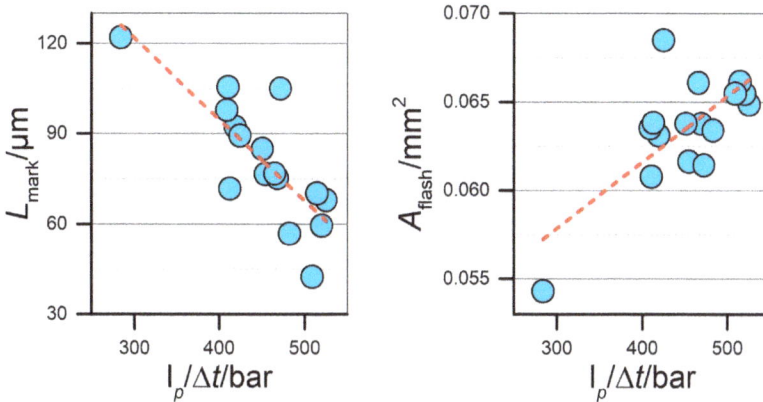

Figure 18. Correlation plots of the two defect size indicators against $I_p/\Delta t$. Red dashed lines represent linear trends. Each point represents the average of the five DoE replicates.

4. Conclusions

The present paper aimed at applying and validating a new optimization concept based on the product and process fingerprints of the µIM process of a micro medical component. The best product and process fingerprint candidates were identified by considering two characteristics: sensitivity to process variations and correlation to part quality. Optical metrology and process monitoring using no external sensors were, for the first time in reported literature, successfully combined to provide an in-line optimization strategy in µIM.

The following conclusions are drawn from the study:

- The variation of µIM process settings had a relevant impact on the quality of the micro component. In particular, the flash area and the length of the gate mark showed the largest sensitivity.
- Varying the four investigated process parameters had an opposite effect on the size of the two defects: an increase in the flash size always came with a decrease in the gate mark size and vice versa. Their simultaneous minimization was therefore not possible to obtain within the investigated process window, posing a great challenge with respect to quality optimization.
- The morphology of the gate mark was deeply influenced by the selected process settings. In particular, a zone of deformation was clearly visible only when moulding with low levels of the parameters, thus significantly increasing the size of the defect.
- The flash area was the measurand with the highest level of correlation to part quality. By measuring the effects of µIM parameters on such indicators, robust conclusions can be made also on three other measurands, namely ODb, L_{mark} and IDb. Therefore, A_{flash} represented the best product fingerprint candidate for the analysed component.
- The indicators extracted from in-line monitored injection and velocity curves were mostly influenced by p_{hold} and v_{inj}. The only one that showed a significant dependence on another parameter, namely T_{mould}, was the mean integral of the pressure during filling, $I_p/\Delta t$. This variable increased when selecting the high levels of the µIM parameters. Being the most sensitive among the investigated process indicators, it was chosen as the best process fingerprint.
- $I_p/\Delta t$ showed a significant correlation with three measurands. In particular, the size of both the defects could be effectively controlled by monitoring the $I_p/\Delta t$ value for each moulding cycle. Such a discovery demonstrated that in-line process optimization in µIM can be carried out by means of a robust monitoring strategy in order to make sure that all the manufactured components have dimensions within the desired range.

Future work will be dedicated to the application of a similar approach to different types of moulded samples such as micro structured and nano structured surfaces. This will extend the validity of this approach to other dimensional ranges and classes of components.

Author Contributions: F.B. conceived and designed the experiments; F.B. performed experiments and measurements; F.B., M.C. and G.T. analysed the data; F.B. wrote the paper; M.C. and G.T. revised the paper.

Acknowledgments: This research work was undertaken in the context of MICROMAN project ("Process Fingerprint for Zero-defect Net-shape MICROMANufacturing", http://www.microman.mek.dtu.dk/). MICROMAN is a European Training Network supported by Horizon 2020, the EU Framework Programme for Research and Innovation (Project ID: 674801).

Conflicts of Interest: The authors declare no conflict of interest.

References

1. Brousseau, E.B.; Dimov, S.S.; Pham, D.T. Some recent advances in multi-material micro- and nano-manufacturing. *Int. J. Adv. Manuf. Technol.* **2010**, *47*, 161–180. [CrossRef]
2. Giboz, J.; Copponnex, T.; Mélé, P. Microinjection molding of thermoplastic polymers: A review. *J. Micromech. Microeng.* **2007**, *17*, R96–R109. [CrossRef]
3. Whiteside, B.; Martyn, M.T.; Coates, P.D.; Greenway, G.; Allen, P.S.; Hornsby, P. Micromoulding: Process measurements, product morphology and properties. *Plast. Rubber Compos.* **2004**, *33*, 11–17. [CrossRef]
4. Yang, C.; Yin, X.-H.; Cheng, G.-M. Microinjection molding of microsystem components: New aspects in improving performance. *J. Micromech. Microeng.* **2013**, *23*, 1–21. [CrossRef]
5. Alting, L.; Kimura, F.; Hansen, H.N.; Bissacco, G. Micro Engineering. *CIRP Ann.-Manuf. Technol.* **2003**, *52*, 635–657. [CrossRef]
6. Sha, B.; Dimov, S.; Griffiths, C.; Packianather, M.S. Micro-injection moulding: Factors affecting the achievable aspect ratios. *Int. J. Adv. Manuf. Technol.* **2007**, *33*, 147–156. [CrossRef]
7. Vera, J.; Brulez, A.-C.; Contraires, E.; Larochette, M.; Trannoy-Orban, N.; Pignon, M.; Mauclair, C.; Valette, S.; Benayoun, S. Factors influencing microinjection molding replication quality. *J. Micromech. Microeng.* **2018**, *28*, 15004. [CrossRef]
8. Surace, R.; Bellantone, V.; Trotta, G.; Fassi, I. Replicating capability investigation of micro features in injection moulding process. *J. Manuf. Process.* **2017**, *28*, 351–361. [CrossRef]
9. Sha, B.; Dimov, S.; Griffiths, C.; Packianather, M.S. Investigation of micro-injection moulding: Factors affecting the replication quality. *J. Mater. Process. Technol.* **2007**, *183*, 284–296. [CrossRef]
10. Tosello, G.; Hansen, H.N.; Gasparin, S. Applications of dimensional micro metrology to the product and process quality control in manufacturing of precision polymer micro components. *CIRP Ann.-Manuf. Technol.* **2009**, *58*, 467–472. [CrossRef]
11. Surace, R.; Bellantone, V.; Trotta, G.; Basile, V.; Modica, F.; Fassi, I. Design and Fabrication of a Polymeric Microfilter for Medical Applications. *J. Micro Nano-Manuf.* **2015**, *4*, 11006. [CrossRef]
12. Masato, D.; Sorgato, M.; Parenti, P.; Annoni, M.; Lucchetta, G. Impact of micro milling strategy on the demolding forces in micro injection molding. *J. Mater. Process. Technol.* **2017**, *246*, 211–223. [CrossRef]
13. Chen, Z.; Turng, L.S. A review of current developments in process and quality control for injection molding. *Adv. Polym. Technol.* **2005**, *24*, 165–182. [CrossRef]
14. Michaeli, W.; Schreiber, A. Online control of the injection molding process based on process variables. *Adv. Polym. Technol.* **2009**, *28*, 65–76. [CrossRef]
15. Speranza, V.; Vietri, U.; Pantani, R. Monitoring of injection moulding of thermoplastics: Adopting pressure transducers to estimate the solidification history and the shrinkage of moulded parts. *Strojniški Vestnik/J. Mech. Eng.* **2013**, *59*, 677–682. [CrossRef]
16. Tsai, K.-M.; Lan, J.-K. Correlation between runner pressure and cavity pressure within injection mold. *Int. J. Adv. Manuf. Technol.* **2015**, 14–23. [CrossRef]
17. Mendibil, X.; Llanos, I.; Urreta, H.; Quintana, I. In process quality control on micro-injection moulding: The role of sensor location. *Int. J. Adv. Manuf. Technol.* **2016**. [CrossRef]
18. Griffiths, C.A.; Dimov, S.; Scholz, S.G.; Hirshy, H.; Tosello, G. Process Factors Influence on Cavity Pressure Behavior in Microinjection Moulding. *J. Manuf. Sci. Eng.* **2011**, *133*, 31007. [CrossRef]

19. Griffiths, C.A.; Dimov, S.S.; Scholz, S.G.; Tosello, G.; Rees, A. Influence of Injection and Cavity Pressure on the Demoulding Force in Micro-Injection Moulding. *J. Manuf. Sci. Eng.* **2014**, *136*, 31014. [CrossRef]
20. Griffiths, C.A.; Dimov, S.S.; Brousseau, E.B. Microinjection moulding: The influence of runner systems on flow behaviour and melt fill of multiple microcavities. *Proc. Inst. Mech. Eng. Part B J. Eng. Manuf.* **2008**, *222*, 1119–1130. [CrossRef]
21. Griffiths, C.A.; Dimov, S.S.; Scholz, S.; Tosello, G. Cavity Air Flow Behavior During Filling in Microinjection Molding. *J. Manuf. Sci. Eng.* **2011**, *133*, 11006. [CrossRef]
22. Zhiltsova, T.V.; Oliveira, M.S.A.; Ferreira, J.A. Integral approach for production of thermoplastics microparts by injection moulding. *J. Mater. Sci.* **2012**, *48*, 81–94. [CrossRef]
23. Baruffi, F.; Calaon, M.; Tosello, G. Effects of micro-injection moulding process parameters on accuracy and precision of thermoplastic elastomer micro rings. *Precis. Eng.* **2018**, *51*, 353–361. [CrossRef]
24. Tosello, G.; Gava, A.; Hansen, H.N.; Lucchetta, G. Study of process parameters effect on the filling phase of micro-injection moulding using weld lines as flow markers. *Int. J. Adv. Manuf. Technol.* **2010**, *47*, 81–97. [CrossRef]
25. Hakimian, E.; Sulong, A.B. Analysis of warpage and shrinkage properties of injection-molded micro gears polymer composites using numerical simulations assisted by the Taguchi method. *Mater. Des.* **2012**, *42*, 62–71. [CrossRef]
26. Ontiveros, S.; Yagüe-Fabra, J.A.; Jiménez, R.; Tosello, G.; Gasparin, S.; Pierobon, A.; Carmignato, S.; Hansen, H.N. Dimensional measurement of micro-moulded parts by computed tomography. *Meas. Sci. Technol.* **2012**, *23*, 125401. [CrossRef]
27. Bellantone, V.; Surace, R.; Trotta, G.; Fassi, I. Replication capability of micro injection moulding process for polymeric parts manufacturing. *Int. J. Adv. Manuf. Technol.* **2012**, *67*, 1407–1421. [CrossRef]
28. Mélé, P.; Giboz, J. Micro-injection molding of thermoplastic polymers: Proposal of a constitutive law as function of the aspect ratios. *J. Appl. Polym. Sci.* **2018**, *135*, 45719. [CrossRef]
29. Annicchiarico, D.; Attia, U.M.; Alcock, J.R. A methodology for shrinkage measurement in micro-injection moulding. *Polym. Test.* **2013**, *32*, 769–777. [CrossRef]
30. Masato, D.; Sorgato, M.; Lucchetta, G. Analysis of the influence of part thickness on the replication of micro-structured surfaces by injection molding. *Mater. Des.* **2016**, *95*, 219–224. [CrossRef]
31. Vella, P.C.; Dimov, S.S.; Brousseau, E.; Whiteside, B.R. A new process chain for producing bulk metallic glass replication masters with micro- and nano-scale features. *Int. J. Adv. Manuf. Technol.* **2014**, *76*, 523–543. [CrossRef]
32. Griffiths, C.A.; Dimov, S.S.; Brousseau, E.B.; Hoyle, R.T. The effects of tool surface quality in micro-injection moulding. *J. Mater. Process. Technol.* **2007**, *189*, 418–427. [CrossRef]
33. Attia, U.M.; Alcock, J.R. Evaluating and controlling process variability in micro-injection moulding. *Int. J. Adv. Manuf. Technol.* **2011**, *52*, 183–194. [CrossRef]
34. Annicchiarico, D.; Attia, U.M.; Alcock, J.R. Part mass and shrinkage in micro injection moulding: Statistical based optimisation using multiple quality criteria. *Polym. Test.* **2013**, *32*, 1079–1087. [CrossRef]
35. Lucchetta, G.; Sorgato, M.; Carmignato, S.; Savio, E. Investigating the technological limits of micro-injection molding in replicating high aspect ratio micro-structured surfaces. *CIRP Ann.-Manuf. Technol.* **2014**, *63*, 521–524. [CrossRef]
36. Masato, D.; Rathore, J.; Sorgato, M.; Carmignato, S.; Lucchetta, G. Analysis of the shrinkage of injection-molded fiber-reinforced thin-wall parts. *Mater. Des.* **2017**, *132*, 496–504. [CrossRef]
37. Baruffi, F.; Parenti, P.; Cacciatore, F.; Annoni, M.; Tosello, G. On the Application of Replica Molding Technology for the Indirect Measurement of Surface and Geometry of Micromilled Components. *Micromachines* **2017**, *8*, 195. [CrossRef]
38. Joint Committee for Guides in Metrology (JCGM). *Evaluation of Measurement Data: Guide to the Expression of Uncertainty in Measurement*; JCGM: Paris, France, 2008.

39. ISO. *15530-3: Geometrical Product Specifications (GPS)—Coordinate Measuring Machines (CMM): Technique for Determining the Uncertainty of Measurement*; ISO: Geneva, Switzerland, 2011.
40. Eladl, A.; Mostafa, R.; Islam, A.; Loaldi, D.; Soltan, H.; Hansen, H.; Tosello, G. Effect of Process Parameters on Flow Length and Flash Formation in Injection Moulding of High Aspect Ratio Polymeric Micro Features. *Micromachines* **2018**, *9*, 58. [CrossRef]

micromachines

MDPI

Article

Product and Process Fingerprint for Nanosecond Pulsed Laser Ablated Superhydrophobic Surface

Yukui Cai [1], Xichun Luo [1,*], Zhanqiang Liu [2,3], Yi Qin [1], Wenlong Chang [1] and Yazhou Sun [4]

[1]   Centre for Precision Manufacturing, DMEM, University of Strathclyde, Glasgow G1 1XJ, UK;
     yukui.cai@strath.ac.uk (Y.C.); qin.yi@strath.ac.uk (Y.Q.); wenlong.chang@strath.ac.uk (W.C.)
[2]   School of Mechanical Engineering, Shandong University, Jinan 250061, China; melius@sdu.edu.cn
[3]   Key Laboratory of High Efficiency and Clean Mechanical Manufacture of MOE/Key National,
     Demonstration Center for Experimental Mechanical Engineering Education, Jinan 250061, China
[4]   School of Mechatronics Engineering, Harbin Institute of Technology, Harbin 150001, China;
     sunyzh@hit.edu.cn
*   Correspondence: xichun.luo@strath.ac.uk; Tel.: +44-(0)-141-574-5280

Received: 7 February 2019; Accepted: 4 March 2019; Published: 7 March 2019

Abstract: Superhydrophobic surfaces have attracted extensive attention over the last few decades. It is mainly due to their capabilities of providing several interesting functions, such as self-cleaning, corrosion resistance, anti-icing and drag reduction. Nanosecond pulsed laser ablation is considered as a promising technique to fabricate superhydrophobic structures. Many pieces of research have proved that machined surface morphology has a significant effect on the hydrophobicity of a specimen. However, few quantitative investigations were conducted to identify effective process parameters and surface characterization parameters for laser-ablated microstructures which are sensitive to the hydrophobicity of the microstructured surface. This paper proposed and reveals for the first time, the concepts of process and product fingerprints for laser ablated superhydrophobic surface through experimental investigation and statistical analysis. The results of correlation analysis showed that a newly proposed dimensionless functional parameter in this paper, R_{hy}, i.e., the average ratio of Rz to Rsm is the most sensitive surface characterization parameter to the water contact angle of the specimen, which can be regarded as the product fingerprint. It also proposes another new process parameter, average laser pulse energy per unit area of the specimen (I_s), as the best process fingerprint which can be used to control the product fingerprint R_{hy}. The threshold value of R_{hy} and I_s are 0.41 and 536 J/mm^2 respectively, which help to ensure the superhydrophobicity (contact angle larger than 150°) of the specimen in the laser ablation process. Therefore, the process and product fingerprints overcome the research challenge of the so-called inverse problem in manufacturing as they can be used to determine the required process parameters and surface topography according to the specification of superhydrophobicity.

Keywords: laser ablation; superhydrophobic surface; process fingerprint; product fingerprint; surface morphology

---

1. Introduction

Superhydrophobic surfaces are defined as those having a water contact angle larger than 150° and sliding angle less than 10°. Artificial superhydrophobic surfaces, created by surface structuring or coating, have received tremendous attention in recent years. It is mainly due to their capabilities of providing several interesting functions, such as self-cleaning, corrosion resistance, anti-icing and drag reduction [1–6]. Surface chemical composition and morphology are two critical factors in determining their hydrophobicity [7–9]. The surface chemical composition affects the intrinsic contact angle, which can be measured by a liquid droplet deposited on a smooth surface. However, in artificial or natural materials, the maximum intrinsic contact angle is only approximately 120° [8,9]. For this reason, more and more structuring technologies have been developed for the fabrication of superhydrophobic surfaces, including wet chemical reaction, lithography, rolling, 3D printing, micro milling and laser ablation [2,6,10–15] etc.

Recently, laser ablation process has been demonstrated as a promising technique to fabricate superhydrophobic structures on varied materials, such as copper, aluminium, steel and glass [15–24]. Yang et al. investigated the wettability transition mechanism of laser ablated aluminium substrate, the results indicated that laser-ablated microstructures had the amplified effects on the hydrophobicity of the specimen [24]. Long et al. reported the effect of the laser pulse energy and width on the morphology of micro/nanostructures on a copper surface. They found that the morphology of the laser ablated structures is more sensitive to the laser pulse energy when nanosecond lasers with long pulse widths are used. Slightly decreasing the laser pulse energy results in the formation of no hierarchical micro- and nanostructures [17,20]. Gregorcic et al. fabricated a 316L stainless steel specimen with a pitch of 50 μm at average pulse power of 0.6 W and 97% pulse overlapping rate and achieved a static contact angle of 153° [18]. Long and Gregorcic both reported that variation of the pitch of channels resulted in completely different surface morphologies—from the highly porous surface to well-separated microchannels, which width and depth depend on laser fluence [18]. Duong Ta et al. concluded that surface roughness could be well controlled by laser power. The arithmetical mean height, Sa increased linearly when laser fluence was higher than 33 J/cm^2. The roughness was around 2 and 7 times larger than that of the untextured surface under fluences of 36 and 48 J/cm^2, respectively [23]. In addition, the effect of laser fluence and line separation on the contact angle of laser structured surfaces were investigated. Experimental results showed that the specimens possess superhydrophobicity has pitches of 50–150 μm and machined at the laser fluence of 36 J/cm^2 [23]. M. Conradi discovered that higher line density resulted in a higher contact angle. However, the average surface roughness Sa increased first then further decreased gradually with the increase of line density [19]. Thus, these researches have indicated that the laser machining parameters would significantly influence the hydrophobicity of the specimens while surface topography is a crucial factor to determine the superhydrophobicity of the specimen. However, there has been little systematic research exploring the correlation between surface topography and hydrophobicity of the specimen. Furtherly, the second challenge is to find out the most effective process parameter and surface characterization parameter for these microstructures which are sensitive to the hydrophobicity of the microstructured surface.

The identification of "product and process fingerprints" of laser ablated surface is a possible solution to solve the above issues. The concept of "product fingerprint" refers to those unique measurable characteristics (e.g., surface characterization parameters) on the laser ablated specimen that, if kept under control and within specifications, will ensure that the specimen possesses superhydrophobicity as required. The product fingerprint must be also sensitive to the variation of process parameters, hence it can be well-controlled by process parameters. For laser ablation process, since the surface characterization parameters are highly related to laser machining parameters, the "Process fingerprint" is defined as a specific process parameter to be controlled in order to maintain the manufacture of the specimen within the specified surface characterization parameters. The product and process fingerprints can be used as an objective function within an optimization tool to assist to determine the required surface topography and process parameters for the superhydrophobic surface.

The purpose of this paper is to reveal the product and process fingerprints for the laser ablation process of superhydrophobic surfaces on 316L stainless steel. A more generalized description can be achieved by linking laser machining parameters, surface characterization parameters and hydrophobicity of the specimen, which is beneficial to precise control of hydrophobicity and simultaneously enhancing its robustness. Therefore, product and process fingerprints are expected to provide a solution to the so-called inverse problem in manufacturing, which means the laser machining parameters and surface characterization parameters can be determined according to the required hydrophobicity, i.e., contact angle. Firstly, analysis of potential process and product fingerprint candidates will be carried out. Then, the most appropriate product fingerprint will be determined from values of Spearman and Kendall rank correlation coefficients according to the experimental results. Thirdly, a new process parameter will be put forward and chosen as the best process fingerprint. Lastly, the correlation between process fingerprint and functional performance, i.e., contact angle will be explored.

2. Analysis of Process and Product Fingerprints

Figure 1 illustrates the concept of process and product fingerprints in the laser ablation process for obtaining the superhydrophobic surface with an array of Gaussian holes of designed geometry. The comparison of all the potential candidates of process and product fingerprints will be discussed in detail later. Most research performed to date has focused on the correlation A; i.e., the effect of laser machining parameters on the contact angle of specimens. However, correlation A is actually composed of correlation B and C. Correlation B refers to the relationship between contact angle and product fingerprint, which is used to explain the underlying mechanism of effect of surface topography on hydrophobicity. Correlation C can describe the relationship between the process fingerprint and product fingerprint, to explore how the process parameters affect the surface topography. Thus, product fingerprint is a bridge to connect process parameters and functional performance-contact angle.

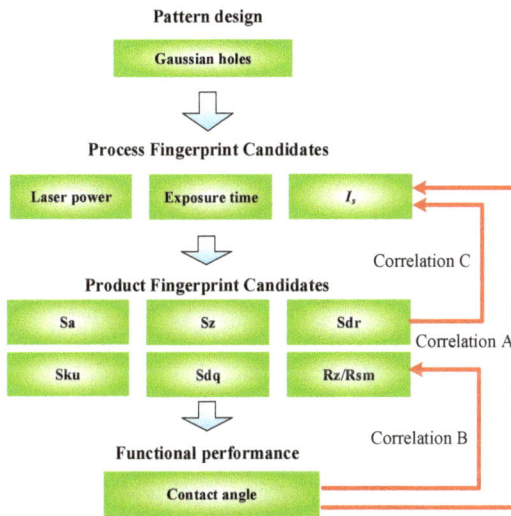

Figure 1. Concept of the process and product fingerprints in laser ablation of the superhydrophobic surface.

2.1. Analysis of Process Fingerprint Candidates: Laser Power, Exposure Time, Laser Pulse Energy Per Unit Area of Specimen

2.1.1. Laser Power (P)

In a nanosecond pulsed laser ablation process, the absorbed energy from the laser pulse melts the stainless steel and heats it to a temperature at which the atoms gain sufficient energy to enter into a gaseous state. Due to the vapour and plasma pressure, the molten materials are partially ejected from the cavity and form surface debris. At the end of a pulse, the heat quickly dissipates into the bulk of the work material and recast layer are formed. Therefore, laser power is a good candidate of process fingerprint as it determines the laser fluence which directly affects the formation of microstructures. The relationship between laser power, pulse repetition rate and peak power can be expressed as:

$$E_p = \frac{P}{f_p} \tag{1}$$

$$P_{peak} = \frac{E_p}{\Delta\tau} \tag{2}$$

where P is laser average power, f_p is pulse repetition rate, E_p is the energy of a single pulse, P_{peak} is the peak power of laser and $\Delta\tau$ is the pulse duration, respectively.

2.1.2. Exposure Time (t)

For substrate with periodic Gaussian holes generated by the laser ablation process, the exposure time t means the machining time for a single Gaussian hole, which determines the number of laser pulses that irradiated the surface. It has a significant effect on the dimension and morphology of Gaussian holes. As shown in Figure 2, the relationship between the number of irradiated pulse N and exposure time t can be expressed as:

$$N = \frac{t}{T} \tag{3}$$

where T is the pulse period.

Laser pulse energy per unit area of the specimen I_s.

I_s means the average laser pulse energy irradiated on a unit area of the specimen. This parameter depends on pulse repetition rate f_p and exposure time t. It can be expressed as:

$$I_s = \frac{t * f_p * E_p \left(\frac{L}{Pitch}\right)^2}{L^2} \tag{4}$$

According to Equation (1), $f_p * E_p = P$, hence Equation (4) can be simplified as:

$$I_s = \frac{t * P}{Pitch^2} \tag{5}$$

where pitch is the distance between adjacent Gaussian holes, and L is the length of the specimen.

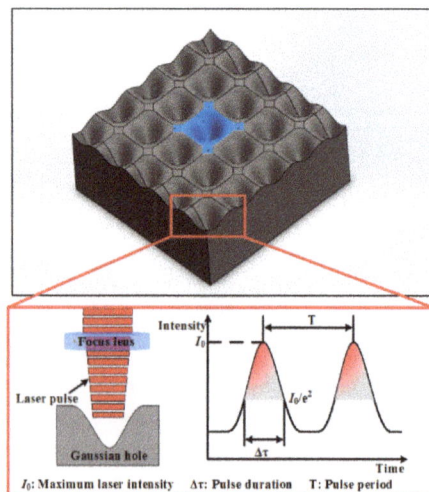

Figure 2. Schematic of periodic Gaussian holes machined by the laser ablation process.

2.2. Analysis of Product Fingerprint Candidates: Sa, Sz, Sku, Sdr, Sdq, R_{hy}

In literature, two typical models have been developed to describe the behavior of a droplet on rough surfaces, i.e., the Wenzel and Cassie-Baxter models [25,26]. According to the Wenzel model, the droplet maintains contact with the structures and penetrates the asperities, and the surface contact area is increased. In addition, the contact angle θ_w can be described as:

$$\cos \theta_w = r \cos \theta \tag{6}$$

$$r = \frac{\text{actual surface area}}{\text{planar area}} \tag{7}$$

where, r is the roughness factor, which defined as the ratio of the actual area of the solid surface to the planar area. θ is the intrinsic contact angle of the material.

Alternatively, according to the Cassie-Baxter model, the droplet is not able to penetrate the microstructure spaces. However, in order to ensure the droplet cannot connect with the bottom of the microstructures, so the sag in height of water droplet between microstructures should be smaller than the depth of microstructures. Moreover, deep microstructures will help to form stable air pockets under the water droplet. Stable air pockets underneath the water droplet help the formation of superhydrophobicity with strong resistance against transition to the Wenzel state. Hence, sufficient depth of microstructure is essential to realize Cassie–Baxter state of the water droplet. The static contact angle θ_{CB} can be expressed as:

$$\cos \theta_{CB} = -1 + f(1 + \cos \theta) \tag{8}$$

$$f = \frac{\text{actual solid and liquid contact area}}{\text{planar area}} \tag{9}$$

where f is the fraction of the solid-liquid contact area.

The above analysis proves that the contact angles obtained in both Wenzel and Cassie-Baxter states are highly related to the vertical and horizontal feature of surface topography. Six surface characterization parameters that most probably correlated with the hydrophobicity of specimens are listed in Table 1. Sa, Sz and Sku are roughness parameters to characterize the height of the surface. Sdr, Sdq, R_{hy} are hybrid parameters which determined from both height and horizontal parameters of the surface. For a rough surface, Sdr means the additional surface area contributed by the texture as compared to the planar definition area. Therefore, 1+Sdr has the same meaning as the roughness factor r in the Wenzel state.

Table 1. Product fingerprint candidates.

Name	Symbol	Meaning
Arithmetical mean height	Sa	The difference in height of each point compared to the arithmetical mean of the surface.
Maximum height	Sz	The sum of the largest peak height value and the largest pit depth value within the defined area.
Kurtosis	Sku	A measure of the sharpness of the roughness profile. Sku < 3: Height distribution is skewed above the mean plane. Sku = 3: Height distribution is normal. (Sharp portions and indented portions co-exist.) Sku > 3: Height distribution is spiked.
Developed interfacial area ratio	Sdr	The percentage of the definition area's additional surface area contributed by the texture as compared to the planar definition area.
Root mean square gradient	Sdq	Root mean square of slopes at all points in the definition area. When a surface has any slope, its Sdq value becomes larger.
Average ratio of Rz to Rsm	R_{hy}	Average ratio of the maximum height of profile (Rz) and mean width of the profile elements (RSm)

Theoretical analysis proved that microstructures should have a high aspect ratio to provide a larger surface area and a smaller separation distance which will help to improve the stabilization of the solid–liquid–air composite interface [27]. However, present functional parameters cannot reflect the aspect ratio of surface asperities. Hence, R_{hy} is proposed for the first time as a dimensionless functional parameter in this research and defined as the average ratio of Rz to Rsm. The subscript "hy" is the short abbreviation of hydrophobicity. The R_{hy} is calculated from the average value of 60 lines that evenly distributed on the structured surface horizontally and vertically. A surface with large R_{hy} can be obtained from a large Rz or smaller Rsm, which means the features of the surface should have a large depth or smaller separation distance (i.e., high density) in the horizontal direction.

3. Experimental Details

Laser machining experiments were carried out on AISI 316L stainless steel by varying the process parameters in order to identify the best product and process fingerprints. All the experiments were carried out on a hybrid ultra-precision machine, as shown in Figure 3. It is equipped with a nanosecond pulsed fiber laser which has a central emission wavelength of 1064 nm. The laser source has a nominal average output power of 20 W and its maximum pulse repetition rate is 200 kHz. For a pulse repetition rate of 20 kHz, the average pulse duration is 100 ns and pulse energy is 1 mJ. The laser machining parameters are listed in Tables 2 and 3. After the laser ablation process, the specimens were cleaned ultrasonically with deionized water, acetone and ethanol successively. Then the prepared specimens were silanized in a vacuum oven using silane reagent (1H, 1H, 2H, 2H-Perfluorooctyltriethoxysilane, 97%, Alfa Aesar Ltd., Ward Hill, MA, USA), at 100 °C for 12 h to reduce their surface free energies.

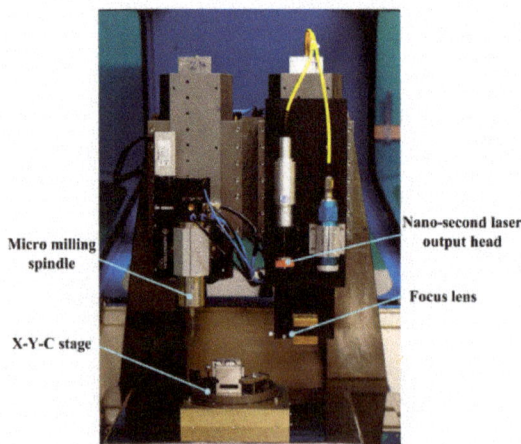

Figure 3. Experimental setup for laser machining trials.

Table 2. The laser machining parameters with varied laser power and pitch.

Pitch (μm)	Laser Power (W)	Pulse Repetition Rate	Feed Rate (mm/min)	Exposure Time (s)	Pattern Types
90	4,6,10,14,20	100K	200	0.4	Gaussian holes
110	4,6,10,14,20	100K	200	0.4	Gaussian holes
130	4,6,10,14,20	100K	200	0.4	Gaussian holes
150	4,6,10,14,20	100K	200	0.4	Gaussian holes

Table 3. The laser machining parameters with varied exposure time and pitch.

Pitch (μm)	Laser Power (W)	Pulse Repetition Rate	Feed Rate (mm/min)	Exposure Time (s)	Pattern Types
70	20	100K	200	0.2,0.4,0.6,1	Gaussian holes
90	20	100K	200	0.2,0.4,0.6,1	Gaussian holes
110	20	100K	200	0.2,0.4,0.6,1	Gaussian holes
130	20	100K	200	0.2,0.4,0.6,1	Gaussian holes
150	20	100K	200	0.2,0.4,0.6,1	Gaussian holes

The surface topography and varied surface characterization parameters of the laser structured surface were measured by a 3D laser scanning confocal microscope (VK-250, Keyence Corporation, Osaka, Japan). The static contact angle on surfaces was measured by a drop shape analyzer (Kruss Ltd., Hamburg, Germany). The selected water droplet volume was 5 μL. For each specimen, the contact angle of the water droplet was measured three times and the average value was adopted.

4. Results and Discussion

4.1. Analysis of Product Fingerprint: Sa, Sz, Sku, Sdr, Sdq, R_{hy}

The investigation of experimental results was carried out to identify the product fingerprint from six candidates related to surface topography. The product fingerprint is the indicator that has the highest level of correlation to contact angle. In this research, the Spearman rank correlation coefficient and Kendall rank correlation coefficient were employed to determine the product fingerprint. Spearman rank correlation coefficient evaluates how strong the correlation between two variables can be defined by a monotonic function. It measures the strength and direction of the monotonic association between two variables, a perfect Spearman correlation of +1 or −1 occurs when each variable is a perfect monotone function of the other [28]. A positive Spearman correlation coefficient corresponds to an increasing monotonic trend between two variables, while a negative value means a decreasing monotonic trend. In addition, Spearman rank correlation coefficient is appropriate for data that is not normally distributed. It can be used to identify a non-linear correlation between two variables. Kendall rank correlation coefficient is a statistic used to measure the ordinal association between two variables [29]. However, unlike the Spearman coefficient, Kendall rank correlation coefficient only considers directional agreement while does not consider the difference between ranks. Therefore, this coefficient is more appropriate for discrete data. This coefficient returns a value of −1 to 1, where 0 is no correlation, 1 is a perfect positive correlation, and −1 is a perfect negative correlation. In most cases, the interpretations of Spearman and Kendall rank correlation coefficients are very similar and thus invariably lead to the same inferences. The above two coefficients were combined to determine the product fingerprint that has the maximum absolute value. The strength of the correlation between the variables can be evaluated by the absolute value of coefficients, as shown in Table 4.

Table 4. Interpretation of the strength of the correlation coefficient.

Value of Coefficient	Correlation Type
1	Perfect correlation
0.81–0.99	Strong correlation
0.71–0.80	Good correlation
0.51–0.70	Weak correlation
0.01–0.50	Poor correlation
0	No correlation

Figure 4 shows scatter plots between the contact angle and the six candidates of product fingerprint. With the increase of Sa, Sz, Sdr, Sdq and R_{hy}, the contact angle shows an increasing trend. It should be noted that a good linear relationship appears between Sz and contact angle, which is similar to the authors' previous study [15]. However, it can be observed that there is no apparent correlation between Sku and contact angle (Figure 4c). As shown in Figure 4d, increasing Sdr from 0.02 to 4.1 leads to contact angle increase rapidly from 89.5° to 159°, but it has a minor impact on the contact angle when Sdr was further increased from 4.1 to 9.8. As Figure 4f indicates, the contact angle increases gradually from 89.5° to 164° with the value of R_{hy} increasing from 0.06 to 0.94.

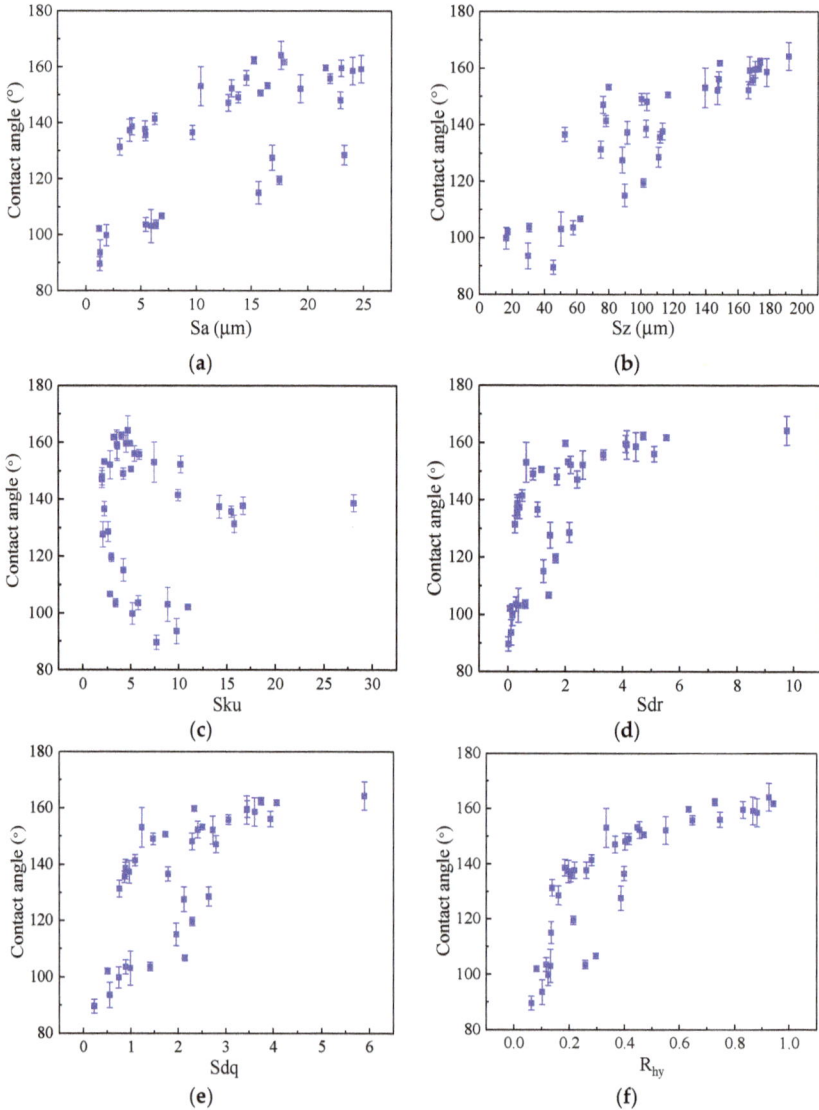

Figure 4. Influence of the product fingerprint candidates on the contact angle for Gaussian hole pattern: (**a**) Sa; (**b**) Sz; (**c**) Sku; (**d**) Sdr; (**e**) Sdq; (**f**) R_{hy}.

Figure 5 shows the variation of Spearman and Kendall rank correlation coefficient between contact angle and candidates of product fingerprint. According to the criterion in Table 4, Sz and R_{hy} both have larger Spearman rank correlation coefficients with the contact angle, which are 0.89 and 0.92

respectively. The Kendall rank correlation coefficient among Sz, R_{hy} and contact angle are 0.74 and 0.76. Thus, the results of Figure 5 suggest that R_{hy} should be determined as the best product fingerprint as it has the maximum Spearman and Kendall rank correlation coefficients.

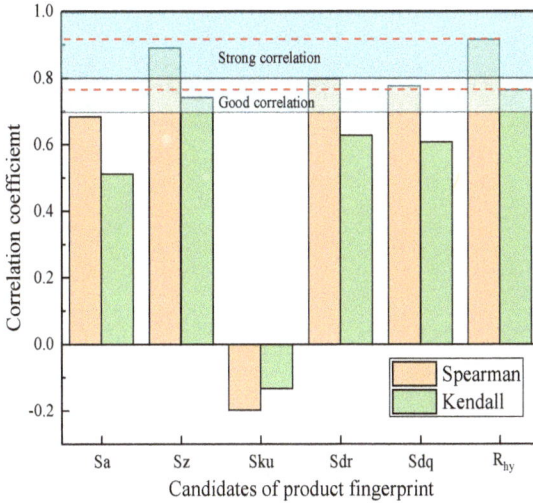

Figure 5. Spearman and Kendall rank correlation coefficient between the contact angle and six candidates of product fingerprint.

According to the results in Figure 4f, an empirical equation was deduced to correlate the experimental R_{hy} and contact angle. The equation is expressed as:

$$\theta_A = a - b*e^{c*R_{hy}} \tag{10}$$

where, θ_A is contact angle; a, b and c are constant values, equal to 164, 105 and −4.9 respectively.

As shown in Figure 6a, the regression curve has good precision to simulate the experimental data. We found that coefficient "a" means the maximum contact angle (164° in this research), the value of "b" is equal to the initial contact angle (105°) of 316L stainless steel after chemical modification. Thus, the contact angle of the specimen is highly related to its maximum contact angle, initial contact angle on a smooth surface and hydrophobicity functional parameter R_{hy}. According to Equation (10), the value of R_{hy} is 0.41 when $\theta_A = 150$. Thus, 0.41 can be regarded as the threshold value of R_{hy} that ensure water contact angle of the specimen higher than 150°.

The dimensionless ratio R_{hy} is the most sensitive candidate parameter for contact angle of the specimen, which can therefore, be regarded as product fingerprint. In literature, many studies proved that a high density of microstructures and smaller period of microstructure will help decrease solid-liquid contact area and increase its hydrophobicity [22,30]. With the increase of R_{hy} from 0.138 to 0.943 (Figure 6b), Rsm decreased from 137.0 μm to 81.8 μm. Therefore, the density of peaks shows a significant increasing trend. Moreover, the depth of microstructures shows an increasing trend, due to average Rz increased from 18.9 μm to 77.2 μm. Therefore, it can be concluded that the superhydrophobicity will benefit from the increase of R_{hy}.

(a)

(b)

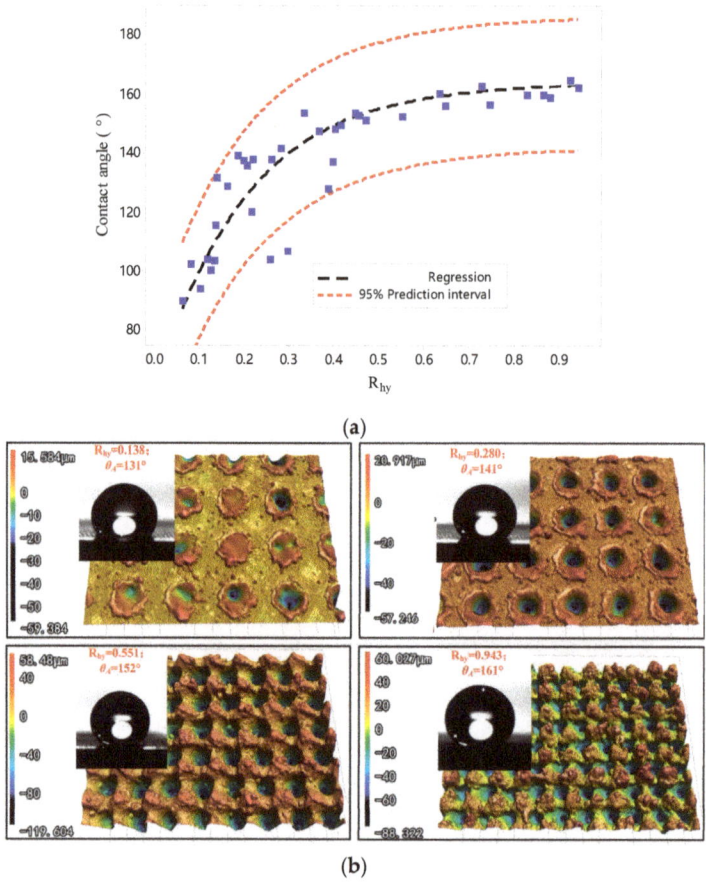

Figure 6. (**a**) Fitted line by exponential function between R_{hy} and contact angle; (**b**) Surface morphology and shape of water drops on specimens with a different value of R_{hy}.

4.2. Analysis of Process Fingerprints: P, t and I_s

The above section proves that R_{hy} is the most appropriate product fingerprint to the laser ablated superhydrophobic structures on 316L stainless steel. In this section, further analysis of the experimental results will be performed to identify the best process fingerprint from the candidates P, t and I_s, i.e., the process fingerprint which has the strongest correlation with R_{hy}. The control of process fingerprints helps to choose appropriate process parameter to obtain a surface with R_{hy} greater than the threshold value ($R_{hy} > 0.41$). The correlation among laser power, pitch of Gaussian hole and R_{hy} is shown in Figure 7. It shows that higher laser power and smaller pitch lead to a higher value of R_{hy}. Laser power and pitch of structures have combined effects on the value of R_{hy}.

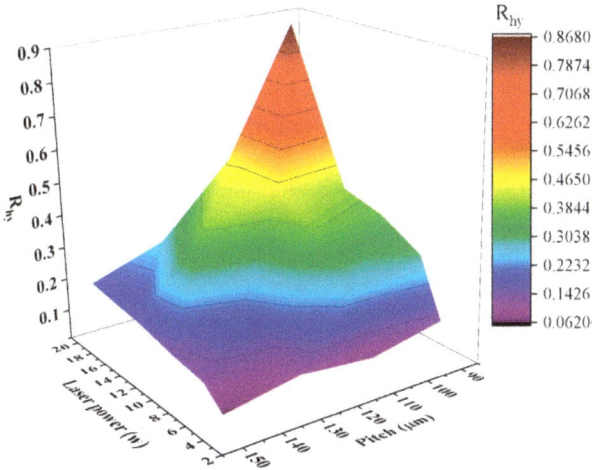

Figure 7. 3D colormap of the product fingerprint (R_{hy}) as a function of laser power and pitch of Gaussian hole.

The effect of exposure time t and pitch of Gaussian holes on the value of R_{hy} is presented in Figure 8. There is no significant linear correlation between exposure time and R_{hy}, but it does not mean exposure time has no effect on R_{hy}. As a whole, it can be found that the value of R_{hy} shows a significant increasing trend with the reduction of pitch from 150 μm to 70 μm.

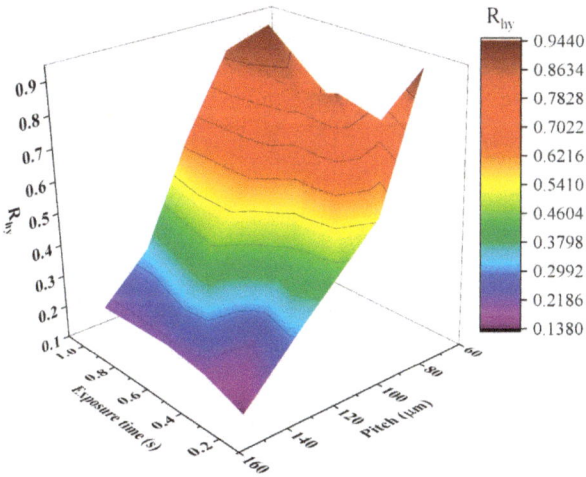

Figure 8. 3D colormap of the product fingerprint (R_{hy}) as a function of exposure time and pitch of Gaussian holes.

The above analysis shows that laser power, pitch and exposure time have a collective influence on R_{hy}. Focusing one of them and ignoring the other two would lead to the determined correlation only effective in certain partial conditions. For instance, the R_{hy} will increase with laser power, but only valid at a precondition of constant pitch and exposure time. Therefore, a comprehensive factor I_s was designed to represent the combined influence of laser power, pitch and exposure time. I_s means the energy intensity that irradiated on the unit area of the specimen and can be calculated by the Equation (5). I_s is proportional to the laser power P and the exposure time t, but inversely proportional to the square of the pitch of the microstructures. Figure 9 reveals that the increasing I_s leads R_{hy} increase rapidly at first, and then level off to become asymptotic to the upper limit. The presence of upper limit means the further increased laser power, exposure time and smaller pitch cannot lead to a further increase of R_{hy}. The correlation between I_s and R_{hy} can be expressed as Equation (11). According to the calculation result, I_s should be greater than 536 J/mm^2 to ensure R_{hy} greater than 0.41, hence the contact angle of the specimen will be larger than 150°.

$$R_{hy} = 0.895 - 0.898 * 0.9985^{I_s} \tag{11}$$

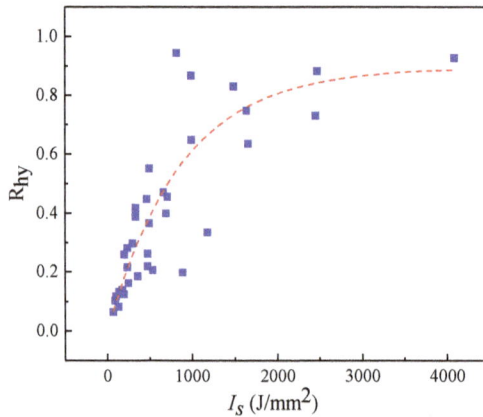

Figure 9. Scatter plots and fitted curve of R_{hy} and I_s.

Therefore, the increased I_s leads to rapidly increase of R_{hy}, the correlation between R_{hy} can be described by the exponential function. I_s is the most sensitive parameters among the investigated three process fingerprint candidates, so it is the best process fingerprint that can be used to control surface morphology, especially the product fingerprint R_{hy}.

4.3. Correlation Between Laser Machining Parameters and Contact Angle

As shown in Figure 10, 3D colormaps are used to display the relationship between laser power, exposure time, pitch of structures and contact angle. To sum up, the greater contact angle benefit from larger laser power and smaller pitch of microstructures except for some outliers.

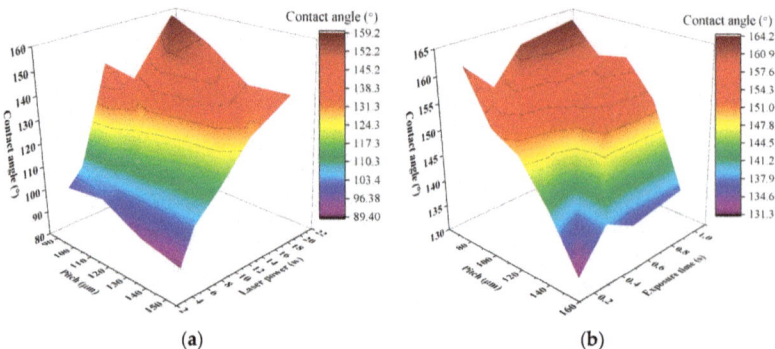

(a)   (b)

Figure 10. (a) 3D colormap of the contact angle as a function of laser power and pitch of microstructures; (b) 3D colormap of the contact angle as a function of exposure time and pitch of microstructures.

Figure 11a shows the scatter diagram and fitted curve between contact angle and I_s. The increasing I_s results in a rapid increase of contact angle at first, and then level off to become asymptotic to the upper limit when I_s greater than 1000 J/mm². The empirical correlation between contact angle and I_s can be expressed by Equation (12). When the value of R_{hy} equals to the threshold value of 0.41, the corresponding I_s is 516.6 J/mm², which is very close to the value of 536 J/mm² obtain from Equation (11). Therefore, I_s should be larger than 536 J/mm² in the laser ablation process, which help ensure the contact angle larger than 150°.

$$\theta_A = a - b * e^{d*I_s} \tag{12}$$

where, θ_A is contact angle, a = 164, b = 105, d = −0.0039. Coefficients of a and b have the same meaning with Equation (10).

The surface morphology and shape of water drops on specimens with a different value of I_s are shown in Figure 11b. With the increase of I_s, the depth and density of structures show a significant increasing trend. Thus, the surface topography and contact angle can be well controlled by choosing the appropriate process parameter I_s.

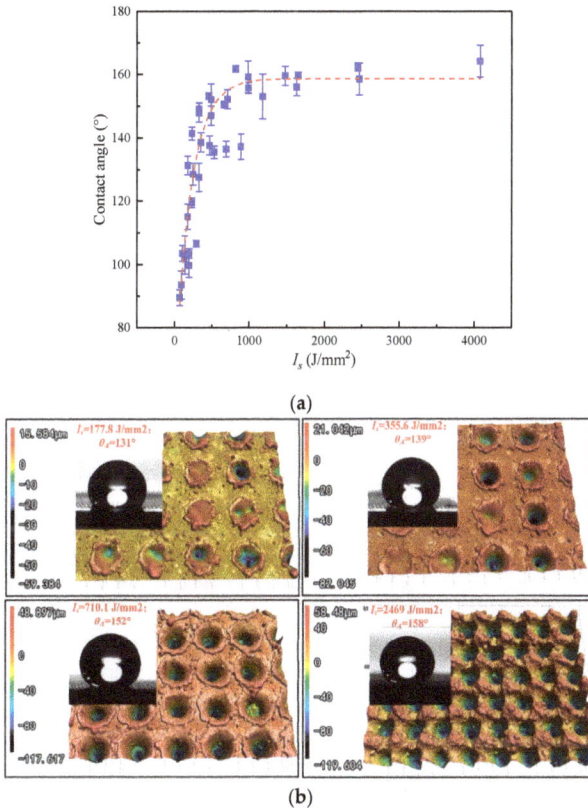

(a)

(b)

Figure 11. (a) Scatter plot and fitted curve between contact angle and I_s; (b) Surface morphology and shape of water drops on specimens with a different value of I_s.

5. Conclusions

In this study, the concepts of product and process fingerprint are put forward for the first time to reveal the correlations among process parameters, surface topography and functional performance,

Micromachines **2019**, *10*, 177

i.e., the contact angle of laser ablated superhydrophobic surface on 316L stainless steel. The most appropriate product fingerprint was determined by the indicators of Spearman and Kendall rank correlation coefficients. Then, the candidate that was most sensitive to product fingerprint was determined as the best process fingerprint. Lastly, the correlation between process fingerprint and functional performance was developed. The conclusions can be drawn as follows:

1. The dimensionless surface functional characterization parameter R_{hy}, i.e., the average ratio of Rz to Rsm is the most sensitive parameter to contact angle of the specimen, which can be regarded as the product fingerprint.

2. Laser pulse energy per unit area on specimen (I_s) represents the combining effect of laser power, exposure time and pitch of structure on surface topography. It is the best process fingerprint that can be used to control the product fingerprint R_{hy}.

3. The increasing I_s leads to the value of R_{hy} increase rapidly at first, and then level off to become asymptotic to the upper limit. A similar trend also can be found between I_s - contact angle and R_{hy} - contact angle. The threshold value of R_{hy} and I_s are 0.41 and 536 J/mm^2 respectively, which help to ensure the superhydrophobicity (contact angle larger than 150°) of the specimen in the laser ablation process.

Author Contributions: Conceptualization, Y.C., X.L. and Y.Q.; methodology, Y.C., Y.S.; writing—original draft preparation, Y.C.; writing—review and editing, X.L., Z.L., W.C. and Y.S.; supervision, X.L., Z.L. and Y.Q.; project administration, X.L. and Y.Q.

Acknowledgments: This research was undertaken in the context of MICROMAN project ("Process Fingerprint for Zero-defect Net-shape MICROMANufacturing", http://www.microman.mek.dtu.dk/). MICROMAN is a European Training Network supported by Horizon 2020, the EU Framework Program for Research and Innovation (Project ID: 674801). The authors would also gratefully acknowledge the financial support from the EPSRC (EP/K018345/1) and the International Cooperation Program of China (grant number 2015DFA70630) for this research.

Conflicts of Interest: The authors declare no conflict of interest.

Data Statement: All data underpinning this publication are openly available from the University of Strathclyde Knowledge Base at https://doi.org/10.15129/06ca19bc-e028-42b7-ad12-0279d57cc940.

References

1. Tong, W.; Xiong, D.; Wang, N.; Yan, C.; Tian, T. Green and timesaving fabrication of a superhydrophobic surface and its application to anti-icing, self-cleaning and oil-water separation. *Surf. Coat. Technol.* **2018**, *352*, 609–618. [CrossRef]

2. Li, S.; Page, K.; Sathasivam, S.; Heale, F.; He, G.; Lu, Y.; Lai, Y.; Chen, G.; Carmalt, C.J.; Parkin, I.P. Efficiently texturing hierarchical superhydrophobic fluoride-free translucent films by AACVD with excellent durability and self-cleaning ability. *J. Mater. Chem. A* **2018**, *6*, 17633–17641. [CrossRef]

3. Anitha, C.; Syed Azim, S.; Mayavan, S. Influence of particle size in fluorine free corrosion resistance superhydrophobic coating—Optimization and stabilization of interface by multiscale roughness. *J. Alloys Compd.* **2018**, *765*, 677–684. [CrossRef]

4. Rastegari, A.; Akhavan, R. The common mechanism of turbulent skin-friction drag reduction with superhydrophobic longitudinal microgrooves and riblets. *J. Fluid Mech.* **2018**, *838*, 68–104. [CrossRef]

5. Xie, X.; Weng, Q.; Luo, Z.; Long, J.; Wei, X. Thermal performance of the flat micro-heat pipe with the wettability gradient surface by laser fabrication. *Int. J. Heat Mass Transf.* **2018**, *125*, 658–669. [CrossRef]

6. Trdan, U.; Hočevar, M.; Gregorčič, P. Transition from superhydrophilic to superhydrophobic state of laser textured stainless steel surface and its effect on corrosion resistance. *Corros. Sci.* **2017**, *123*, 21–26. [CrossRef]

7. Karlsson, M.; Forsberg, P.; Nikolajeff, F. From hydrophilic to superhydrophobic: Fabrication of micrometer-sized nail-head-shaped pillars in diamond. *Langmuir* **2010**, *26*, 889–893. [CrossRef] [PubMed]

8. Nishino, T.; Meguro, M.; Nakamae, K.; Matsushita, M.; Ueda, Y. The Lowest Surface Free Energy Based on −CF$_3$ Alignment. *Langmuir* **1999**, *15*, 4321–4323. [CrossRef]

9. Bell, M.S.; Shahraz, A.; Fichthorn, K.A.; Borhan, A. Effects of Hierarchical Surface Roughness on Droplet Contact Angle. *Langmuir* **2015**, *31*, 6752–6762. [CrossRef] [PubMed]

10. Ta, D.V.; Dunn, A.; Wasley, T.J.; Kay, R.W.; Stringer, J.; Smith, P.J.; Connaughton, C.; Shephard, J.D. Nanosecond laser textured superhydrophobic metallic surfaces and their chemical sensing applications. *Appl. Surf. Sci.* **2015**, *357*, 248–254. [CrossRef]

11. Ngo, C.V.; Chun, D.M. Fast wettability transition from hydrophilic to superhydrophobic laser-textured stainless steel surfaces under low-temperature annealing. *Appl. Surf. Sci.* **2017**, *409*, 232–240. [CrossRef]

12. Ngo, C.V.; Chun, D.M. Effect of Heat Treatment Temperature on the Wettability Transition from Hydrophilic to Superhydrophobic on Laser-Ablated Metallic Surfaces. *Adv. Eng. Mater.* **2018**, *20*, 1–11. [CrossRef]

13. Kim, J.H.; Mirzaei, A.; Kim, H.W.; Kim, S.S. Facile fabrication of superhydrophobic surfaces from austenitic stainless steel (AISI 304) by chemical etching. *Appl. Surf. Sci.* **2018**, *439*, 598–604. [CrossRef]

14. Brinksmeier, E.; Reese, S.; Klink, A.; Langenhorst, L.; Lübben, T.; Meinke, M.; Meyer, D.; Riemer, O.; Sölter, J. Underlying Mechanisms for Developing Process Signatures in Manufacturing. *Nanomanuf. Metrol.* **2018**, *1*, 193–208. [CrossRef]

15. Cai, Y.; Chang, W.; Luo, X.; Sousa, A.M.L.; Lau, K.H.A.; Qin, Y. Superhydrophobic structures on 316L stainless steel surfaces machined by nanosecond pulsed laser. *Precis. Eng.* **2018**, *52*, 266–275. [CrossRef]

16. Yang, C.; Tartaglino, U.; Persson, B.N.J. Influence of surface roughness on superhydrophobicity. *Phys. Rev. Lett.* **2006**, *97*, 1–4. [CrossRef] [PubMed]

17. Long, J.; Cao, Z.; Lin, C.; Zhou, C.; He, Z.; Xie, X. Formation mechanism of hierarchical Micro- and nanostructures on copper induced by low-cost nanosecond lasers. *Appl. Surf. Sci.* **2019**, *464*, 412–421. [CrossRef]

18. Gregorčič, P.; Šetina-Batič, B.; Hočevar, M. Controlling the stainless steel surface wettability by nanosecond direct laser texturing at high fluences. *Appl. Phys. A Mater. Sci. Process.* **2017**, *123*, 1–8. [CrossRef]

19. Conradi, M.; Drnovšek, A.; Gregorčič, P. Wettability and friction control of a stainless steel surface by combining nanosecond laser texturing and adsorption of superhydrophobic nanosilica particles. *Sci. Rep.* **2018**, *8*, 2–10. [CrossRef] [PubMed]

20. Long, J.; He, Z.; Zhou, C.; Xie, X.; Cao, Z.; Zhou, P.; Zhu, Y.; Hong, W.; Zhou, Z. Hierarchical micro- and nanostructures induced by nanosecond laser on copper for superhydrophobicity, ultralow water adhesion and frost resistance. *Mater. Des.* **2018**, *155*, 185–193. [CrossRef]

21. Ngo, C.V.; Chun, D.M. Control of laser-ablated aluminum surface wettability to superhydrophobic or superhydrophilic through simple heat treatment or water boiling post-processing. *Appl. Surf. Sci.* **2018**, *435*, 974–982. [CrossRef]

22. Park, K.C.; Choi, H.J.; Chang, C.H.; Cohen, R.E.; McKinley, G.H.; Barbastathis, G. Nanotextured silica surfaces with robust superhydrophobicity and omnidirectional broadband supertransmissivity. *ACS Nano* **2012**, *6*, 3789–3799. [CrossRef] [PubMed]

23. Ta, V.D.; Dunn, A.; Wasley, T.J.; Li, J.; Kay, R.W.; Stringer, J.; Smith, P.J.; Esenturk, E.; Connaughton, C.; Shephard, J.D. Laser textured superhydrophobic surfaces and their applications for homogeneous spot deposition. *Appl. Surf. Sci.* **2016**, *365*, 153–159. [CrossRef]

24. Yang, Z.; Liu, X.; Tian, Y. Insights into the wettability transition of nanosecond laser ablated surface under ambient air exposure. *J. Colloid Interface Sci.* **2019**, *533*, 268–277. [CrossRef] [PubMed]

25. Cassie, B.D. Wettability of porous surfaces. *Trans. Faraday Soc.* **1944**, *40*, 546–551. [CrossRef]

26. Wenzel, R.N. Resistance of Solid Surfaces To Wetting By Water. *Ind. Eng. Chem.* **1936**, *28*, 988–994. [CrossRef]

27. Nosonovsky, M.; Bhushan, B. Roughness optimization for biomimetic superhydrophobic surfaces. *Microsyst. Technol.* **2005**, *11*, 535–549. [CrossRef]

28. Spearman's Rank Correlation Coefficient. Available online: https://en.wikipedia.org/wiki/Spearman% 27s_rank_correlation_coefficient (accessed on 21 January 2019).

29. Kendall Rank Correlation Coefficient. Available online: https://en.wikipedia.org/wiki/Kendall_rank_ correlation_coefficient (accessed on 21 January 2019).

30. Celia, E.; Darmanin, T.; Taffin de Givenchy, E.; Amigoni, S.; Guittard, F. Recent advances in designing superhydrophobic surfaces. *J. Colloid Interface Sci.* **2013**, *402*, 1–18. [CrossRef] [PubMed]

micromachines

MDPI

Article

Process Understanding of Plasma Electrolytic Polishing through Multiphysics Simulation and Inline Metrology

Igor Danilov [1],*, Matthias Hackert-Oschätzchen [1], Mike Zinecker [1], Gunnar Meichsner [2], Jan Edelmann [2] and Andreas Schubert [1,2]

[1] Professorship Micromanufacturing Technology, Faculty of Mechanical Engineering, Chemnitz University of Technology, 09107 Chemnitz, Germany; matthias.hackert@mb.tu-chemnitz.de (M.H.-O.); mike.zinecker@mb.tu-chemnitz.de (M.Z.); andreas.schubert@mb.tu-chemnitz.de (A.S.)
[2] Fraunhofer Institute for Machine Tools and Forming Technology, 09126 Chemnitz, Germany; Gunnar.Meichsner@iwu.fraunhofer.de (G.M.); jan.edelmann@iwu.fraunhofer.de (J.E.)
* Correspondence: igor.danilov@mb.tu-chemnitz.de; Tel.: +49(0)-371-5313-3665

Received: 27 February 2019; Accepted: 22 March 2019; Published: 26 March 2019

Abstract: Currently, the demand for surface treatment methods like plasma electrolytic polishing (PeP)—a special case of electrochemical machining—is increasing. This paper provides a literature review on the fundamental mechanisms of the plasma electrolytic polishing process and discusses simulated and experimental results. The simulation shows and describes a modelling approach of the polishing effect during the PeP process. Based on the simulation results, it can be assumed that PeP can be simulated as an electrochemical machining process and that the simulation can be used for roughness and processing time predictions. The simulation results exhibit correlations with the experimentally-achieved approximation for roughness decrease. The experimental part demonstrates the results of the PeP processing for different times. The results for different types of roughness show that roughness decreases exponentially. Additionally, a current efficiency calculation was made. Based on the experimental results, it can be assumed that PeP is a special electrochemical machining process with low passivation.

Keywords: plasma-electrolytic polishing; PeP; surface modification; finishing; electro chemical machining; ECM

1. Introduction and Literature Review

1.1. Introduction

Plasma electrolytic polishing (PeP) is a special case of electrochemical machining [1] which requires high voltage and uses an environmentally-friendly aqueous solutions of salts. A summary of the PeP process is shown in Figure 1, where the workpiece is an anode connected to a DC energy source.

PeP is a technology that is used as a finishing surface treatment for precision metallic parts because of low achievable roughness ($Ra < 0.02$ µm) and low removal rates [2,3]. Nowadays state-of-the-art method for polishing is mechanical polishing. In comparison to mechanical polishing, in PeP, the whole workpiece surface can be polished in a few minutes, as can complex free forms [2,4,5].

In the extant literature, there is a lot of information on solutions used for polishing different metal alloys (e.g., steels, aluminium, titan and others) [5–12] and process parameters (e.g., temperature, electrolyte concentration, voltage, etc.). For example, 3–6% ammonium sulphate solutions are widely used for polishing steel workpieces; the applied voltage is between 200 V and 350 V; a combination of ammonium fluoride and potassium fluoride is common for polishing titanium parts, etc. For metals, the common temperature range is 70–90 °C [5,10,12].

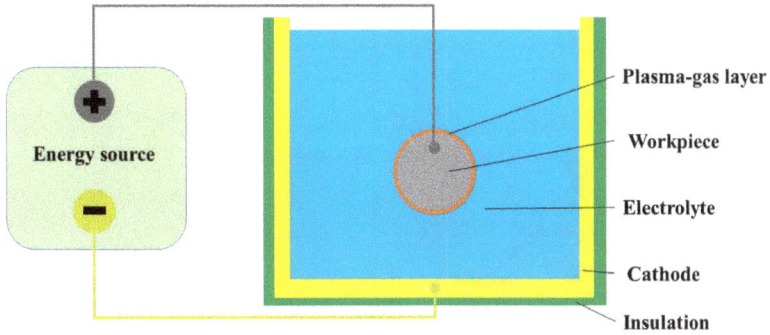

Figure 1. Principle scheme of PeP.

Figure 2 shows a typical current vs. voltage plot. The first section, "ab", is a conventional electrolysis process that can be described by classical electrochemistry [5]. Section "bc" is a transient or switching mode, when a plasma-gas layer periodically occurs on the anode [5]. The section "cd" is an electrolytic plasma mode [6] when the plasma-gas layer is stable and polishing is possible. At section "de", the plasma-gas layer becomes unstable. At sections "bc", "cd" and "de", an increase in voltage leads to a decrease in current because of the increase in the thickness of the plasma-gas layer [13].

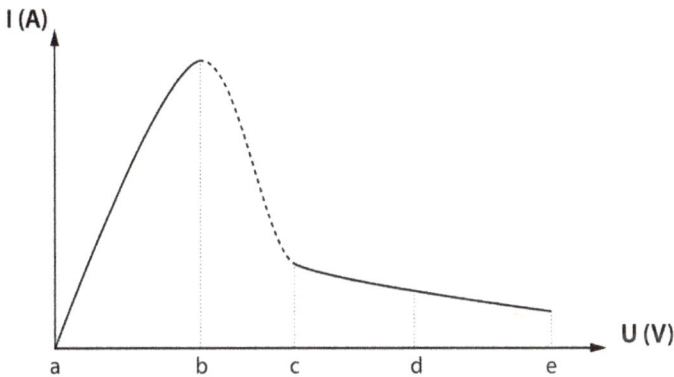

Figure 2. Schematic current–voltage characteristic according to [5].

However, the minimum achievable roughness is limited by the initial roughness of the workpiece. Mukaeva [14] has shown that the surface roughness parameter *Ra* can be approximated by the following parametric dependence on time *t*:

$$Ra = A \cdot \exp(-t/\tau) + C \tag{1}$$

where:

A – max decrease in roughness;
τ – time constant;
C – min achievable roughness;
t – processing time;

All of these variables can be obtained with experimental data.

The parameters used by Mukaeva for the experiments are provided in Table 1. The results are shown in Figure 3. The roughness parameter *Ra* is plotted as a function of time for three different voltage values: 250 V, 300 V, 350 V. From Figure 3, it can be seen that with higher temperatures, the achievable roughness is greater; this can be explained as follows. Higher temperatures increase the thickness of the plasma-gas layer, leading to a decrease in current [3,13].

Table 1. Experimental parameters [14].

Parameter	Value
Voltage	250 V, 300 V, 350 V
Electrolyte	Ammonium sulfate
Electrolyte concentration (wt %)	5%
Electrolyte temperature	70 °C, 80 °C, 90 °C
Samples material	Stainless steel (1.4021)
Initial roughness (*Ra*)	0.63 μm, 0.32 μm

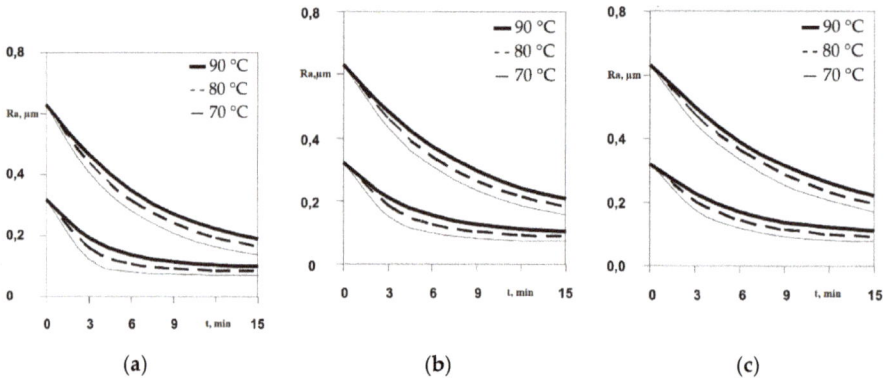

Figure 3. Roughness as function of time, (**a**) 250 V, (**b**) 300V, (**c**) 350 V [14].

Some authors have demonstrated that surface roughness depends on the immersion depth of the sample: the deeper the sample was immersed, the lower the achieved surface roughness [3,15,16]. This effect can be explained by the fact that the plasma-gas layer is not uniform along the workpiece height. The thickness of the plasma-gas layer closer to the electrolyte-air interface is higher [15,16]. This happens because the gas bubbles move towards the electrolyte-air interface.

Some information about PeP and workpiece geometry was provided by Kulikov et al. [5]. Their experiments show that products with small cavities, when the depth is less than the diameter of the bore, can be polished. Inside the deep bores, the walls were not polished when the depth exceeds the bore diameter. Nevertheless, a large number of deep bores, cavities and cracks leads to an increase in the current, and also can lead to interruptions in the polishing process.

A possible solution for the problem with polishing difficult geometries, small bores and cavities is the usage of an electrolyte jet [4,5,17–22]. Ablyaz et al. [17] and Novoselov et al. [22] demonstrated the possibility of polishing an object with a complex geometry with a free electrolyte jet. In the works of Alekseev et al. [20] and Cornelsen et al. [4], the polishing of the inner surfaces of pipes was shown.

Regarding the process, there are still two big gaps in our understanding of the PeP process: the detailed description of conductivity of the plasma-gas layer and the polishing mechanism of PeP. The information provided below is an overview of existing theories on the process.

1.2. Electrical Conductivity of the Plasma-Gas Layer

There are several theories of the nature of the electrical conductivity of the plasma-gas layer. One of the first to have observed the plasma-gas layer within the anode process at the high voltages was Kellogg [23]. In his work, Kellogg hypothesised that the plasma-gas layer is "primarily a water-vapour film". He observed some sparks inside plasma-gas layer during the process and assumed that conduction could be explained by the ionisation of the gas within the plasma-gas layer, caused by a high electric field.

A lot of papers have examined this of or a similar mechanism of conduction [10,14,16,24–26]. A high electric field causes ionisation on the vapour film that surrounds workpiece, and the formation of plasma. So, the plasma-gas layer consists of ions from the electrolyte and ions from the workpiece surface. In some papers, ions from the electrolyte and metal ions play the main role in the conduction of the plasma-gas layer [10,16,25].

Others assume that ions have an important, but not a principal, role in conduction. According to them, the values of current densities, which were observed in experiments, cannot be provided only by ions. In this work, ions contribute to the release of electrons that provide the necessary conductivity [14,24,26].

Vaňa et al. [27] mentioned that the plasma-gas layer is mainly water steam. The water steam layer is ionized due to high voltage. This leads to the formation of electric current which flows in the form of a glow discharge.

Another theory consists of the appearance of so-called electrolyte bridges [3,28,29]. In this theory, the thickness of the plasma-gas layer is not constant. The high electric field and the fact that the plasma-gas layer is not homogeneous lead to the appearance of zones with small thicknesses of the plasma-gas layer (Figure 4a). Then, under the ponderomotive force, the electrolyte continues to shift toward the workpiece surface, so the thickness of the plasma-gas layer gets smaller (Figure 4b). Getting closer to the workpiece surface, the electric field is higher, so the ponderomotive force is also higher. When the distance between the electrolyte and the surface of workpiece is around a couple of microns, additional, small "electrolyte bridges" appear (Figure 4c). When all this small bridges touch the surface, they create a current impulse and quickly boil because of Joule heating (Figure 4d,e). Sinkevitch et al. [3,28] compare this with explosive boiling. In this theory, the whole current in the system is a superposition of a large number of impulses from "electrolyte bridges". Also, because of the small explosions of the bridges, the plasma-gas layer vibrates during the process. Sinkevitch also mentions that the current in the PeP process is a combination of a constant and a high-frequency component [3,28]. Duradji observed two components in the current during the polishing process [26].

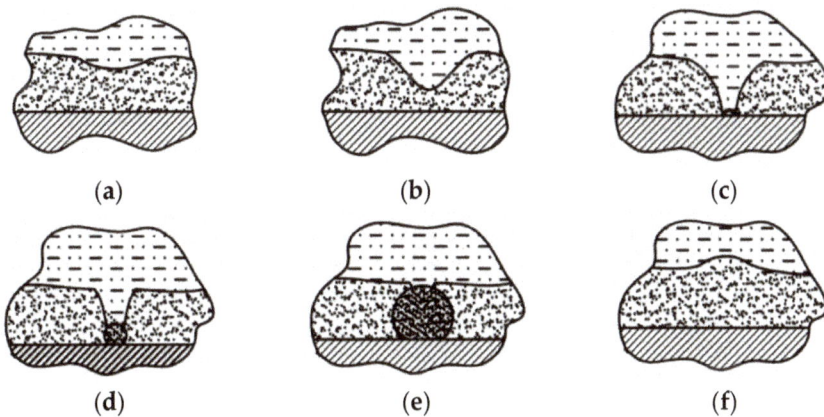

Figure 4. Electrolyte bridge formation scheme [28].

Another theory, called the 'streamer theory', is based on streamer discharges [4,30,31]. A schematic drawing is provided in Figure 5. In this figure (1) is electrolyte, (2) photoionization, (3) ions, (4) electrons, (5) workpiece surface, (6) secondary electrons, (7) avalanche head, (8) the streamer, (9) plasma channel, and (10) a gas explosion. Because of the high electric field, electrons have enough speed and energy to ionize molecules and release other electrons from them (Figure 5a). These electrons strike other molecules, forming some form of electron avalanches (Figure 5b). Then, the flow of electrons and ions forms a conductive channel that connects the workpiece surface with the electrolyte (Figure 5c–f)).

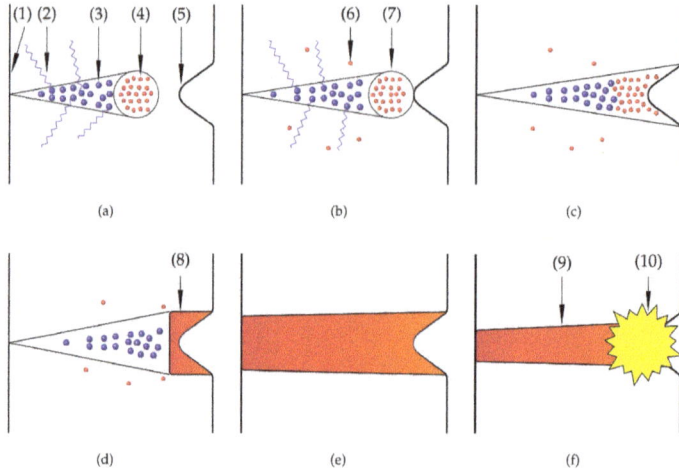

Figure 5. Schematic drawing of the formation of a plasma discharge according to streamer theory [4,30].

In summary, none of the presented models fully describes the conduction in plasma-gas layer.

1.3. Mechanism during Plasma Electrolytic Polishing

In the literature, there are many theories about the mechanisms of polishing during the PeP process. Many papers mention melting the workpiece surface as the main, or as an important, part of the polishing mechanism [4,5,14,25,27,29–32].

In the streamer theory, mentioned above, melting is considered a primary polishing mechanism [4,30–32]. In this theory, ions and electrons move toward the workpiece surface through a conductive channel, i.e., a streamer, and start to interact with the workpiece. This interaction leads to an increase in temperature of both the workpiece surface and the conductive channel, and melting. A further increase in temperature leads to an explosion in the channel. The explosion removes the melted metal.

Plotnikov et al. [25] suggested a mechanism which is a combination of melting and oxidation. At the beginning, a new gas bubble is formed; this is caused by the high temperature of the workpiece. Because of the high electric field, the gas inside the bubble gets ionized and turns into a high temperature plasma which starts to melt the oxide layer on the workpiece surface. The plasma bubble expands due to the high temperature, thereby creating a shock wave. The shock wave partially reflects from the interface between the plasma-gas layer and the electrolyte. The reflected wave compresses the gas bubble, causing it to collapse. When the bubble collapses, a process similar to cavitation occurs. Ions get into the formed void. Then, ions start to react with the workpiece surface, leading to the formation of an oxide film. According to this model, the polishing process takes place when the rate of the oxide layer formation is comparable to the rate of its removal by high-temperature plasma. Gas bubbles are present on the entire workpiece surface, so the polishing takes place everywhere. However, the removal rate for the peaks is much greater than that for the cavities.

Vaňa et al. [27] mentioned that the plasma-gas layer is an ionized medium in which glow discharges are present.. Discharges melt the workpiece surface first at the points where the thickness of the plasma-gas layer is lower. This leads to the rapid removal of the peaks on the workpiece surface, and to a smoothing effect. Because each plasma discharge removes the same amount of material (S1 = S2), the thickness of removed layer h2 is less than that of h1. According to Vaňa, this leads to a slow down of the removal rate over time. A schematic representation of the polishing process can be seen in Figure 6. However, it should be noted that better gloss can be achieved only by using the proper electrolyte. This may indicate the important role of electrochemical reactions in the polishing process.

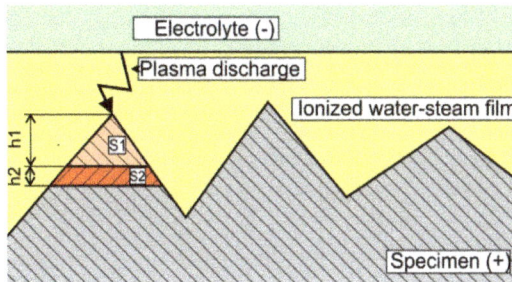

Figure 6. Schematic representation of the polishing process according to Vaňa et al. [27].

Another approach considers PeP as an electrochemical process [3,5,10,29,33–37].

According to Kalenchukova et al. [10], there is no melting during the PeP process; the main polishing mechanism is electrochemical dissolution. The plasma-gas layer (named steam-gas shell in Figure 7) has a different thickness on the peaks (h2) and in the cavities (h1), leading to a higher current density on the peaks, and consequently, to a higher removal rate. Higher removal of the peaks leads to a rapid decrease in roughness.

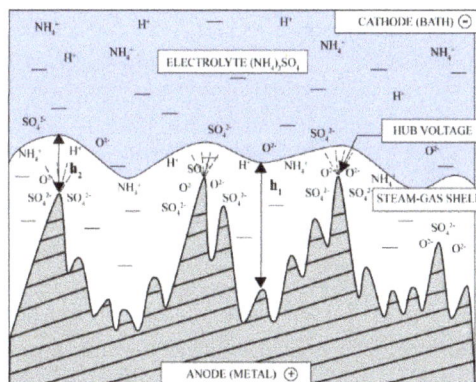

Figure 7. The mechanism of electrolytic-plasma polishing according to Kalenchukova et al. [10].

Smyslov et al. [36] describe the polishing mechanism as a superposition of anodizing and simultaneous chemical etching of the formed oxide layer. Etching the peaks is faster, and the oxide layer there is thinner than that in the cavities; this leads to a decrease in roughness.

Parfenov et al. [34] have studied the current efficiency of the PeP and electrolysis processes. It was concluded that PeP is mainly an electrochemical process, with a current efficiency of about 30%. It was stated that there is no melting during the PeP process, despite the presence of discharges in the plasma-gas layer. These discharges do not cause melting or the removal of the workpiece surface. Based on data obtained from a comparison of PeP and electrolysis by Parfenov et al. [34], it can be concluded that the appearance of a plasma-gas layer leads to changes in some electrochemical reactions. By traditional electrolysis, the formation of oxygen can be observed on the anode surface.

Electrochemical machining (ECM) with the same current density as in PeP process has a current efficiency of about 9% [34]. The low current efficiency of the ECM process can be explained by the passivation of the workpiece surface by oxygen. In the case of PeP, because of a stable plasma-gas layer, it can no longer be due to oxygen formation on the workpiece surface. It can therefore be assumed that oxygen formation still takes place, albeit on the interface between the plasma-gas layer and the electrolyte. A similar assumption was made by Kellogg [23].

Sinkevitch et al. [3,35,37] assumed that the PeP process can be considered as an anodic dissolution that can be described by the mechanism of complexation through a series of sequential or sequential and parallel stages.

Volenko et al. [29] assumed PeP as a superposition of physicochemical, thermal, electrical and hydrodynamic processes. According to Volenko, the plasma-gas layer is a dielectric. The conduction in the plasma-gas layer is described by the electrolytic bridges model, as mentioned above. In this paper, polishing is a combination of electrochemical removal and electrical discharge machining (EDM). The same mechanism is mentioned by Saushkin et al. [14].

Kulikov et al. [5] mentioned that the polishing mechanism is not fully understood, but assumed that polishing can occur because of a combination of electrochemical dissolution and melting by the discharges.

Alekseev et al. [16] describe the polishing mechanism as a combination of discharges, ion sputtering and chemical sputtering.

2. Multiphysics Simulation of Plasma Electrolytic Polishing

To simulate the PeP process, a model was developed. Based on a literature review and on experimental experience, the simulation in this paper is based on the assumption that the main mechanism of PeP for stainless steel is electrochemical. The coupling scheme of the model is provided in Figure 8. The model set up and calculation were made in COMSOL Multiphysics®. Electric Currents and Deformed Geometry interfaces were chosen for this model to simulate the current and electric potential and the polishing effect during the PeP process. The developed model is used to simulate the PeP process after the appearance of a stable plasma-gas layer.

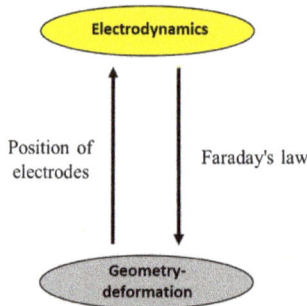

Figure 8. Coupling scheme of the multiphysical model [38].

2.1. Geometry

The model geometry and boundary conditions are provided in Figure 9. The model set up was based on the principle scheme shown in Figure 1. The bath with the electrolyte had dimensions of 20 cm × 20 cm. The workpiece was a disc with a mounting bore, which was completely immersed in the bath to a depth of 5 cm. A plasma-gas layer surrounded the workpiece from the beginning of the simulation.

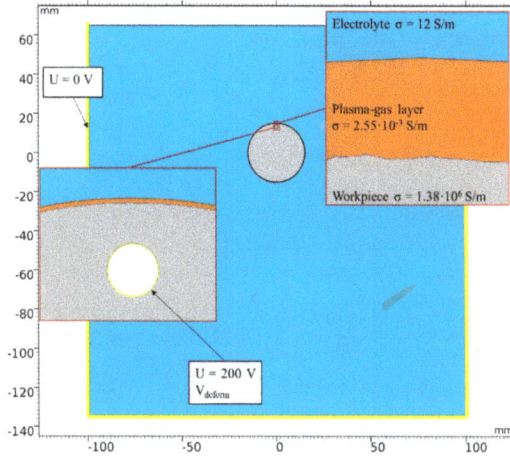

Figure 9. Modell geometry and boundary & domain conditions.

The model had 3 domains: an electrolyte, plasma-gas layer and the workpiece. The side and bottom boundaries of the model were grounded. A voltage of 200 V was applied to the boundaries of the mounting bore.

To simulate the polishing effect of PeP and to analyse the current density distribution on a surface, the workpiece surface profile was generated randomly in COMSOL Multiphysics® using the Spatial Frequencies method [39] and the following equations:

$$y = \sin(2\pi s)\left(15 + A \sum_{m=-N}^{N} \left(m^2\right)^{\frac{-b}{2}} g1(m)\cos(2\pi ms + u1(m))\right), \tag{2}$$

$$x = \cos(2\pi s)\left(15 + A \sum_{m=-N}^{N} \left(m^2\right)^{\frac{-b}{2}} g1(m)\cos(2\pi ms + u1(m))\right), \tag{3}$$

The parameters that were used for this are provided in Table 2. The initial roughness in the model was 2.49 μm. Electrical conductivity of the electrolyte domain was set 120 mS/cm. This value corresponded to an ammonium sulphate solution with a concentration of 50 g/L at 75 °C [13]. This electrolyte is common for polishing stainless steels. Steel 1.4301 was chosen as the material for the anode.

Table 2. Parameters for Spatial Frequencies method.

Parameter	Description	Value
N	Spatial frequency resolution	2000
b	Spectral exponent	0.5
A	Scale parameter in y coordinate	0.001
s	Phase coefficient	from 0 to 1
g1	Gaussian random function	
u1	Uniform random function	

Electrical conductivity of the plasma-gas layer was calculated based on the assumption that almost all voltage would drop in the plasma-gas layer. Using experimental data and data provided in the existing literature, it was possible to calculate first electric field, and then, the electrical conductivity. For this calculation, a thickness of 150 μm for the plasma-gas layer was chosen, based on the literature [2–4].

Then, based on the chosen thickness of the plasma-gas layer and a voltage of 200 V that is used in the model, the electric field can be calculated as follows:

$$E = \frac{V}{dh} = \frac{200 \text{ V}}{0.015 \text{ cm}} = 13333 \text{ V/cm.} \tag{4}$$

This corresponds with the range mentioned in extended literature; common values of the electric field are 10^4–10^5 V/cm [2–4,14]. The current density can be calculated with following equation: $j_n = \sigma \cdot E$. Using the current density and the electric field, the electrical conductivity can be calculated. Taking the average j_n based on experimental data from Rajput et al. [13] for 200 V of 0.3399 A/cm^2 and the above calculated electric field, the electrical conductivity of the plasma-gas layer can be calculated as follows:

$$\sigma = \frac{j_n}{E} = \frac{0.3399 \text{ A/cm}^2}{13333 \text{ V/cm}} = 2.55 \cdot 10^{-2} \text{mS/cm.} \tag{5}$$

Other simulation parameters are provided in Table 3.

Table 3. Simulation parameters.

Parameter	Value
Voltage	200 V
Anode conductivity	1.38×10^7 mS/cm
Electrolyte conductivity	120 mS/cm
Plasma-gas layer conductivity	2.55×10^{-2} mS/cm
Plasma-gas layer thickness	0.15 mm
Anode relative permittivity	1
Electrolyte relative permittivity	55
Plasma-gas layer relative permittivity	1

2.2. Model Mesh

A visualisation of the model mesh is provided in Figure 10. The complete mesh consists of 213,870 domain elements and 9247 boundary elements. All parameters used for meshing are provided in Table 4. The finest mesh is realised near the anode surface where the removal takes place.

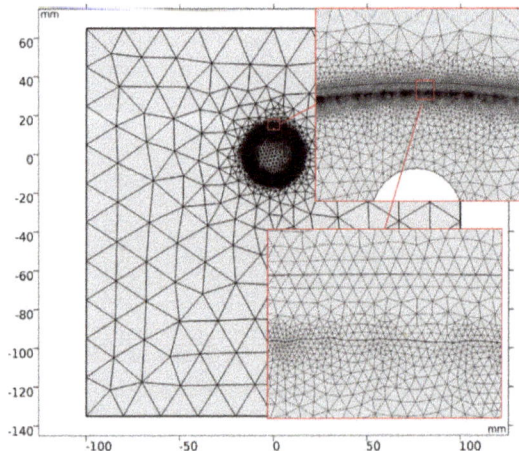

Figure 10. Visualisation of the model mesh.

Table 4. Parameters for mesh.

Parameter	Electrolyte and Plasma-Gas Layer	Workpiece
Maximum element size	20 mm	20 mm
Minimum element size	0.005 mm	0.005 mm
Maximum element growth rate	1.5	1.2
Curvature factor	0.2	0.2
Resolution of narrow regions	1	1

The mesh deformation is calculated according to equation below:

$$V_{\text{deform}} = K \cdot (-j_n),$$ (6)

where:

K is the removal coefficient, and j_n is the normal current density. K is calculated from experimental data from Rajput et al. [13] and based on the average removal speed in a one-dimensional direction and on the average current density. For example, the average removal speed and current density for 200 V can be used to determine the next removal coefficient:

$$K = \frac{MRR}{j_n} = \frac{5.24 \times 10^{-8} \text{ m/s}}{3398.69 \text{ A/m}^2} = 1.54 \times 10^{-11} \text{m}^3/(\text{A} \cdot \text{s}),$$ (7)

It was assumed for this model that voltage would only have an influence on the thickness of the plasma-gas layer. In this case, it may be concluded that the MRR is primarily dependent upon the current density. So, the removal coefficient K can be described as a function of the current density. The removal coefficient K as a function of current density can be seen in Figure 11. The black dots on the graph represent data obtained from experimental data from Rajput et al. [13]. It was assumed that for a current density equal to and less than zero, a metal removal process does not occur. A linear approximation was also applied for current density values exceeding 0.34 A/cm^2. A removal simulation was made for 300 s machining time.

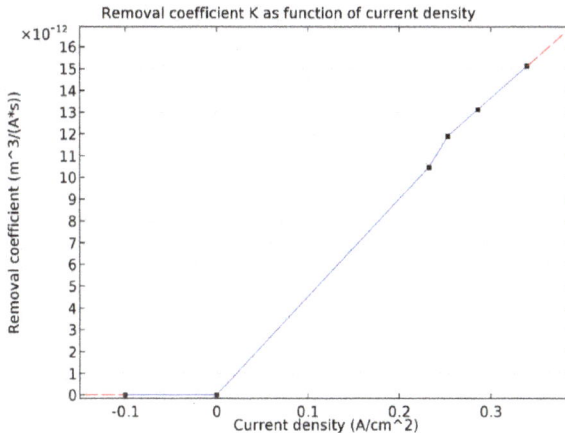

Figure 11. Removal coefficient K as function of current density.

2.3. Simulation Results

The results of the modelling of electric potential can be seen in Figures 12 and 13. It can be seen that almost the whole voltage drops in the plasma-gas layer. This was expected from the experimental data. This result allowed us to assume that the plasma-gas layer could be considered as a special electrochemical cell, where the interface between the plasma-gas layer and the electrolyte acts as cathode.

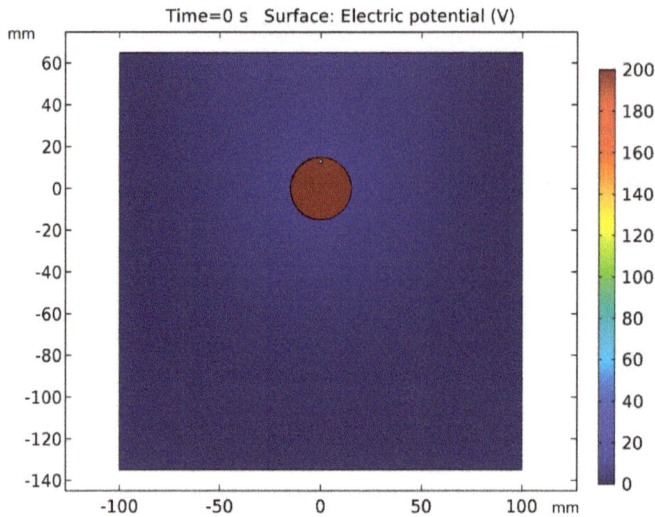

Figure 12. Electric potential.

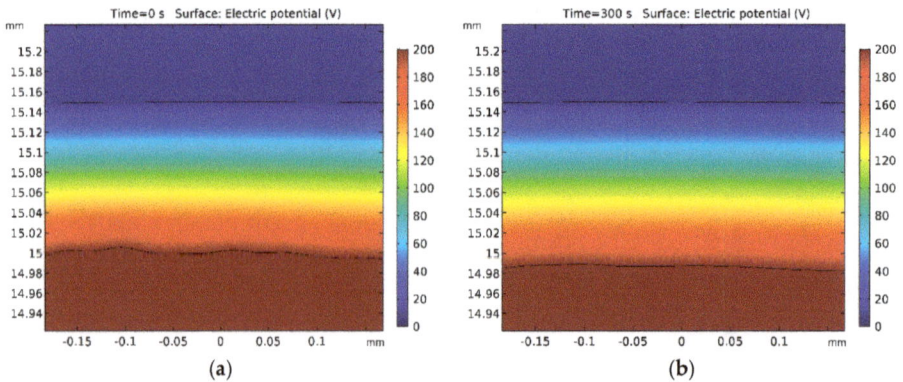

Figure 13. Detailed view of workpiece surface and electric potential (**a**) at $t = 0$ s, (**b**) at $t = 300$ s.

Figure 14 shows the surface profile before and after 300 s polishing. In Figure 15, the normal current density on the anode surface at the beginning of the process and after 300 s polishing is provided. It can be seen that despite the fact that the overall shape of the surface is retained, the peaks were visibly removed.

Comparing Figures 14 and 15, it can be concluded that the normal current density is mainly influenced by the shape of the surface. This leads to higher current densities on the surface profile peaks and lower ones in the cavities. Because of the electrochemical mechanism of the process, a higher current density on the peaks, and consequentially, a higher removal rate than in the cavities, leads to a polishing effect on the workpiece surface. In Figure 15, it can also be seen that the current density in the deeper cavities increases with the processing time; this can be explained by the decrease in peak heights, and therefore, a more even current distribution on the surface.

Figure 16 presents a comparison of the average current density in the model and in the experiment of Rajput et al. [13].

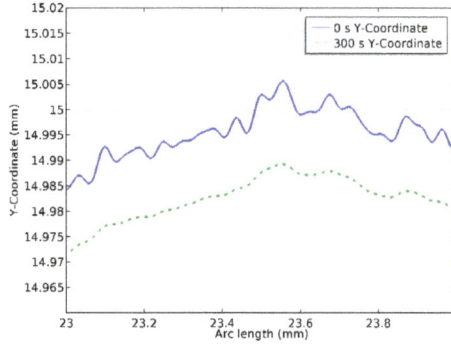

Figure 14. Detailed view of surface profile at 0 s and 300 s.

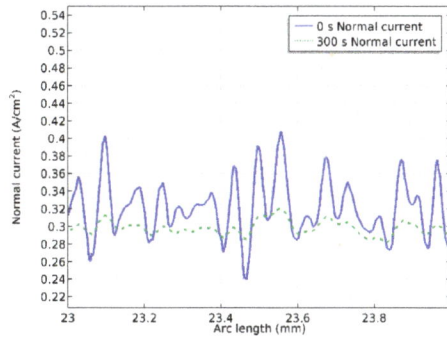

Figure 15. Detailed view of normal current density at 0 s and 300 s.

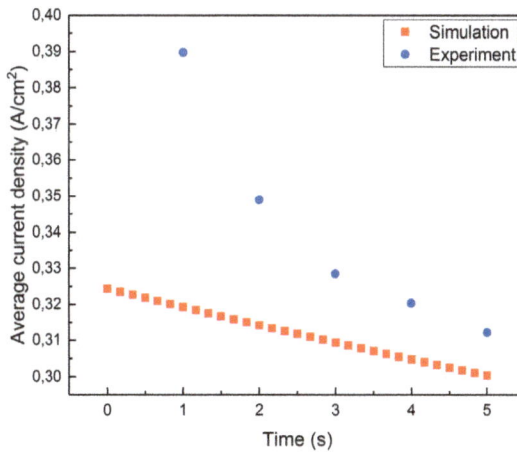

Figure 16. Average current density in model (red) and experiment from Rajput (blue) [13].

The time average current density in the model is 0.312 A/cm^2, compared to 0.340 A/cm^2 in the experiment from Rajput [13]. It can be seen that the average current density in the model is systematically lower than that in the experiment. This can be explained by the fact that the thermodynamic effects are not taken into account in the model. In the real process of PeP, a large amount of the current is used to heat the anode and electrolyte and to evaporate the electrolyte. Because at the beginning of the process, there is no plasma-gas layer, the current density is maximal.

Presumably, a lot of energy is used at the beginning of the process to form the plasma-gas layer. Then, part of the energy is required to stabilise it. In the model, plasma exists from the beginning and remains stable throughout the simulation time. So, no energy is used in the formation and stabilisation of the plasma-gas layer.

To analyse the polishing effect, the roughness parameter *Ra* was calculated. The equation for *Ra* was developed based on the following formula [38]:

$$Ra = \frac{1}{l} \int_0^l |h(x)| dx, \tag{8}$$

where:

l is the evaluation length and $h(x)$ represents deviations from the mean line at position *x*.

$$h(x) = |y - \bar{y}|, \tag{9}$$

To calculate this in COMSOL, the following component couplings were used: intop1 - integration over boundaries 13 and 14, and aveop1 – average over the sample boundaries.

Because the workpiece is a disc, it is necessary to make some changes to this equation. First of all, *x* in this case is changed to the radius, *r*, of the disc. So, the deviations from the mean line were calculated using the following equation:

$$h(x) = |r - \bar{r}|, \tag{10}$$

where $\bar{r}$ is an overage radius, calculated with aveop1.

Then, for the disc, *l* is the circumference. The following equation was used:

$$l = 2\pi\bar{r}, \tag{11}$$

Applying everything to equation (8):

$$Ra = \frac{1}{2\pi\bar{r}} \int |r - \bar{r}| dl, \tag{12}$$

The results of this calculation are presented in Figure 17. It can be seen that the roughness decreases according to exponential decay (1) from Mukaeva [14]. According to this equation, the minimal achievable roughness Ra in this model has a value of 1.67 µm.

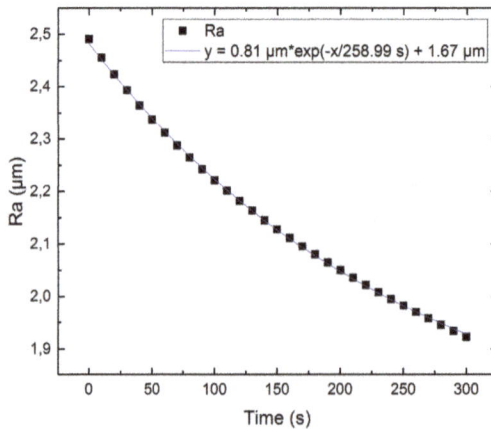

Figure 17. Selected results for *Ra* as function of time with fit curve.

3. Inline Metrology in Plasma Electrolytic Polishing

Inline metrology systems are important to ensure product quality control during processing. In the case of plasma polishing, the most important parameter of the final product is the surface roughness. To ensure the stability of the polishing process, it is necessary to control the current density and temperature.

According to the existing literature, common value of average current density range from 0.1 A/cm^2 to 0.5 A/cm^2 [1–3,5,7,16,34,40]. Thus, knowing the initial sample area and monitoring the current during processing, the process can be controlled. If the current density values are too large or too small, this may mean that the process is not stable; it could indicate, for example, that the plasma-gas layer is unstable or that it has collapsed. In this case, the end result may exceed the requirements. In addition, based on the experimental data, it can be assumed that the current can be analysed to gain information about the surface roughness.

Temperature control is important because the temperature directly affects the formation of the plasma-gas layer, and thus, the current during the process. At higher temperatures, the plasma-gas layer thickness increases, which leads to a decrease in current density. Also, if the temperature is too high, it can lead to the destruction of the chemical components of the electrolyte and/or to their evaporation.

A PeP prototype lab system was developed at Chemnitz University of Technology. A summary of the system can be seen in Figure 18. The setup has one axis in the z-direction. Different clamping systems for the sample can be mounted. The selected power supply Keysight N8762A makes it possible to set the voltage up to 600 V and the current up to 8.5 A, and includes a built-in current measurement system.

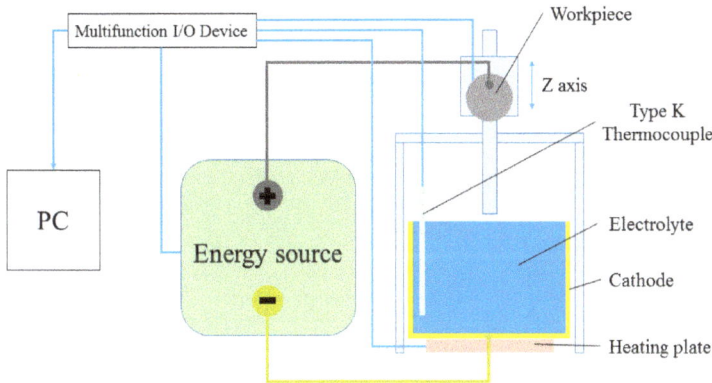

Figure 18. Scheme of the PeP prototype system.

The temperature was measured by thermocoupling using a multifunction I/O device NI USB-6215. The whole system was controlled by PC with controlling software which was developed in LabVIEW. Before starting, a heating plate heated the electrolyte to a pre-set temperature.

A disc with a 30 mm diameter was selected as the sample for experiment. Each disc had a bore with a 3.8 mm diameter for mounting. A simple hook was used as a sample holder. The discs were made with different initial roughnesses on both sides. All samples were immersed at the same depth. The measured current values were corrected for the current value obtained when the holder hook was immersed without a sample. The initial parameters for experiment are given in Table 5.

The experiment was undertaken to measure changes in mass and roughness. Surface roughness was measured with 3D Laserscanning-Microscope Keyence VK-9700. *Sa* and *Ra* roughnesses were chosen as the main parameters for the measurement. The mass of the samples was measured with precision balances Sartorius ME36S.

Sa and *Ra* roughnesses were measured before and after polishing at different positions on the sample surface. Two positions were chosen for measurement: one on the top and one on the bottom of the sample on each side.

Table 5. Initial parameters.

Parameter	Value
Voltage	250 V
Pre-set temperature of electrolyte	75 °C
Electrolyte salt	Ammonium sulphate
Electrolyte salt concentration	5% of mass
Samples material	steel 1.4301 (AISI 304)
Initial roughness *Sa*	(0.15 ± 0.02) μm
	(0.63 ± 0.08) μm
Initial roughness *Ra*	(0.10 ± 0.02) μm
	(0.14 ± 0.02) μm

Experimental Results

The typical surface of the disc before and after polishing can be seen in Figures 19 and 20 respectively. In Figure 19, the pattern of the surface after turning can be seen. In Figure 20, this pattern is not visually observed on the polished surface. At the same time, dark spots can be observed on the entire polished surface. Some of them are peaks of up to 5 μm. Presumably, these may be undissolved inclusions of carbon or other elements from the composition of the steel.

Figure 19. Sample surface before polishing, $Ra = 0.080$ μm, $Sa = 0.124$ μm.

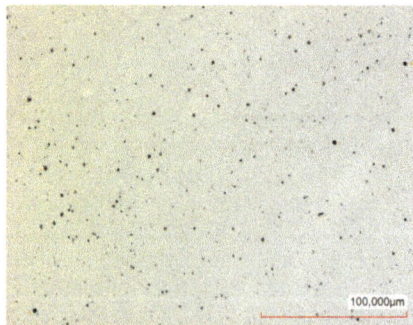

Figure 20. Sample surface after 5 min polishing, $Ra = 0.023$ μm, $Sa = 0.045$ μm.

Figure 21 shows an example of a monitored temperature of the electrolyte as a function of time. This figure demonstrates a typical increase in temperature during polishing of a single sample. In this example, it increases from about 75 °C to approximately 85 °C after 100 s, and then increases only slightly more in the last 200 s of the process.

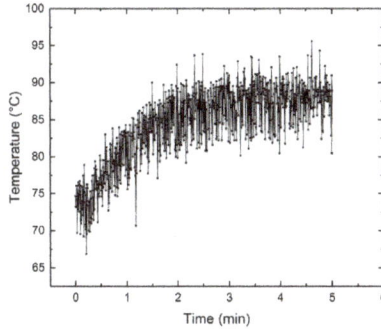

Figure 21. Monitored temperature as a function of time.

Figure 22 displays an example of current density as function of time. This figure shows a typical decrease in current density during the polishing of a single sample. There are two possible reasons for the observed decrease; firstly, the decrease in current density can be explained by the increase in temperature [14]. At a higher temperature, the thickness of the plasma-gas layer increases, leading to a decrease in current. An alternative possible reason may be the decrease in roughness. Because of the roughness, the current density distribution is not even. The current density is focused mainly on the peaks, so it is higher there. Consequently, the removal rate for the peaks will be also higher, and they will be removed more rapidly. This leads to a fast decrease in the heights of these peaks and, because of this, in roughness. The decrease in the heights of the peaks leads to more even distribution of the current density and a decrease in the current density over time.

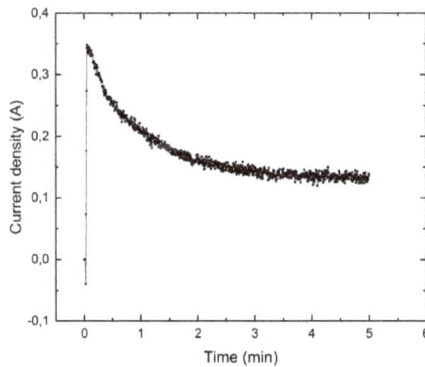

Figure 22. Current density as function of time.

Figure 23 shows the average surface roughness, Sa, at both sides of the samples as a function of time with the exponential decay fit function (1). The results for average Ra roughness are presented in Figure 24.

It can be seen that the exponential decay fit works for Sa roughness, as well as for Ra. The minimum achievable roughness, Sa, for initial roughnesses of 0.15 µm and 0.63 µm, are 0.066 µm and 0.301 µm, respectively. The minimum achievable roughness, Ra, for initial roughnesses of 0.10 µm and 0.14 µm, are 0.032 µm and 0.045 µm, respectively.

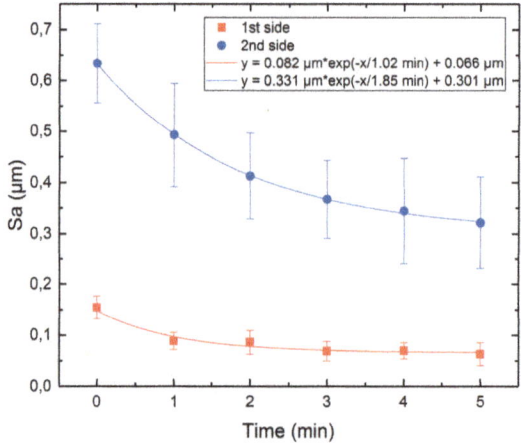

Figure 23. Surface roughness *Sa* as function of time.

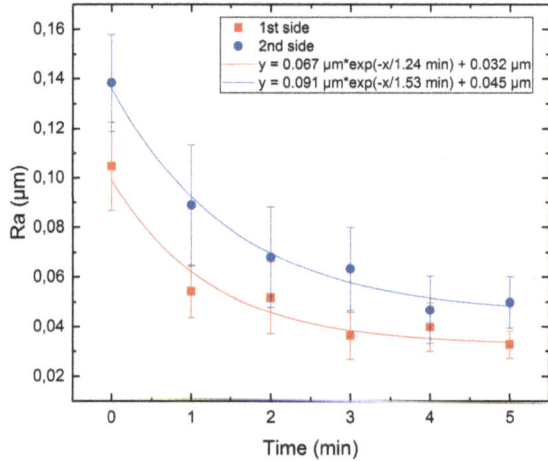

Figure 24. Surface roughness *Ra* as function of time.

Also, the roughness along the sample height was studied. The results are provided in Table 6. On both sample sides, roughnesses *Ra* and *Sa* change differently, depending on the position along the height. It can be seen that the decrease in roughness for the bottom of the samples is bigger; this can be explained by the continuous gas formation which occurred during the process. The plasma-gas layer formed in the first seconds of the process and remained stable until the end of processing, mainly due to the evaporation of the electrolyte [10,14,16,24–26]. However, the gas moved up along the sample height; this led to an increase in the thickness of the plasma-gas layer from the lowest point of the sample to the highest. Because of this, the current density increased from the top to the bottom of the sample. Taking into account the assumption that PeP is mainly an electrochemical process, it may be concluded that a higher current leads to a higher removal rate. Based on the data obtained in the simulation of the PeP process, we can conclude that the removal rate for the peaks is higher; this also leads to a decrease in the roughness. Thus, a higher current density at the bottom of the sample leads to a faster reduction of roughness.

Table 6. *Sa* and *Ra* minimum achievable roughness.

Parameter	Initial Roughness	Minimum Achievable Roughness	
		Top	Bottom
Sa	0.15 μm	0.083 μm	0.045 μm
	0.63 μm	0.414 μm	0.221 μm
Ra	0.10 μm	0.033 μm	0.028 μm
	0.14 μm	0.055 μm	0.032 μm

The samples were weighed before and after polishing to calculate the material removal rate and current efficiency. The result of the material removal rate (MRR) calculation is provided in Figure 25, in which the average MRR decrease for the longer processing times are given. This may be due to the decrease in current over time. This leads to a decrease in charge, and therefore, to less removal.

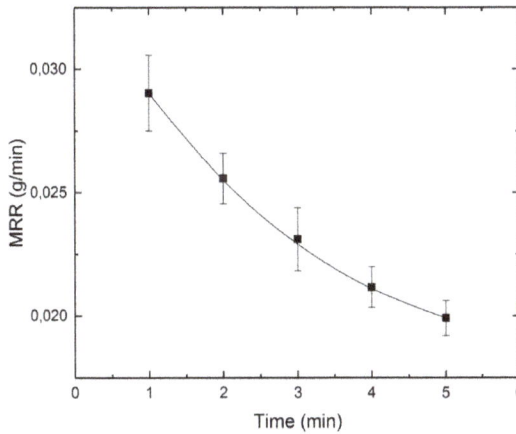

Figure 25. Average MRR as function of time.

The current efficiency was calculated to evaluate the process, and based on the assumption that the PeP process can be considered an electrochemical one. The calculation was based on equations (13) - (15). The parameters used for the calculation can be seen in Table 7.

$$m_{spec} = \frac{1}{F} \sum_{i=1}^{n} \frac{c_i \cdot M_i}{z_i},$$ (13)

where:

Table 7. Parameters for stainless steel 1.4301.

Chemical Element	Fe	Cr	Ni	N	Mn	Si	C
Mass fraction c in %	68.8	18	10	0.1	2	1	0.1
Valence z	3	6	2	3	2	4	4
Molar Mass M in g/mol	55.85	51.996	58.7	14.007	54.94	28.09	12.01

c_i is the mass fraction, M_i is the Molar Mass, and z_i is the valence

$$m_{eff} = \frac{m_{rem}}{Q},$$ (14)

where:

m_{rem} is the real mass removal and Q is the exchanged electric charge

$$\eta = \frac{m_{eff}}{m_{spec}},$$ (15)

The results of the calculation can be seen in Figure 26. This figure shows the current efficiency of a process for five different processing times. The processing time for an average current efficiency is around 59%. This value is similar for electrochemical machining, but for the same current density values of ECM, current efficiency can be lower due to passivation. As mentioned above, Kellogg [23] and Parfenov et al. [34] made the assumption that when the plasma-gas layer is stable, oxygen formation no longer occurs on the workpiece surface, but rather that oxygen formation still takes place but on the interface between plasma-gas layer and electrolyte.

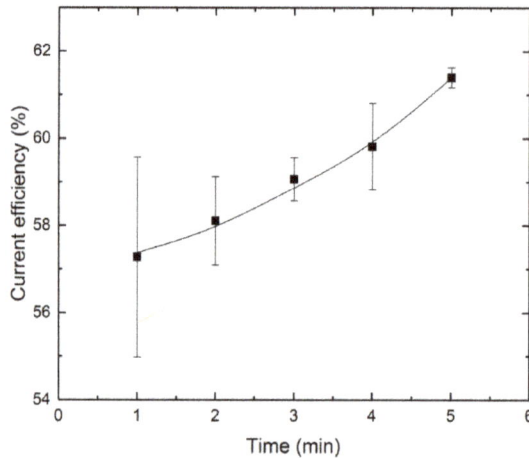

Figure 26. Current efficiency as function of time.

It also can be seen in Figure 26 that the current efficiency increases over time. This can be explained by the change in current density over time, as shown in Figure 22. It can be assumed that a decrease in current leads to a lower level oxygen formation, and therefore to lower passivation.

4. Conclusions

This paper provides a review of the literature on the possible mechanisms of the plasma-gas layer conductivity, and on the possible material removal mechanisms during plasma electrolytic polishing. The most popular theories about conductivity in the plasma-gas layer are based on the assumption that the plasma-gas layer is a highly-ionized medium in which different ions and electrons provide conductivity. The two most popular theories about the removal mechanism are electrochemical removal and removal because of melting.

Based on our simulation and experimental results, the following conclusions may be drawn:

- The main voltage drop in PeP occurs in the plasma-gas layer. The current distribution inside plasma-gas layer is determined by the workpiece surface. Thus, a higher current density on the peaks leads to a faster removal of the peaks and to a reduction of roughness.
- PeP can be simulated as an electrochemical machining process. Moreover, future simulations can be undertaken considering only the plasma-gas layer where the interface with the electrolyte acts as a cathode and the workpiece surface as the anode.
- PeP processing time and Ra roughness can be predicted with the simulation.
- Changes in roughness depend on the position along the sample height.

- *Sa* roughness can be fitted with exponential decay fit, as well as with *Ra* roughness.
- The changes in roughness for both *Sa* and *Ra* depend on the position along the sample's height. The decrease in roughness for the bottom of the sample is bigger.
- The current efficiency of PeP is comparable to that of ECM, but the current density is lower.
- Passivation in PeP and oxygen formation should be different from a typical electrochemical process because of the plasma-gas layer.

Future experiments comparing PeP and ECM with the same electrolyte and same current density value are planned, in order to compare the resulting current efficiencies and surface topologies. Furthermore, it will be important to conduct a study of the residual stresses after applying PeP in the future.

Author Contributions: Conceptualization, I.D., M.H.-O., M.Z., G.M., J.E. and A.S.; formal analysis, I.D., M.H.-O., M.Z., G.M., J.E. and A.S.; writing—original draft preparation, I.D., M.H.-O., M.Z. and G.M.; writing—review and editing J.E. and A.S.; supervision, M.H.-O., M.Z., G.M., J.E. and A.S.; funding acquisition, M.H.-O., M.Z. and A.S.

Funding: This research was funded by Horizon 2020, the EU Framework Programme for Research and Innovation (Project ID: 674801).

Acknowledgments: This research work was undertaken in the context of MICROMAN project ("Process Fingerprint for Zero-defect Net-shape MICROMANufacturing"). MICROMAN is a European Training Network supported by Horizon 2020, the EU Framework Programme for Research and Innovation (Project ID: 674801).

Conflicts of Interest: The authors declare no conflict of interest.

References

1. Nestler, K.; Böttger-Hiller, F.; Adamitzki, W.; Glowa, G.; Zeidler, H.; Schubert, A. Plasma Electrolytic Polishing-An Overview of Applied Technologies and Current Challenges to Extend the Polishable Material Range. *Procedia CIRP* **2016**, *42*, 503–507. [CrossRef]

2. Zeidler, H.; Boettger-Hiller, F.; Edelmann, J.; Schubert, A. Surface Finish Machining of Medical Parts Using Plasma Electrolytic Polishing. *Procedia CIRP* **2016**, *49*, 83–87. [CrossRef]

3. Синькевич, Ю.В.; Шелег, В.К.; Янковский, И.Н.; Беляев, Г.Я. Электроимпульсное Полирование Сплавов на Основе Железа, Хрома и Никеля (*Electropulse Polishing of Alloys Based on Iron, Chromium and Nickel*); Белорусский Национальный Технический Университет: Minsk, Belarus, 2014; ISBN 978-985-550-516-8.

4. Cornelsen, M.; Deutsch, C.; Seitz, H. Electrolytic Plasma Polishing of Pipe Inner Surfaces. *Metals* **2017**, *8*, 12. [CrossRef]

5. Куликов, И.С.; Ващенко, С.В.; Каменев, А.Я. Электролитно-Плазменная Обработка Материалов (*Electrolytic-Plasma Treatment of Materials*); Беларуская Навука: Минск, Belarus, 2010; ISBN 978-985-08-1215-5.

6. Алексеев, Ю.Г.; Паршуто, А.Э.; Нисс, В.С.; Королев, А.Ю. Способ Электролитно-Плазменной Обработки Стального Изделия (The Method of Electrolyte-Plasma Treatment of Steel Products). BY Patent 21103, 4 January 2012.

7. Duradji, V.N.; Kaputkin, D.E.; Duradji, A.Y. Aluminum treatment in the electrolytic plasma during the anodic process. *J. Eng. Sci. Technol. Rev.* **2017**, *10*, 81–84. [CrossRef]

8. Смыслов, А.М.; Таминдаров, Д.Р.; Мингажев, А.Д.; Смыслова, М.К.; Самаркина, А.Б. Способ Электролитно-Плазменного Полирования Деталей из Титановых Сплавов (Method of Electrolyte-Plasma Grinding Parts Made from Titanium Alloys). RU Patent 2495967, 3 July 2012.

9. Valiev, R.I.; Khafizov, A.A.; Shakirov, Y.I.; Sushchikova, A.N. Polishing and deburring of machine parts in plasma of glow discharge between solid and liquid electrodes. *Mater. Sci. Eng.* **2015**, *86*, 012026. [CrossRef]

10. Kalenchukova, V.O.; Nagula, P.K.; Tretinnikov, D.L. About changes in the chemical composition of the electrolyte in the process of electrolytic-plasma treatment of materials. *Mater. Methods Technol.* **2015**, *9*, 404–413.

11. Kashapov, L.N.; Kashapov, N.F.; Kashapov, R.N. Investigation of the influence of plasma-electrolytic processing on the surface of austenitic chromium-nickel steels. *J. Phys. Conf. Ser.* **2013**, *479*, 012003. [CrossRef]

12. Дураджи, В.Н.; Капуткин, Д.Е. Способ Электролитно-Плазменной Обработки Поверхности Металлов (Method of electrolytic-plasma treatment of metal surface). RU Patent 2550393, 27 May 2014.

13. Rajput, A.S.; Zeidler, H.; Schubert, A. Analysis of voltage and current during the Plasma electrolytic Polishing of stainless steel. In Proceedings of the 17th International Conference European Society Precision Engineering Nanotechnology, EUSPEN 2017, Hannover, Germany, 29 May–2 June 2017; pp. 2–3, ISBN 9780995775107.

14. Мукаева, В.Р. Управление Технологическим Процессом Электролитно-плазменного Полирования на Основе Контроля Шероховатости пов. по Импедансным Спектрам (Management of Processing Procedure of Electrolytic-Plasma Polishing on the Base of Control of Surface Roughness by Impedance Spectra). Ph.D. Thesis, Ufa State Aviation Technical University, Ufa, Russia, 2014.

15. Кревсун, Э.П.; Куликов, И.С.; Каменев, А.Я.; Ермаков, В.Л. Устройство для Электролитно-Плазменного Полирования Металлического Изделия (Device for Electrolytic-Plasma Polishing of Metal Product). BY Patent 13937, 4 December 2008.

16. Алексеев, Ю.Г.; Кособуцкий, А.А.; Королев, А.Ю.; Нисс, В.С.; Кучерявый, В.Д.; Повжик, А.А. Особенности процессов размерной обработки металлических изделий электролитно-плазменным методом (Features of the processes of dimensional processing of metal products by electrolytic-plasma method). *Литье и Металлургия* **2005**, *4*, 188–195.

17. Ablyaz, T.R.; Muratov, K.R.; Radkevich, M.M.; Ushomirskaya, L.A.; Zarubin, D.A. Electrolytic Plasma Surface Polishing of Complex Components Produced by Selective Laser Melting. *Russ. Eng. Res.* **2018**, *38*, 491–492. [CrossRef]

18. Попов, А.И.; Тюхтяев, М.И.; Радкевич, М.М.; Новиков, В.И. Анализ тепловых явлений при струйной фокусированной электролитно-плазменной обработке (The analysis of thermal phenomena occuring under jet focused electrolytic plasma processing). *Научно-Технические Ведомости Санкт-Петербургского Государственного Политехнического Университета* **2017**, *4*, 141–150. [CrossRef]

19. Попов, А.И.; Радкевич, М.М.; Кудрявцев, В.Н.; Захаров, С.В.; Кузьмичев, И.С. Установка для Электролитно-Плазменной Обработки Турбинных Лопаток (Plant for Electrolyte-Plasma Treatment of Turbine Blades). RU Patent 2623555, 24 May 2016.

20. Алексеев, Ю.Г.; Королев, А.Ю.; Нисс, В.С.; Паршуто, А.Э. Электролитно-плазменная обработка внутренних поверхностей трубчатых изделий russian(Electrolytic-Plasma Treatment of Inner Surface of Tubular Products). *Наука и Техника* **2016**, *15*, 61–68.

21. Кургузов, С.А.; Залетов, Ю.Д.; Косматов, В.И.; Гусева, О.С.; Шевцова, И.Н. Электролитно-плазменная очистка поверхности стального металлопроката (Electrolytic and plasma cleaning of the surface of steel rolled metal product). *Электротехнические Системы и Комплексы* **2016**, *2*, 48–51.

22. Новоселов, М.В.; Шиллинг, Н.Г.; Рудавин, А.А.; Радкевич, М.М.; Попов, А.И. Оценка возможности полирования нержавеющих сталей струйной электролитно-плазменной обработкой (Assessment of a possibility polishing of stainless steels jet electrolytic and plasma processing). *Вестник ПНИПУ* **2018**, *20*, 94–102. [CrossRef]

23. Kellogg, H.H. Anode Effect in Aqueous Electrolysis. *J. Electrochem. Soc.* **1950**, *97*, 133. [CrossRef]

24. Гирговьев, А.И.; Ширяева, С.О.; Морозов, В.В. О некоторых закономерностях формирования электрического тока в окрестности опущенного в электролит нагретого электрода (About some regularities of the electric current formation in the vicinity of a heated electrode lowered into the electrolyte). *Электронная Обработка Материалов* **2004**, *5*, 16–20.

25. Плотников, Н.В.; Смыслов, А.М.; Таминдаров, Д.Р. К вопросу о модели электролитно-плазменного полирования поверхности (To a question on model of electrolytic-plasma polishing). *Вестник УГАТУ* **2013**, *17*, 90–95.

26. Дураджи, В.Н. Особенности установления электрогидродинамического режима, используемого для полирования металлов в электролитной плазме (Features of the establishment of the electrohydrodynamic regime used for polishing metals in electrolytic plasma). *Металлообработка* **2013**, *3*, 35–39.

27. Vana, D.; Podhorsky, S.; Hurajt, M.; Hanzen, V. Surface Properties of the Stainless Steel X10 CrNi 18/10 after Aplication of Plasma Polishing in Electrolyte. *Int. J. Mod. Eng. Res.* **2013**, *3*, 788–792.

28. Синькевич, Ю.В. Концепт. модель коммутационного механизма электрич. проводимости парогазовой оболочки в режиме электроимпульсного полирования (Conceptual Model of Commutation Mechanism for Electric Conductivity of Vapor-Gas Envelope in Electro-Impulse Polishing Mode). *Наука и Техника* **2016**, *15*, 407–414. [CrossRef]

29. Воленко, А.П.; Бойченко, О.В.; Чиркунова, Н.В. Электролитно-плазменная обработка металлических изделий (Electrolyte-plasma treatment of metals). *Вектор Науки ТГУ* **2012**, *4*, 144–147.

30. Wang, J.; Suo, L.C.; Guan, L.L.; Fu, Y.L. Analytical Study on Mechanism of Electrolysis and Plasma Polishing. *Adv. Mater. Res.* **2012**, *472–475*, 350–353. [CrossRef]

31. Wang, J.; Zong, X.; Liu, J.; Feng, S. Influence of Voltage on Electrolysis and Plasma Polishing. In Proceedings of the 2017 International Conference on Manufacturing Engineering and Intelligent Materials (ICMEIM 2017), Guangzhou, China, 25–26 February 2017; Volume 100, no. Icmeim. pp. 10–15, ISBN 978-94-6252-317-3. [CrossRef]

32. Wang, J.; Suo, L.C.; Fu, Y.L.; Guan, L.L. Study on Material Removal Rate of Electrolysis and Plasma Polishing. In Proceedings of the 2012 IEEE International Conference on Information and Automation, Shenyang, China, 6–8 June 2012; pp. 917–922, ISBN 9783037853702. [CrossRef]

33. Парфенов, Е.В.; Ерохин, А.Л.; Невьянцева, Р.Р.; Мукаева, В.Р.; Горбатков, М.В. Управление электролитно-плазменными и электрохимическими технологическими процессами на основе контроля состояния объекта методом импедансной спектроскопии (Control of electrolytic-plasma and electrochemical processes based on control of condition of object by impedance spectroscopy). In Proceedings of the Xii Всероссийское Совещание по Проблемам Управления Вспу-2014, Moscow, Russia, 16–19 June 2014; pp. 4348–4359.

34. Parfenov, E.V.; Farrakhov, R.G.; Mukaeva, V.R.; Gusarov, A.V.; Nevyantseva, R.R.; Yerokhin, A. Electric field effect on surface layer removal during electrolytic plasma polishing. *Surf. Coat. Technol.* **2016**, *307*, 1329–1340. [CrossRef]

35. Иванова, Н.П.; Синькевич, Ю.В.; Шелег, В.К.; Янковский, И.Н. Механизм анодного растворения коррозионностойких и конструкц. углерод. сталей в условиях электроимпульсного полирования (The mechanism of anodic dissolution of corrosion-resistant and structural carbon steels under conditions of electropulse polishing). *Наука и Техника* **2013**, *1*, 24–30.

36. Смыслов, А.М.; Смыслова, М.К.; Мингажев, А.Д.; Селиванов, К.С. Многоэтапная электролитно-плазменная обработка изделий из титана и титановых сплавов (Multistage electrolytic-plasma processing of products from titanium and titanium alloys). *Вестник УГАТУ* **2009**, *13*, 141–145.

37. Иванова, Н.П.; Синькевич, Ю.В.; Шелег, В.К.; Янковский, И.Н. Исследование морфологии и химического состава электроимпульсно полированной поверхности углеродистых и коррозионностойких сталей (Study of the morphology and chemical composition of electropulse polished surface of carbon and corrosion-resistant steels). *Наука и Техника* **2012**, *6*, 3–10.

38. Danilov, I.; Hackert-Oschätzchen, M.; Schaarschmidt, I.; Zinecker, M.; Schubert, A. Transient Simulation of the Removal Process in Plasma Electrolytic Polishing of Stainless Steel. In Proceedings of the COMSOL Conference 2018, Lausanne, Switzerland, 22–24 October 2018; Available online: https://www.comsol.com/paper/download/573171/danilov_paper.pdf (accessed on 10 December 2018).

39. Sjodin, B. How to Generate Random Surfaces in COMSOL Multiphysics® | COMSOL Blog. Available online: https://www.comsol.com/blogs/how-to-generate-random-surfaces-in-comsol-multiphysics/ (accessed on 24 May 2018).

40. Böttger-Hiller, F.; Nestler, K.; Zeidler, H.; Glowa, G.; Lampke, T. Plasma electrolytic polishing of metalized carbon fibers. *AIMS Mater. Sci.* **2016**, *3*, 260–269. [CrossRef]

micromachines

MDPI

Article

On the Process and Product Fingerprints for Electro Sinter Forging (ESF)

Emanuele Cannella [1,2,*], **Chris Valentin Nielsen** [1] **and Niels Bay** [1]

[1] Department of Mechanical Engineering, Technical University of Denmark, DK-2800 Kongens Lyngby, Denmark; cvni@mek.dtu.dk (C.V.N.); nbay@mek.dtu.dk (N.B.)
[2] IPU, Instituttet for Produktudvikling, DK-2800 Kongens Lyngby, Denmark
* Correspondence: emcann@mek.dtu.dk; Tel.: +45-4525-6286

Received: 25 February 2019; Accepted: 25 March 2019; Published: 27 March 2019

Abstract: Electro sinter forging (ESF) represents an innovative manufacturing process dealing with high electrical currents. Classified in the category of electrical current assisted sintering (ECAS) processes, the main principle is that Joule heating is generated inside the compacted powder, while the electrical current is flowing. The process is optimized through the analysis of the main process parameters, namely the electrical current density, sintering time, and compaction pressure, which are also evaluated as process fingerprints. The analysis was conducted on commercially pure titanium powder. Small discs and rings were manufactured for testing. The influence of the process parameters was analysed in terms of the final material properties. The relative density, microstructures, hardness, and tensile and compressive strengths were analysed concerning their validity as product fingerprints. Microstructural analyses revealed whether the samples were sintered or if melting had occurred. Mechanical properties were correlated to the process parameters depending on the material. The different sample shapes showed similar trends in terms of the density and microstructures as a function of the process parameters.

Keywords: Electro sinter forging; resistance sintering; electrical current; fingerprints

1. Introduction

Among several manufacturing processes, sintering represents a well-known solution based on powder consolidation via thermal heating. Ceramic objects were the first examples of sintered components, while metal sintering is relatively new. The first literature containing sintering is dated from 1829 [1]. In that case, platinum was attempted. Since that time, researchers have mostly focused on new materials and technologies. Conventional sintering includes two steps, namely (i) compaction of the sintering powder with the generation of the green body, and (ii) heating in a sintering furnace. The material increases its density by thermal energy, enabling a reduction of the free-surface energy as described by the Gibbs equation [2]. There are three main contributions in achieving the energy reduction, i.e., external pressure, particles geometry, and chemical reactions [3]. The main advantage of sintering is the possibility of manufacturing complex shapes with a large reduction of scrap material. Furthermore, the internal structure is less affected by the process nature, as it appears in conventional hot and cold forging. However, the high temperatures involved in sintering can generate some problems concerning creep, grain growth, and oxidation [4]. To reduce these effects, new sintering technologies have focused on reducing the time inside the furnace. Hot pressing (HP), where the two main steps are carried out simultaneously, represents an alternative to conventional sintering. The green body is heated inside the sintering die while the compaction pressure is maintained during the whole process. The thermo-mechanical field enables an increased densification rate [5]. Higher densities are achieved because the micro-porosities and elastic spring back are limited by the pressure action [6] and increased compact strength [7]. Later developments introduced a new heating approach based on an electrical current and Joule heating. To achieve high temperatures in a very short time, the electrical current flows through the green compact and/or the die. The pressure is maintained during the process as in HP. In general terms, it is identified as field-assisted sintering technology (FAST) [8]. As described by Grasso et al. [9], the term, "field", may identify different sources, e.g., mechanical, electrical, or electro-magnetic sources. Thus, electrical current assisted sintering (ECAS) is suggested as a subgroup to identify those processes where the electrical current is used for heating. The classification is done as a function of the electrical current waveforms and sources, loading methods, discharge time, and maximum peak current. In the case that electrically non-conductive powders are used, the electrical sintering provides the heating of the die. The approach is therefore comparable to HP, but the heating rate is increased by utilizing Joule heating. One of the most representative and oldest names is spark plasma sintering (SPS). The die and electrodes are usually made of graphite, enabling the current to flow within both the die and green compact, in which case it is conductive. The high density achieved is justified by the idea of plasma formation. However, no proof of plasma has been shown experimentally [10]. In the case of conductive powders, the electrical current flowing inside the green compact enables direct heating of the samples. The achieved high density is therefore a consequence of the complex electro-thermo-mechanical field, generating several phenomena among the particles, e.g., dielectric breakdown of the oxide layers and pinch-effects [11]. Alternating current, direct current, and capacitor discharge technologies are used [12]. The different current profiles result in different heating rates, involving "fast" processes, above 0.1 s, and "ultrafast" processes, below 0.1 s. The electrical current density is therefore set with lower values in the "fast" case, below 10 A/mm^2, than in the "ultrafast" case, above 100 A/mm^2 [9]. For example, flash sintering enables such reduced times by using high capacitor discharges [13]. The pressure can be kept constant or altered while processing, which is the case of electro sinter forging (ESF) [14].

Generally, the main process parameters are the electrical current, time, and compaction pressure. Several conductive powders have successfully been sintered with those processes, e.g., gold [15], copper [16], titanium [17], and melt spun magnetic powders [18]. However, the process limits are related to the material and machine properties. This requires a careful trade-off of the process parameters in connection with the final product characteristics, defined as process and product fingerprints.

By applying the "ultrafast" ESF principle, the present research focuses on the analysis of the effects derived from different process settings, defined as process fingerprint candidates, for ESF. The final quality of the components is affected by the electrical current, pressure, and time. The process window is limited by the required energy to ensure adequate consolidation of the particles and yet is sufficiently low enough to avoid any melting of the material. In the case of an alternating current, the electrical sine wave influences the results, because of the high current amplitudes. Furthermore, programming the alternating current machine is more complicated than the direct current machine. All sintered parts are analysed in terms of density. Mechanical properties are evaluated in terms of hardness and tensile and compressive strengths. Micrographs are made to identify and compare the achieved microstructures. The validity of these properties as product fingerprints is investigated based on their empirical connections with the process parameters. Fingerprints are found when unique input/output correlations are established.

2. Materials and Methods

Disc- and ring-shaped specimens were manufactured during the experiments (Figure 1). An axisymmetric design facilitated the investigation of the sintering process by limiting the problems related to thermal gradients. Rings were produced to analyse the results obtained with an additional tool contact surface, which influences the obtained properties. The tested material was commercial pure titanium (GoodFellow Inc., Huntingdon, UK), which is electrically conductive and has a melting temperature of 1668 °C [19]. A neodymium alloy, namely NdFeBCo (Magnequench, Tianjin, China), suitable for permanent magnets, was investigated in parallel research by the present authors (ongoing work by the present authors). The titanium raw powder is shown in Figure 2 by scanning electron microscope (SEM) backscattered electron detector (BSD) images. The sintered discs have an external diameter of 10 mm and a thickness of 3 mm. The rings were designed to have a comparable cross-sectional area and therefore similar electrical current density with equal nominal electrical values. They have an outer diameter of 13 mm, an internal diameter of 6.7 mm, and a 3 mm thickness.

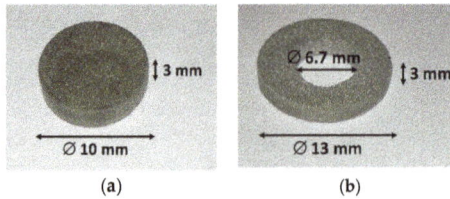

(a) (b)

Figure 1. Examples of sintered titanium samples by (**a**) a disc and (**b**) a ring.

Figure 2. Scanning electron microscope (SEM) backscattered electron detector (BSD) image of the titanium powder used for sintering.

The ESF experiments were carried out on two different resistance welding machines (Figure 3), i.e., middle frequency direct current (MFDC) machine from Expert (Expert Maschinenbau GmbH, Lorsch, Germany) and an alternating current (AC) machine from Tecna (Tecna, Castel S. Pietro Terme, Italy), in order to test different current profiles with comparable levels of total electrical energy. The two machines are equipped with different mechanical systems. The MFDC machine is hydraulically operated and has disc-springs for the follow-up of the compaction force, while the AC machine is pneumatically operated.

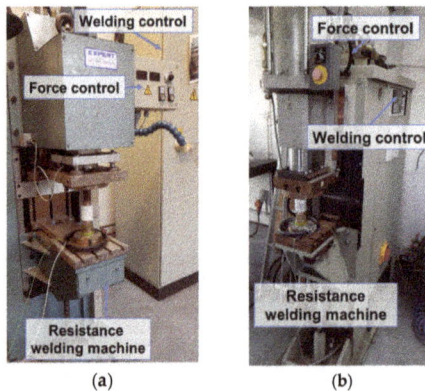

(a) (b)

Figure 3. Resistance welding machines used for sintering, (**a**) middle frequency direct current (MFDC) and (**b**) alternating current (AC) resistance welders.

The tool system was made by an electrically insulated die made of aluminium oxide and conductive punches/electrodes made of copper (Figure 4a). The two different tool designs shown in Figure 4b,c were used for discs and rings, respectively. The disc tool design consists of a cylindrical die with an inner hole. The ring tool design has an additional, removable alumina mandrel radially aligned by the two hollow electrodes. The mandrel prevents inwards flow of the ring specimen during ESF. After each ESF, the mandrel is removed and the ring is ejected from the die.

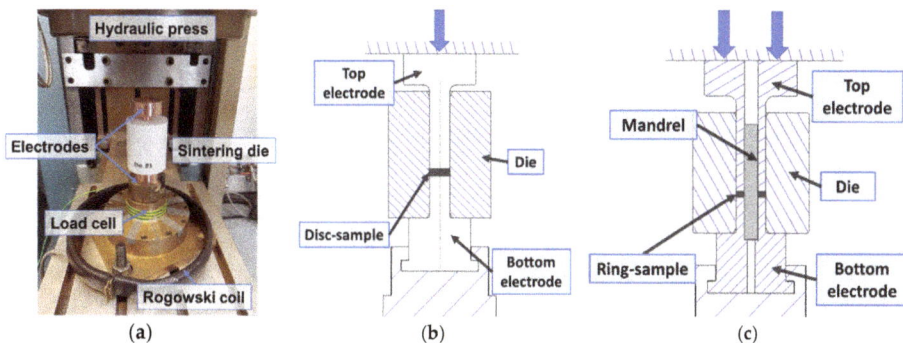

Figure 4. Pictures showing the (**a**) tool system used for electro sinter forging, (**b**) tool design for the disc, and (**c**) tool design for the ring.

3. Process and Product Fingerprints

Process and product fingerprints are considered as unique inputs and outputs for a manufacturing process. The definition of fingerprints is important to control and monitor the ESF process and to guarantee product quality. The following sections present specific process and product fingerprints for ESF of the discs and rings.

3.1. Process Fingerprints

Two different resistance welding machines were used. They differ in the generated current profiles, which are direct (Figure 5a) and alternating electrical currents (Figure 5b).

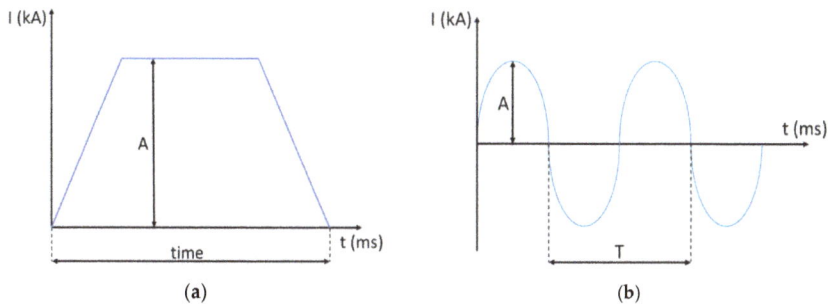

Figure 5. Typical electrical profiles in the case of (**a**) direct and (**b**) alternating currents.

In the case of the MFDC (Figure 5a), the nominal value, A, is set on the machine, while on the AC machine, the RMS value is entered. RMS indicates the root-mean-square value of the electrical current profile, which in the case of a 180° conduction angle, would be estimated as:

$$RMS = \frac{A}{\sqrt{2}} \qquad (1)$$

where *A* is the current amplitude (Figure 5b). For smaller conduction angles, the relation becomes more complicated because the wave form changes from the ideal sinusoidal form. The conduction angle represents the amount of time in which the current flows. Figure 6 shows the current profile for two experiments running with different conduction angles, 75% and 100%, but the same *RMS*. Figure 6 shows how the required amplitude for a given *RMS* depends on the conduction angle.

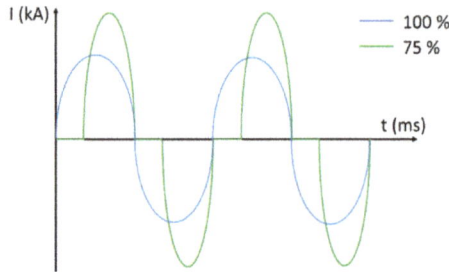

Figure 6. Illustrated examples of alternating current waves for different conduction angles, 75% and 100%.

The monitoring of the current was performed in-line by using a Rogowski coil (Tecna) as shown in Figure 4a. Data were saved for post processing and the electrical current was converted into an average current density. The load was in-line measured by using a piezoelectric load cell (Kistler, Winterthur, Switzerland) placed inside a brass cage as shown in Figure 4.

In-line monitoring of both the electrical current and load is important to verify that the process is performed correctly. If the electrical resistance of the powder compact is too large, no current is delivered due to the machine safety control, and sintering does not take place. This information was immediately available from the measured process parameters as shown in Figure 7. Figure 7a shows a typical diagram of the compaction pressure and the current density for a successful case, whereas Figure 7b shows an example in which the current is not delivered properly. In the successful case (Figure 7a), there is also a characteristic pattern between the pressure and the current density. The compaction pressure decreases when the current raises due to softening of the material and stabilizes shortly after the mechanical follow up from the machine.

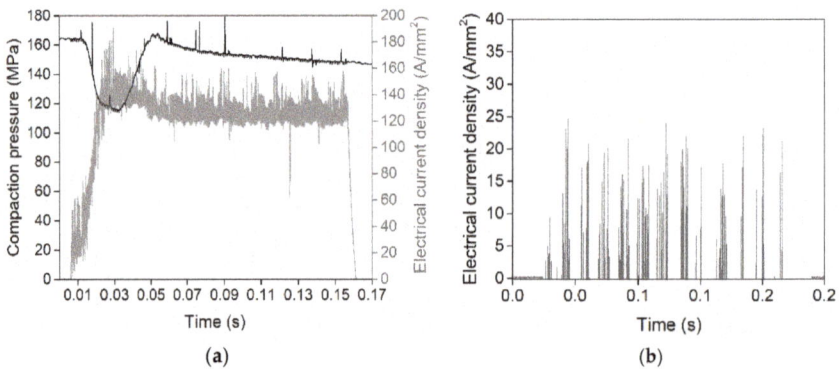

Figure 7. Typical MFDC diagrams when the sintering process is (a) correctly carried out and (b) aborted.

The voltage and temperature are parameters that can be considered as process fingerprints. Measuring the voltage may help in monitoring the density achieved while sintering. The model considers the variation of the electrical resistance of the material due to the increasing densification of the sample. According to Montes et al. [20], the electrical resistivity, ρ_e, of a metal powder aggregate, at room temperature, can be expressed as:

$$\rho_e = \rho_0(1 - \theta_r)^{-r} \tag{2}$$

where ρ_0 is the electrical resistivity at bulk density and room temperature, θ_r is the instantaneous porosity of the compact, and r is a resistivity exponent defined as:

$$r = 1 + (1 - \theta_m)^{4/5} \tag{3}$$

where θ_m is the starting porosity before sintering. The electrical resistivity, ρ_t, is also influenced by the temperature according to the following formula [21]:

$$\rho_t = \rho_e[1 + \alpha(T - T_0)] \tag{4}$$

where ρ_e is the aforementioned electrical resistivity at room temperature for a metal powder aggregate, α is the temperature coefficient of resistance, and T and T_0 are the instantaneous and room temperatures, respectively. Knowing the temperature also helps in understanding the process limits in terms of the maximum electrical energy. The maximum temperature should not go beyond the melting point, since the subsequent rapid cooling results in a dendritic structure, which is unsuitable for some applications due to generations of micro-voids.

In the present work, both the voltage and temperature were difficult to measure because of technological and process limits. For the voltage, attempts resulted in an increasing noise being generated in the data acquisition system, which affected the electrical current measurements. Temperature was an issue for the process characteristic itself. Very short sintering times and a closed die configuration made the conventional instruments, e.g., thermocouples and pyrometers, unsuitable for the purpose. Attempts were made by numerical simulations [22].

3.2. Product Fingerprints

To achieve properties as good as those in conventional manufacturing processes, the final density must be close to the bulk one. The internal porosities influence the achieved mechanical strength and increase the risk of crack propagation. However, for some advanced applications, e.g., biomedical products, a certain porosity is preferred [23]. For this reason, density is the most fundamental product fingerprint. The estimation of density was based on individual measurements of mass and volume. The mass was easily estimated by using a precision balance (Sartorius GmbH, Göttingen, Germany), while the volume was analysed by Archimedes' method [24] and 3D volume reconstruction. The 3D volume was estimated by using metrology software (GOM GmbH, Braunschweig, Germany). Compatible density estimations were achieved by the authors [25]. Although volume reconstruction requires a long measuring time, the sample was not influenced by the uncertainty produced by Archimedes' method in the case of highly porous samples, where liquid penetration affects the result by underestimating volumes and therefore overestimating densities. Precautions were taken by embedding the porous specimens in a waterproof resin.

The average porosity can be estimated by:

$$\theta = 1 - \frac{d}{d_0} \tag{5}$$

where d and d_0 are the sample and bulk densities, respectively. A detailed porosity analysis was achieved by cutting the specimen at a diametrical cross-section. After being ground and polished, the surface was investigated by optical analysis (Figure 8). An image analysis software was used to compute the porosity for the investigated area, by counting the number of black and white pixels representing porosity and material, respectively. Computed tomography is also a powerful instrument for porosity investigation [26].

Figure 8. Example of a diametrical disc cross-section ready for local porosity investigation. This sample is made of titanium and sintered by MFDC. The compaction pressure is 170 MPa, the sintering time is 150 ms, and the electrical current density is 178 A/mm^2.

The microstructure evaluation gives information regarding the sintering effect on the powder particles. The appearance of difference phases can vary as a function of the sintering parameters. In the case of ESF, the analysis of microstructures gave indications of the possible melting of the material. If the melting temperature was not reached, the micrographs clearly showed the bond interface between the particles. When high electrical currents were applied, the high temperatures exceeded the melting point of the sintered material. The achieved structures were typical of a melted material, as will be shown in Section 4.2. Microstructures were also used to investigate possible contaminations and oxidations of the material.

Mechanical properties were investigated in terms of hardness, tensile, and compressive strengths. Indirect tensile tests (IDT) [27] were conducted for the discs and compression tests were carried out for both the discs and the rings. A 600 kN hydraulic press (Mohr & Federhaff AG, Manheim, Germany) was used for both approaches. As shown in Figure 9a, the IDT requires compression of the sample perpendicular to its centre axis to generate a tensile fracture at the centre of the disc specimen by tension perpendicular to the compression direction. Linear elastic behaviour is assumed until fracture occurs in the IDT. Compression tests were performed as illustrated in Figure 9b.

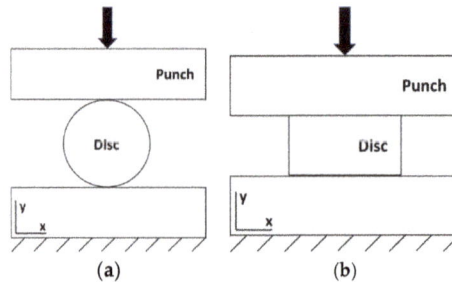

Figure 9. Schematic setups of the (**a**) indirect tensile test and (**b**) compression test.

Hardness tests were performed according to the Vickers method given by ISO 6507 [28]. A pyramidal indenter was applied on the polished surface. Micro Vickers equipment was used for the purpose, with standard HV0.1 (100 g) and 50× optical lens (Olympus, Tokyo, Japan).

In addition, further product fingerprints can be selected based on the specific product. Geometrical and dimensional features can be important fingerprints to be monitored and related to the process settings. For example, in the case of axisymmetric components designed for rotational applications, e.g. rotors, the eccentricity is a natural product fingerprint.

4. Results

4.1. Density Analysis

Titanium discs were sintered on both the MFDC and AC machines to evaluate the influence of the applied current profile. In both cases, an evaluation of the parameters with the most influence and an optimization of the process parameters were carried out. Figure 10 shows such a study based on the AC machine. The relative densities of the obtained samples are shown for various process settings, and in agreement with Joule's law of heating, the electrical current was found to be the most influencing process parameter. The parameter ranges were (i) 61–185 A/mm^2 *RMS* for the electrical current density, (ii) 36–156 MPa for the compaction pressure, and iii) 120–200 ms for the sintering time, which is entered as the number of cycles with a 20 ms period (50 Hz). The error bars in Figure 10 represent the measurement uncertainty estimated by taking into account the resolution of the balance and 3D scanner for measuring the mass and volume, respectively. The error propagation formula was applied based on the density formula and a coverage factor, k = 2, at a confidence level of 95%, according to the ISO GUM [29]. Figure 10a shows the obtained relative density as a function of the electrical current density. The density increases with the increasing current until around 140 A/mm^2, where it stabilizes. Internal melting of the samples occurred for the highest current densities, as shown by the micrographs in Section 4.2. Until 60 MPa, the compaction pressure increased the density (Figure 10b). Hereafter, the compaction pressure did not influence the density. The sintering time did not affect the final density in the range of 120–200 ms (Figure 10c).

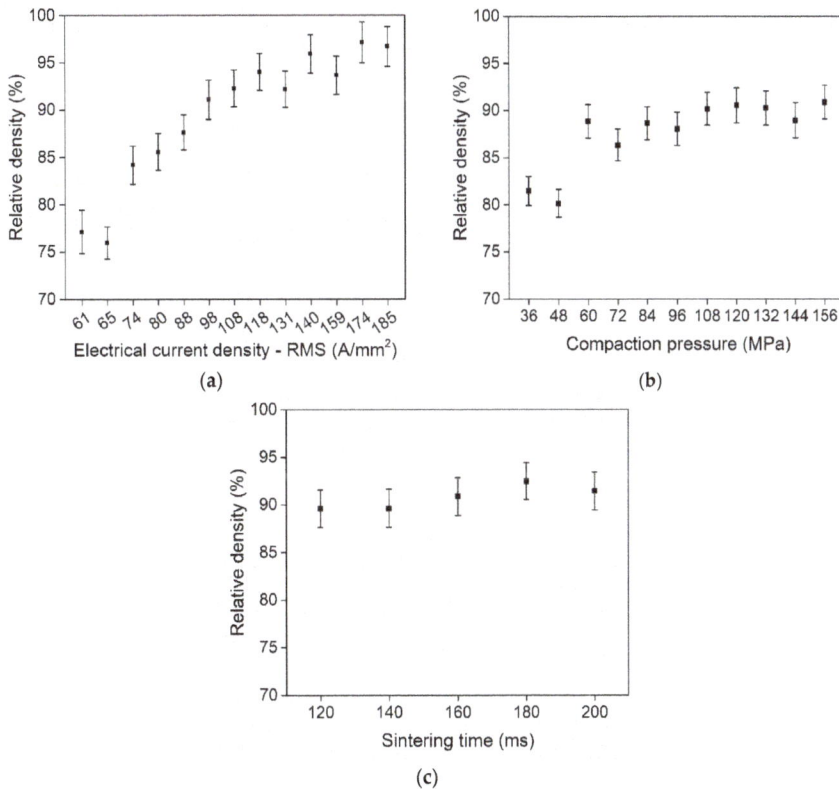

Figure 10. Density variation for titanium discs sintered in the AC machine as a function of the (**a**) electrical current density (84 MPa, 100 ms), (**b**) compaction pressure (98 A/mm^2, 100 ms), and (**c**) sintering time (98 A/mm^2, 94 MPa).

Similar trends were obtained when sintering titanium (ongoing work by the present authors) and magnetic (ongoing work by the present authors) discs in the MFDC machine. As a general conclusion, sintering was possible in both of the machines. However, the MFDC is more controllable than the

AC, which is more prone to melting the samples due to the high current amplitudes, at the same electrical energies.

The titanium rings were only sintered in the MFDC machine to avoid further complications from the AC current profile. Figure 11 shows the obtained relative densities for both the disc and ring-shaped samples as a function of the applied current density. The error bars again show the measurement uncertainties, which are calculated as described above in relation to Figure 10. As noted in Section 2, the axial cross-section of the disc and the ring were chosen to have comparable areas. The heights are also the same, and therefore, the initial resistance and expected current density are also comparable. The results showed how the contact area between the rod and compacted powder introduced larger thermal gradients, resulting in a lower density in the ring than in the disc for the same current density. This is more pronounced at high electrical currents, where the thermal gradients are steeper. Furthermore, the cross-sectional analyses in Section 4.2 show the poorer distribution of porosities in the ring samples due to the additional thermal gradient on the inside of the ring.

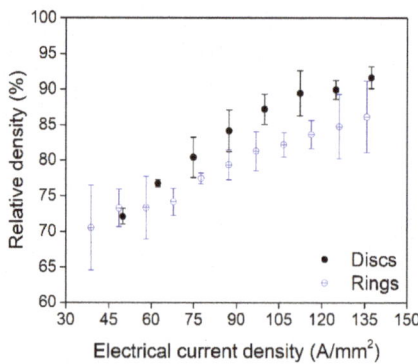

Figure 11. Comparison between the achieved densities for titanium discs and rings produced in the MFDC machine as a function of the electrical current density (150 ± 20 MPa, 150 ms).

4.2. Microstructures

Microstructural observations give an important overview of the particle bonding effect after sintering. The sintering limit can be defined as the maximum total energy that can be applied before melting. Depending on the application, melting can be accepted or not. However, in the case of melting, closed porosities as found in Figure 12, can be generated and therefore influence the final properties of the material.

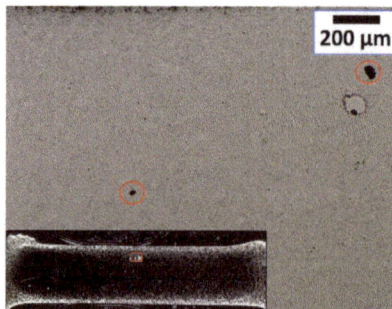

Figure 12. Micrograph showing examples of closed porosities for a titanium sample sintered by MFDC at 115 A/mm^2 for 150 ms under a 170 MPa compaction pressure.

Furthermore, it is possible to note how the most heated zone is the core of the sample (Figure 13). This can be explained by the cooling of the tools on the surface and the larger electrical resistance in the core before sintering. The reason for the latter is the compaction itself, which is not homogeneous. The result is a fully dense core surrounded by porous material.

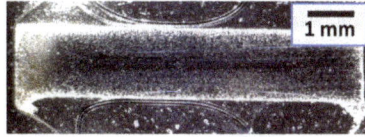

Figure 13. Optical picture showing the diametrical section of a titanium disc sintered by MFDC at 165 A/mm^2 for 150 ms under a 170 MPa compaction pressure. An internal fully dense core (dark) and a porous outer part (bright) can be seen.

Micrographs of samples sintered with the AC machine were compared with the samples sintered with MFDC. The results highlighted the difficulty in controlling the alternating current in terms of energy peaks due to the high amplitudes. The appearance of melted material was found at 76–89 A/mm^2 in the case of AC, while 115 A/mm^2 could be applied before melting with the MFDC. In the pictures shown in Figure 14, the comparison between titanium samples made by MFDC and AC is shown. At 76 A/mm^2, MFDC samples (Figure 14a) have a typical sintered microstructure with bonded particles. Conversely, in Figure 14b, dendritic structures typical of a melted and rapidly solidified material are shown in a sample made by AC. In the latter case, the high amplitudes resulted in current density peaks of 188 A/mm^2, which is 2.5 times the *RMS* value and the corresponding current density when using MFDC.

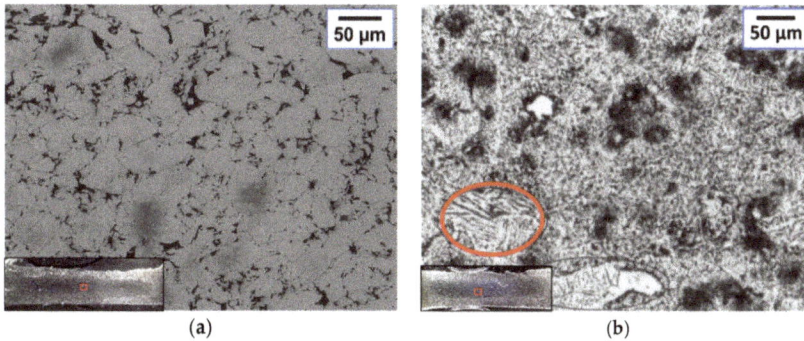

| (a) | (b) |

Figure 14. Micrographs showing the achieved microstructures for titanium samples sintered at (a) 76 A/mm^2 MFDC for 150 ms at a compaction pressure of 170 MPa and (b) 76 A/mm^2 *RMS* AC for 100 ms at a compaction pressure of 82 MPa.

When sintering titanium rings, the structures and density become asymmetric. This was observed when samples were diametrically cut and embedded into resin (Figure 15). The achieved structures suffered from thermal and pressure gradients created between the two sections of the same diametrical cut. The main reason is the inhomogeneous starting density of the green body after the initial compaction. The central mandrel constituted an obstacle to the reallocation of particles during compaction. Furthermore, according to the theoretical model described by Equations (2) and (3), an increase in electrical resistivity is seen with an increasing porosity. This provoked inhomogeneous Joule heating, inducing a thermal gradient. Figure 16 shows the influence on the microstructure in one example. In Figure 15e, the geometry is also affected, showing a well-shaped rectangular section on one side (right) and a slightly deformed one on the other side (left).

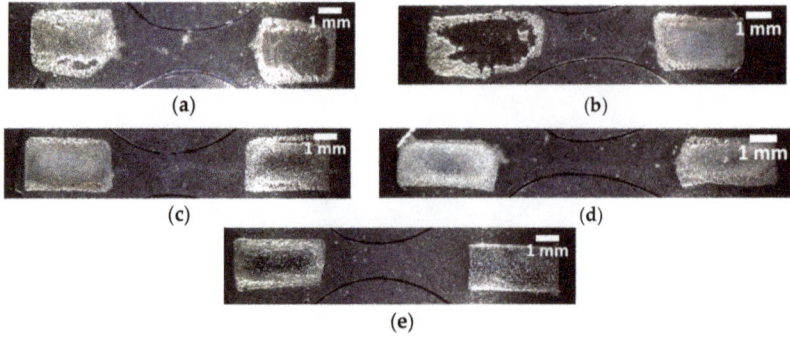

Figure 15. Optical images showing the diametrical sections obtained for sintered titanium rings at 133 MPa, 150 ms and (**a**) 60, (**b**) 82, (**c**) 103, (**d**) 123, and (**e**) 144 (A/mm^2).

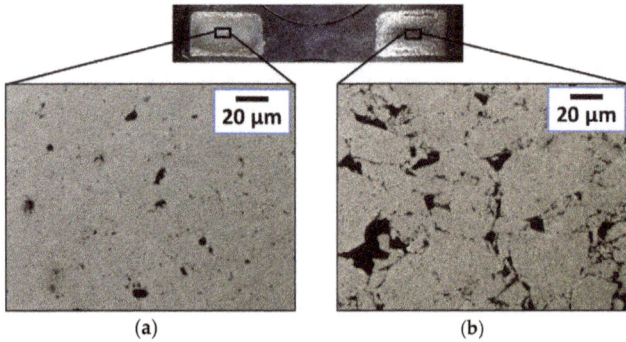

Figure 16. Optical micrographs showing the detailed microstructures achieved for a titanium ring sintered at 133 MPa, 150ms, and 103 A/mm^2. The differences between the two sides are clear by having (**a**) a dense core with micro-voids and (**b**) bonded particles not completely densified.

Magnetic rings broke during ejection due to their brittleness. The structure was not strong enough to withstand ejection. The problems experienced when sintering rings may be reduced by limiting thermal dissipation to the sintering tools. Pre-heating of the die and electrodes could also be investigated.

4.3. Mechanical Properties

Hardness, tensile, and compressive tests were performed to evaluate the mechanical quality of the sintered samples.

The hardness measurements are shown in Figure 17 for titanium discs sintered by AC. Seven indentations were performed at the core with about a three diameters distance to avoid any influence from previous indentations. The average values and standard uncertainties were estimated for each analysed sample. The expanded uncertainty was evaluated with a coverage factor of $k = 2$ at a confidence level of 95%, according to the ISO GUM [29]. Figure 17 shows how the hardness is influenced by the achieved microstructures. The bonded microstructures seen for low relative densities (<80%) showed a lower average hardness than in the case of high relative densities (>80%).

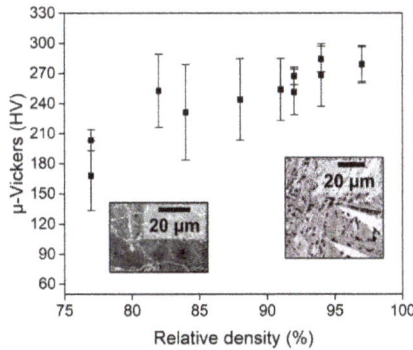

Figure 17. Hardness measured for titanium samples sintered by AC as a function of the relative density. Each point represents a different sample, and corresponding error bars refer to measurement uncertainties. Microstructures are distinguished for low and high densities.

Compressive and indirect tensile tests were done on the titanium samples sintered by AC. For the compressive tests, the samples exhibited a double brittle/ductile behaviour, corresponding to low and high densities. A high degree of plasticity was noticed with high densities, above 80%. Low densities showed evident cracks at the end of the compression, as shown in Figure 18a. The indirect tensile tests showed the same double nature, being both brittle and ductile. In this case, low density samples broke by brittle fracture. The IDT was estimated as 30 ± 10 MPa. High density samples fractured by shearing, and the IDT principle was therefore not applied due to the deviation from linear elastic deformation until fracture, which is a requirement for the test [30]. Titanium discs sintered by MFDC show similar mechanical trends as a function of the obtained relative density (ongoing work by the present authors).

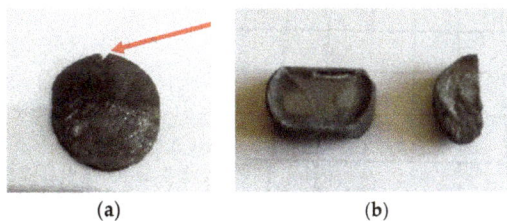

(a)   (b)

Figure 18. Titanium samples sintered by AC and tested by (**a**) compression with the red arrow identifying a fracture and (**b**) indirect tensile testing. The sample in (**a**) has a relative density of 71% and the samples in (**b**) have relative densities >80% (left) and <80% (right) with corresponding ductile and brittle behaviour, respectively.

5. Conclusions

An analysis of the main parameters led to the definition of the process and product fingerprints for ESF being identified, which corresponds to the most influencing process parameters and final sample properties. The process fingerprints identified were the current and compaction pressure profiles. In terms of current input, both an alternating current (AC) and middle-frequency direct current (MFDC) were analysed. ESF was possible with both machines. However, the AC machine had the disadvantage of overheating the sample, with high current peaks required to deliver a given *RMS* current.

The product fingerprints identified were the density, microstructure, and mechanical properties in terms of the hardness and strength. Density is a main parameter and increases with increasing process energy, but melting should be avoided to minimize undesired effects, such as closed porosities, oxidation, and possible grain growth. The increase in density was greatly influenced by higher values of currents. The electrical current was therefore considered as the most valuable process fingerprint for ESF. Microstructure analysis revealed whether samples were involved in particle bonding or melting.

By testing a different geometry, namely a ring, the process was affected more by the compaction itself, generating different grades of density in symmetric regions of the rings. This was provoked by the different temperature achieved because of the non-uniform porosity distribution, involving different values of electrical resistances and therefore not the same Joule heat value. Future work should focus on force sensors as a process fingerprint to monitor the uniformity of the compaction pressure while the green body is formed.

Additional product fingerprints can be defined for specific purposes, e.g., eccentricity would be a key product fingerprint for applications in rotors or other equipment involving high-speed rotation. General dimensional and geometrical features can also be selected. Future work should focus on a statistical analysis to establish robust correlations between process fingerprints and specific product fingerprints for given applications.

Author Contributions: E.C. and C.V.N. conceived and designed the experiments; E.C. performed experiments and measurements; E.C. and C.V.N analysed the data; E.C. wrote the paper; C.V.N. and N.B. revised the paper.

Funding: This research received no external funding.

Acknowledgments: This research work was undertaken in the context of MICROMAN project ("Process Fingerprint for Zerodefect Net-shape MICROMANufacturing", http://www.microman.mek.dtu.dk/). MICROMAN is a European Training Network supported by Horizon 2020, the EU Framework Programme for Research and Innovation (Project ID: 674801).

Conflicts of Interest: The authors declare no conflict of interest.

References

1. German, R.M. History of sintering: Empirical phase. *Powder Metall.* **2013**, *56*, 117–123. [CrossRef]
2. Castro, R.H.R. *Sintering*; Castro, R., van Benthem, K., Eds.; Engineering Materials; Springer: Berlin/Heidelberg, Germany, 2013; Volume 35, ISBN 978-3-642-31008-9.
3. Kong, L.B.; Huang, Y.; Que, W.; Zhang, T.; Li, S.; Zhang, J.; Dong, Z.; Tang, D. Sintering and Densification (I)—Conventional Sintering Technologies. In *Transparent Ceramics*; Springer: Cham, Switzerland, 2015; Volume 37, pp. 291–394. ISBN 3319189565.
4. Panigrahi, B.B.; Godkhindi, M.M.; Das, K.; Mukunda, P.G.; Ramakrishnan, P. Sintering kinetics of micrometric titanium powder. *Mater. Sci. Eng. A* **2005**, *396*, 255–262. [CrossRef]
5. Atkinson, H.V.; Davies, S. Fundamental aspects of hot isostatic pressing: An overview. *Metall. Mater. Trans. A Phys. Metall. Mater. Sci.* **2000**, *31*, 2981–3000. [CrossRef]
6. Garay, J.E. Current-Activated, Pressure-Assisted Densification of Materials. *Annu. Rev. Mater. Res.* **2010**, *40*, 445–468. [CrossRef]
7. Xiao, Z.Y.; Ke, M.Y.; Fang, L.; Shao, M.; Li, Y.Y. Die wall lubricated warm compacting and sintering behaviors of pre-mixed Fe-Ni-Cu-Mo-C powders. *J. Mater. Process. Technol.* **2009**, *209*, 4527–4530. [CrossRef]
8. Guillon, O.; Gonzalez-Julian, J.; Dargatz, B.; Kessel, T.; Schierning, G.; Räthel, J.; Herrmann, M. Field-Assisted Sintering Technology/Spark Plasma Sintering: Mechanisms, Materials, and Technology Developments. *Adv. Eng. Mater.* **2014**, *16*, 830–849. [CrossRef]
9. Grasso, S.; Sakka, Y.; Maizza, G. Electric current activated/assisted sintering (ECAS): A review of patents 1906–2008. *Sci. Technol. Adv. Mater.* **2009**, *10*, 053001. [CrossRef] [PubMed]
10. Bonifacio, C.S.; Holland, T.B.; van Benthem, K. Evidence of surface cleaning during electric field assisted sintering. *Scr. Mater.* **2013**, *69*, 769–772. [CrossRef]
11. Vanmeensel, K.; Laptev, A.; Huang, S.G.; Vleugels, J.; Van der Biest, O. The Role of the Electric Current and Field during Pulsed Electric Current Sintering. In *Ceramics and Composites Processing Methods*; John Wiley & Sons, Inc.: Hoboken, NJ, USA, 2012; pp. 43–73. ISBN 111817660X.
12. Anselmi-Tamburini, U.; Groza, J.R. Critical assessment: Electrical field/current application—A revolution in materials processing/sintering? *Mater. Sci. Technol.* **2017**, *33*, 1855–1862. [CrossRef]

13. Yu, M.; Grasso, S.; Mckinnon, R.; Saunders, T.; Reece, M.J. Review of flash sintering: Materials, mechanisms and modelling. *Adv. Appl. Ceram.* **2017**, *116*, 24–60. [CrossRef]

14. Fais, A. Processing characteristics and parameters in capacitor discharge sintering. *J. Mater. Process. Technol.* **2010**, *210*, 2223–2230. [CrossRef]

15. Forno, I.; Actis Grande, M.; Fais, A. On the application of Electro-sinter-forging to the sintering of high-karatage gold powders. *Gold Bull.* **2015**, *48*, 127–133. [CrossRef]

16. Fais, A.; Scardi, P. Capacitor discharge sintering of nanocrystalline copper. *Zeitschrift fur Krist. Suppl.* **2008**, *27*, 37–44. [CrossRef]

17. Montes, J.M.; Rodríguez, J.A.; Cuevas, F.G.; Cintas, J. Consolidation by electrical resistance sintering of Ti powder. *J. Mater. Sci.* **2011**, *46*, 5197–5207. [CrossRef]

18. Castle, E.; Sheridan, R.; Zhou, W.; Grasso, S.; Walton, A.; Reece, M.J. High coercivity, anisotropic, heavy rare earth-free Nd-Fe-B by Flash Spark Plasma Sintering. *Sci. Rep.* **2017**, *7*, 1–12. [CrossRef] [PubMed]

19. Brandes, E.A.; Brook, G.B. *Smithells Metals Reference Book: Seventh Edition*; Elsevier: Amsterdam, The Netherlands, 2013; ISBN 9780080517308.

20. Montes, J.M.; Cuevas, F.G.; Cintas, J. Electrical resistivity of a titanium powder mass. *Granul. Matter* **2011**, *13*, 439–446. [CrossRef]

21. Weiner, L.; Chiotti, P.; Wilhelm, H.A. *Temperature Dependence of Electrical Resistivity of Metals*; Ames Laboratory ISC Technical Reports; Iowa State University: Ames, IA, USA, 1952.

22. Pavia, A.; Durand, L.; Ajustron, F.; Bley, V.; Chevallier, G.; Peigney, A.; Estournès, C. Electro-thermal measurements and finite element method simulations of a spark plasma sintering device. *J. Mater. Process. Technol.* **2013**, *213*, 1327–1336. [CrossRef]

23. Dutta, B.; (Sam) Froes, F.H. The additive manufacturing (AM) of titanium alloys. In *Titanium Powder Metallurgy*; Elsevier: Amsterdam, The Netherlands, 2015; pp. 447–468. ISBN 978-0-12-800054-0.

24. *ASTM International Standard Test Methods for Density of Compacted or Sintered Powder Metallurgy (PM) Products Using Archimedes' Principle*; Astm B962-13; ASTM International: West Conshohocken, PA, USA, 2013. [CrossRef]

25. Cannella, E.; Nielsen, C.V.; Bay, N. Process parameter influence on Electro-sinter-forging (ESF) of titanium discs. In Proceedings of the 18th International Conference of the european Society for Precision Engineering and Nanotechnology (euspen 18), Venice, Italy, 4–8 June 2018; pp. 315–316.

26. Cannella, E.; Nielsen, E.K.; Stolfi, A. Designing a Tool System for Lowering Friction during the Ejection of In-Die Sintered Micro Gears. *Micromachines* **2017**, *8*, 1–15. [CrossRef] [PubMed]

27. Fahad, M.K. Stresses and failure in the diametral compression test. *J. Mater. Sci.* **1996**, *31*, 3723–3729. [CrossRef]

28. *ISO BS EN 6507-1:2018—Metallic Materials—Vickers Hardness Test—Part 1: Test Method*; ISO: Geneva, Switzerland, 2018.

29. *ISO/IEC Guide 98-3: 2008 Uncertainty of Measurement—Part 3: Guide to the Expression of Uncertainty in Measurement (GUM:1995)*; ISO: Geneva, Switzerland, 2008.

30. Procopio, A.T.; Zavaliangos, A.; Cunningham, J.C. Analysis of the diametrical compression test and the applicability to plastically deforming materials. *J. Mater. Sci.* **2003**, *38*, 3629–3639. [CrossRef]

Article

Process Fingerprint in Micro-EDM Drilling

Mattia Bellotti, Jun Qian and Dominiek Reynaerts *

Department of Mechanical Engineering, KU Leuven, Member Flanders Make, 3001 Leuven, Belgium;
mattia.bellotti@kuleuven.be (M.B.); jun.qian@kuleuven.be (J.Q.)
* Correspondence: dominiek.reynaerts@kuleuven.be; Tel.: +32-16-32-2640

Received: 7 March 2019; Accepted: 8 April 2019; Published: 11 April 2019

Abstract: The micro electrical discharge machining (micro-EDM) process is extensively used in aerospace, automotive, and biomedical industries for drilling small holes in difficult-to-machine materials. However, due to the complexity of the electrical discharge phenomena, optimization of the processing parameters and quality control are time-consuming operations. In order to shorten these operations, this study investigates the applicability of a process fingerprint approach in micro-EDM drilling. This approach is based on the monitoring of a few selected physical quantities, which can be controlled in-line to maximize the drilling speed and meet the manufacturing tolerance. A Design of Experiments (DoE) is used to investigate the sensitivity of four selected physical quantities to variations in the processing parameters. Pearson's correlation is used to evaluate the correlation of these quantities to some main performance and hole quality characteristics. Based on the experimental results, the potential of the process fingerprint approach in micro-EDM drilling is discussed. The results of this research provide a foundation for future in-line process optimization and quality control techniques based on machine learning.

Keywords: electrical discharge machining; micro drilling; process monitoring; quality control

---

1. Introduction

Micro electrical discharge machining (micro-EDM) drilling is a well-established non-contact thermal process for making small holes in electrically conductive materials, such as cooling holes in turbine blades and diesel injector nozzles [1,2]. In this process, the removal of material occurs through sequences of high-frequency electrical discharges within an electrically insulated gap between two electrodes. Deionized water or hydrocarbon oil are commonly applied as dielectric medium. Tubular tools of brass or copper material are used to increase the drilling speed [3,4].

Due to the complexity involved in the electrical discharging process, optimization of the processing parameters in micro-EDM drilling is an iterative and time-consuming process, which is often based on manual experimentation and user experience. Likewise, the complexity of the discharging phenomena does not facilitate the establishment of correlations between the applied processing parameters and the final hole quality. It follows that significant post-processing metrology efforts are often required for quality control and tolerance verification of micro-EDMed holes. Recent advancements in machine learning techniques offer new solutions for shortening and automating process optimization and allow for the creation of technology databases to reduce post-process metrology [5,6]. However, effective and reliable approaches for correlating the processing parameters and the relevant outputs in terms of efficiency of the drilling process and hole quality are needed for these purposes.

Traditionally, a so-called 'direct approach' has been used for optimizing the processing parameters and predicting the hole quality in micro-EDM drilling. As shown in Figure 1, this approach focuses on establishing direct correlations between the processing parameters and the performance and quality characteristics. Extensive research has been conducted in recent years to establish such correlations.

For instance, Lin et al. [7] used a response surface method to model the effects of different processing parameters on the material removal rate (MRR), tool wear rate (TWR), and diameter overcut (DOC) in micro-EDM of carbon tool steel. Jung et al. [8] and Ay et al. [9] used gray relational analysis for multiple performance optimization when processing stainless steel and Inconel 718 as a workpiece material. D'Urso et al. [10] defined two process windows representing the TWR and MRR as a function of the hole depth, considering various processing parameters and electrode materials. Jahan et al. [11] studied the influence of the process-energy parameters on the MRR and taper ratio (TR) in micro-EDM drilling of tungsten carbide. Suganthi et al. [12] used adaptive neuro-fuzzy inference system and artificial neural networks to predict the MRR, TWR, and surface roughness (SR) from the processing parameter settings. Although using direct correlations could be an effective solution for optimizing the processing parameters, no real-time feedback can be obtained during the drilling process to check whether any variation from the process conditions, under which the correlations were established, is occurring. Therefore, post-processing measurements are required to check the hole quality. Accurately performing these measurements could be tedious or, in some cases, not even possible. (e.g., small hole size or accessibility problems).

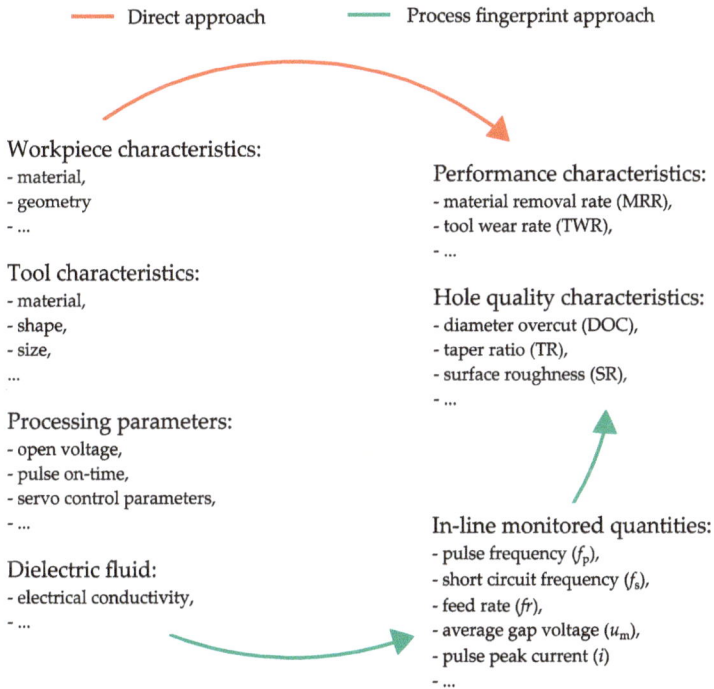

Figure 1. Following the process fingerprint approach, in-line monitored quantities can be used for optimizing the processing parameters and predicting the hole quality. This could reduce or even omit the post-processing metrology efforts required when following the direct approach.

An alternative to the direct approach could be the process fingerprint approach, which attempts to find correlations between the performance and quality characteristics and measurable physical quantities that can be monitored and controlled in-line (Figure 1). In this context, the term 'process fingerprint' refers to what is left on the workpiece after the manufacturing process, considering the mechanism and dynamics of the material removal process. The physical quantities, which are strongly correlated to the process fingerprint and are in-line monitored and controlled, can be denoted as indicators for the process fingerprint. The process fingerprint has previously also been referred to as 'process signature' in the specific case of surface integrity or surface modification in manufacturing processes [13,14]. For example, Sealy et al. [15] identified energy process signatures for the surface integrity problem in hard milling. Klink [16] introduced the concept of process signatures from the point of view of part functionality and surface modification in electrochemical machining (ECM) and EDM.

Following the process fingerprint approach, real-time optimization algorithms and decision-making systems can be implemented to maximize the drilling rate and meet the manufacturing tolerance by keeping the in-line monitored quantities within a desired range. Hence, the time for reaching the optimal setting of processing parameters could be shortened and post-processing metrology efforts could be drastically reduced or even omitted. The applicability of the process fingerprint approach to some micro-manufacturing processes has been proved, such as for micro-milling [17] and micro-injection moulding [18]. However, despite its unique advantages, the approach has not currently been applied in micro-EDM drilling.

In order to shorten the process optimization time and reduce post-process metrology efforts in micro-EDM drilling, this paper explores the applicability of the process fingerprint approach by quantitatively studying the correlations between some selected physical quantities, and the main performance and hole quality characteristics. Four quantities that are conventionally used for in-line monitoring of the micro-EDM process are investigated as potential indicators of the process fingerprint. A Design of Experiments (DoE) is used as a screening methodology to analyze the sensibility of these quantities to variations in the processing parameters. A correlation study is performed to identify the quantities that are mostly correlated to the main outputs in terms of drilling efficiency (MRR and TWR) and hole quality (DOC and TR). The most suitable indicators to be considered for the process fingerprint are subsequently identified and discussed.

2. Materials and Methods

2.1. Experimental Setup

Micro-EDM drilling experiments were carried out on a SARIX® SX-100-HPM (Sarix SA, Sant'Antonino, Switzerland) machine tool, which was equipped with the latest SARIX® PULSAR pulse generator. A tool guiding system was used to reduce the run out of the tool when approaching the workpiece (Figure 2). This system consists of a ceramic tool guide and a guide holder. The tool guide was positioned at a distance of approximately 2 mm above the top surface of the workpiece.

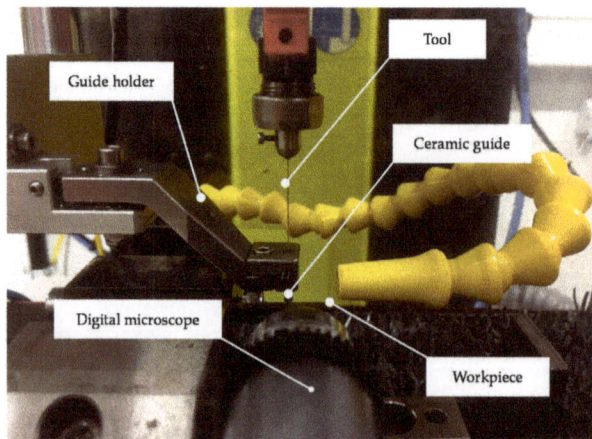

Figure 2. Experimental setup.

Ti6Al4V was chosen as the workpiece material, since it is an alloy widely-used in biomedical and aerospace industries due to its excellent biocompatibility, high corrosion resistance, and good mechanical properties at high temperatures [19–22]. Commercial brass tubes provided by SARIX® (outer diameter: 350 μm, inner diameter: 130 μm) were used as the tool electrode. Hydrocarbon oil (HEDMA® 111) was applied as dielectric liquid. A combination of side and internal flushing was used. The flushing pressures were set to 0, 2, and 4 MPa respectively.

Series of through-holes were drilled into Ti6Al4V small plates of 2 mm thickness and 6 mm width. The plates were fixed as cantilevers into the workpiece holder to allow the tool to be fed through the hole outlet after the breakthrough. A digital microscope (Dino-Lite® Edge AM4115ZT, AnMo Electronics Corporation, Hsinchu, Taiwan) was used to monitor the drilling process and interrupt the process as soon as a breakthrough was observed.

2.2. Process Monitoring

During the experiments, five variables were monitored in-line by the monitoring system embedded in the power generator of the SARIX® machine. The main advantage of using this embedded system is that no external sensors or data processing systems are required. This considerably simplifies potential future implementations of the monitoring operations performed in this research into an industrial production environment.

The monitored variables were (i) n_p: The number of normal discharge pulses, (ii) n_s: The number of short circuits, (iii) u_m: The average gap voltage, (iv) Δz: The z-axis displacement of the ram of the machine tool, and (v) t: The drilling time. In particular, n_p, n_s, and u_m were recorded through the pulse generator system, Δz was monitored by means of the z-axis encoder of the machine tool, and t was recorded using the clock of the computer of the control unit.

These five variables were synchronized and simultaneously updated at a rate of 100 Hz by the control unit of the machine, whereby a cumulative sum of each variable was stored. In order to monitor the variables at regular intervals, the drilling process was divided into steps of 50 μm in depth. This value refers to the nominal drilling depth as read by the z-axis encoder of the machine. Therefore, it does not correspond to the depth of the hole being drilled because of the longitudinal wear of the tool. At the end of each step, four quantities were computed in-line from the cumulative sums of n_p, n_s, u_m, Δz and t, which were subsequently zeroed. These quantities are:

- Pulse frequency (f_p): The amount of normal discharge pulses per time unit,
- Short circuit frequency (f_s): The amount of short circuits and arcs per time unit,
- Feed rate (fr): The speed at which the tool electrode is advanced into the workpiece, and
- Differential gap voltage (Δu): The difference between the open voltage and average gap voltage.

Figure 3 provides a summary of the operations performed to compute these four quantities, which are considered as potential indicators for the process fingerprint in this research. While f_p, f_s, and fr are quantities that are conventionally monitored in micro-EDM operations for various purposes such as tool wear compensation [23,24] and breakthrough detection [25,26]. For the purposes of this study, the meaning of Δu deserves a more detailed explanation. This Δu quantity varies with the number of normal and abnormal discharge pulses, depending whether the abnormal discharges are short circuits or open circuits. In particular, Δu is expected to approach zero when the discharging process is mostly in the state of open circuit, while it increases with the amount of normal discharge pulses or short circuits. Since short circuits and arcs are characterized by a lower discharge voltage than normal discharge pulses [27], the increase of Δu is larger at the occurrence of short circuits and arcs rather than normal discharge pulses.

At the end of each drilling experiment, a comma-separated value (.csv) text file including the monitored values of f_p, f_s, fr, and Δu at every 50 μm drilling step was automatically generated. The text files were later post-processed in order to identify the most suitable indicators for the process fingerprint.

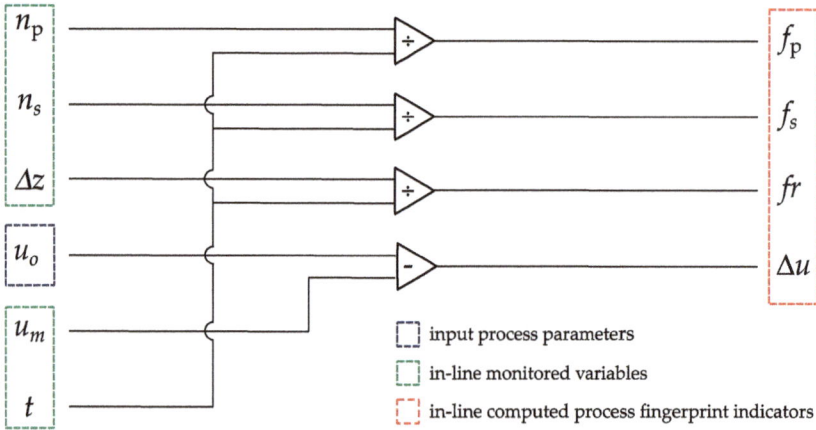

Figure 3. In-line calculation of the four quantities that are investigated as potential indicators for the process fingerprint of the micro electrical discharge machining (micro-EDM) drilling process (n_p: Number of normal discharge pulses, n_s: Number of short circuits and arcs, u_o: Open voltage, u_m: Average gap voltage, Δz: Nominal drilling depth, t: Drilling time, f_p: Frequency of normal discharge pulses, f_s: Frequency of short circuits and arcs, fr: Feed rate, Δu: Differential voltage).

2.3. Process Fingerprint Analysis

A post-process analysis was carried out to evaluate the correlation of the potential indicators for the process fingerprint to the performance and hole quality characteristics. A three-step analysis process was followed. First, the sensitivity of each indicator to changes of the processing parameters was investigated. Secondly, the correlation of each indicator to the MRR and TWR was evaluated. Thirdly, the correlations to the DOC and TR were computed. The Pearson's correlation coefficient [28] was chosen to evaluate all correlations since it is a widely-used metric to evaluate the strength of the statistical relationship between two variables [29,30].

First, a DoE approach was adopted to investigate the sensitivity of the potential indicators for the process fingerprint to changes to the processing parameters. In particular, four processing parameters were varied in a two-level full factorial design. Four repetitions were carried out for each experiment, resulting in a total of 64 experimental runs. The selected processing parameters were the (A) pulse off-time, (B) open voltage, (C) servo adjustment factor, and (D) reference gap voltage. The servo adjustment factor is the proportional gain factor of the servo control loop. The processing parameters were chosen in order to include changes related either to the cycle time of the power supply system (A), to the energy input per discharge (B), or to the settings of the servo feed control system (C,D). The levels of the processing parameters were set so as to cover a wide process window. The selection was carried out according to the machine vendor's recommendation and to previously reported experimental research involving a similar combination of tool and workpiece materials [4,22,25]. Table 1 is a summary of the experimental factors and their levels.

Table 1. Design of Experiments (DoE): factors and levels.

Factor	Process Parameter	Unit	Level −1	Level +1
A	Pulse off-time	μs	5	20
B	Open voltage	V	80	160
C	Servo adjustment factor	–	20	80
D	Reference voltage	V	30	70

The following processing parameters, not included in the DoE, were kept unchanged during the experiments: Pulse on-time – 4 μs, energy index – 301, current index – 70, tool rotation – 750 rpm, tool polarity – negative. According to the machine vendor, the energy index determines the energy and shape of the discharge pulses. The energy index 301 provides triangular pulses of medium-high energy content. When this energy index is selected, the current index can be used to regulate the pulse peak current.

Secondly, the correlation of each indicator to the MRR and TWR was analyzed in order to investigate the possibility of using the indicators of the process fingerprint for the purpose of optimizing the processing parameters. The MRR was calculated as the ratio between the volume of material removed from the workpiece (V_w) and the total machining time (T), approximating the through-holes to a conical frustum:

$$\text{MRR} = \frac{V_w}{T} = \frac{\frac{\pi}{12} H (D_{in}^2 + D_{out}^2 + D_{in} D_{out})}{T} \tag{1}$$

where H is the thickness of the workpiece, and D_{in} and D_{out} are the inlet and outlet diameters of the holes, respectively, which were measured using a Werth VideoCheck HA coordinate measuring machine in optical mode. Similarly, the TWR was computed as the ratio between the volume of material removed from the tool (V_t) and the total machining time (T). To compute such volume, the nominal inner (d_t) and outer (D_t) diameters of the tool were used, while the longitudinal wear was calculated by subtracting the workpiece thickness (H) to the monitored value of the drilling depth at the occurrence of breakthrough (Z_b) as shown in Equation (2).

$$\text{TWR} = \frac{V_t}{T} = \frac{\frac{\pi}{4} (D_t^2 - d_t^2)(Z_b - H)}{T} \tag{2}$$

Lastly, the DOC and TR were used to evaluate the quality of the drilled holes, since these are the two parameters that are mostly considered for assessing the geometrical characteristics of micro-EDMed holes [4,5,31–34]. The DOC was computed as:

$$\text{DOC} = D_{out} - D_t \tag{3}$$

while the TR was calculated as:

$$\text{TR} = \frac{D_{out} - D_{in}}{H} \tag{4}$$

The Pearson's correlation coefficient (r_p) was then calculated for all possible combinations between the indicators for the process fingerprint and the performance and hole quality characteristics. The r_p coefficient was computed as [28]:

$$r_p = \frac{\sum_{i=1}^{64} (x_i - \bar{x})(y_i - \bar{y})}{\sqrt{\sum_{i=1}^{64} (x_i - \bar{x})^2} \sqrt{\sum_{i=1}^{64} (y_i - \bar{y})^2}} \tag{5}$$

where x and y are the vectors containing the datasets of the two quantities that are correlated, and $\bar{x}$ and $\bar{y}$ are their respective mean values. The average values of the indicators for the process fingerprint in each of the 64 experimental runs were considered. The value of r_p can vary between -1 and $+1$, which correspond to a perfect negative correlation and a perfect positive correlation, respectively. A value equal to 0 indicates the absence of a correlation between two considered data sets. Therefore, the closer the value of r_p to -1 or $+1$, the stronger the linear correlation between the two data sets.

3. Results and Discussion

3.1. Sensitivity to Changes to the Processing Parameters

Figure 4 shows the data sets resulting from the 64 experimental runs. These data sets were used to analyze the sensitivity of the four potential indicators for the process fingerprint to changes to the processing parameters. For each indicator, the main effects plot and the Pareto chart of the effects are provided in Figure 5. The main effects plots can be used for analyzing the overall influence of each processing parameters on the indicators, while the Pareto charts show which effects are significant. This is of particular interest in this study since the indicators for the process fingerprint should be sensitive to variations to the process conditions, especially for the purpose of optimizing the processing parameters.

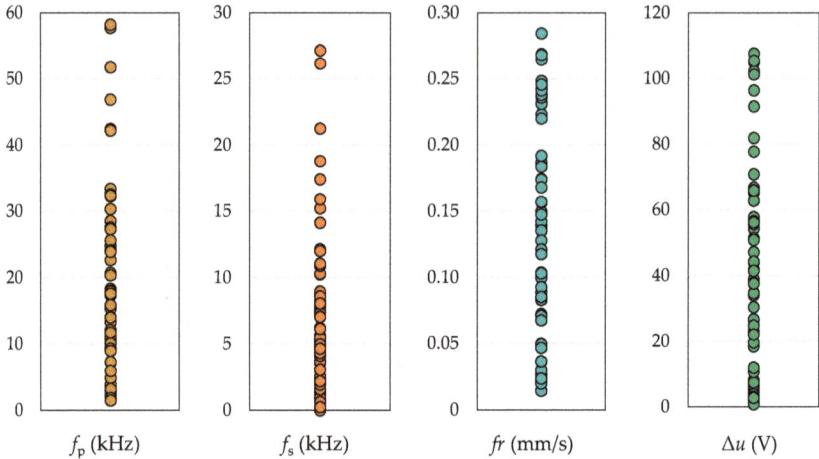

Figure 4. Plots of the data sets for the four potential indicators for the process fingerprint. Each data point corresponds to one of the 64 experimental runs.

From the main effect plots it can be seen that f_p increased when a shorter pulse off-time, a higher energy amount per discharge pulse (i.e., higher open voltage), or more aggressive settings of the servo control system (i.e., higher servo adjustment factor or lower reference voltage) were applied. These effects are in line with previously reported results [24,31]. Therefore, they are not further discussed. Similar responses can be seen for f_s and fr, even though in both cases the effects of one factor (the open voltage for f_s, and the pulse off-time for fr) were considerably less important than the ones of the other three factors. On the contrary, Δu shows an increasing trend with the pulse off-time. This effect can be explained by the fact that a longer pulse off-time reduced the occurrence of both normal pulses and short circuits, thus increasing the amount of time spent in open-circuit state.

The Pareto charts reveal that all factors and second-order interactions, besides the one between the open voltage and reference voltage, had a significant influence on f_p. This was deduced from the fact that the standardized effects were above the significance level. Moreover, f_p displayed similar standardized effects for the four factors. This means that f_p varied rather uniformly when changing different processing parameters. This is an appreciable characteristic for a process fingerprint indicator. Regarding the other three indicators, the Pareto charts show that f_s, fr, and Δu were all sensible to changes to the single processing parameters, but not to most of the second-order interactions. In particular, fr was sensible to only one second-order interaction, i.e., the interaction between the open voltage and reference gap voltage. Furthermore, unlike f_p, it can be noticed that f_s, fr, and Δu did not display a limited variability of the standardized effects. For instance, Δu was highly sensible to changes to the open voltage, and fr to changes to the servo control parameters.

Based on the Pareto charts, it can be concluded that f_p, f_s, fr, and Δu could be suitable quantities to be considered as indicators for the process fingerprint of the micro-EDM drilling process. This conclusion is drawn since the four indicators were at least sensible to changes to the single processing parameters. Nevertheless, thanks to more uniform sensitivity properties to changes to the processing parameters, f_p was the quantity that shows the highest potential.

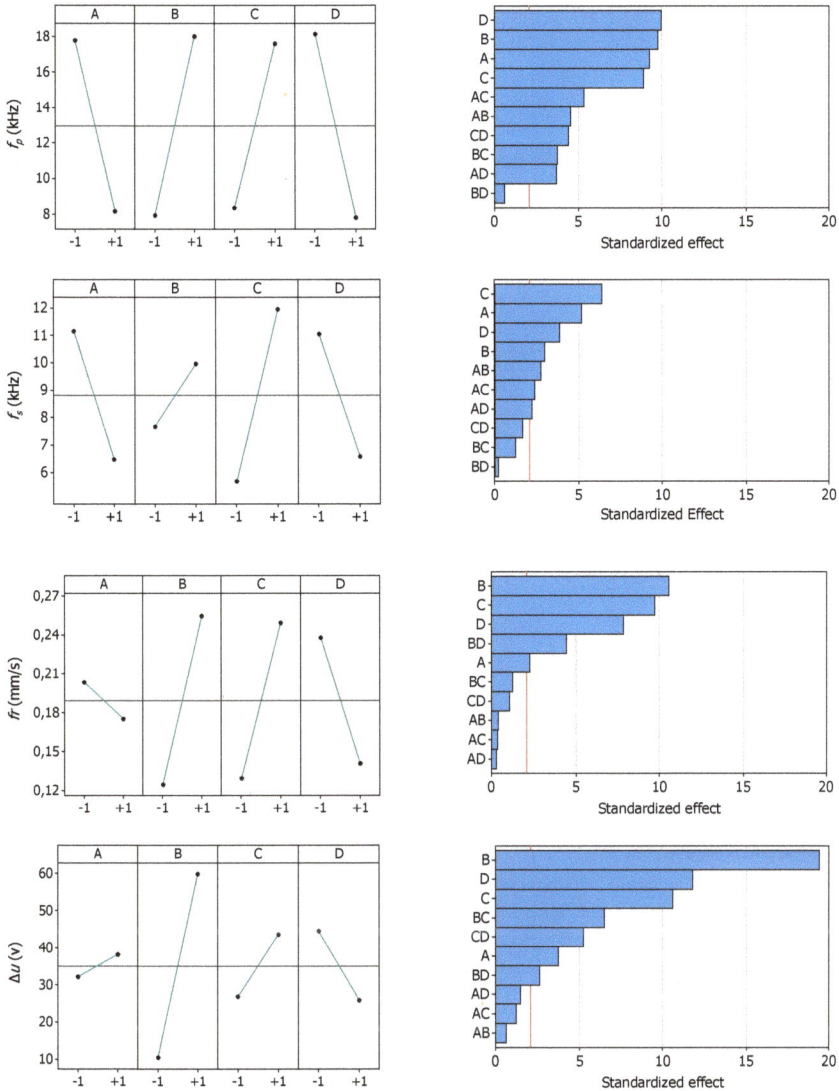

Figure 5. Influence of the four factors on the four potential indicators for the process fingerprint. The main effects plots (left column) and Pareto charts of the standardized effects (right column) are shown. The red vertical line in the Pareto charts corresponds to the significance level at 95% confidence level.

3.2. Correlation with Performance Characteristics

The correlation coefficients of the potential indicators for the process fingerprint to the MRR and TWR are shown in Figure 6. It is evident that the correlation coefficient of f_s was significantly lower than the others. The reason for this can be found in the fact short circuits do not contribute to material removal in a predictable manner, as do normal discharge pulses. Short circuits are normally considered to be harmful to the drilling speed [35]. This explains why the correlation coefficients of f_s were negative, while the correlations coefficients of f_p were positive. Besides f_s, the other three indicators displayed positive and relatively good correlations with the MRR and TWR. This indicates that maximization of f_p, fr, or Δu could be a viable way to increase the MRR, while minimization of one of these quantities could be pursued to reduce the TWR. The fact that the values of the r_p coefficients of fr were higher than the ones of f_p and Δu suggests that fr could be the most suitable quantity to be monitored in-line for optimizing the processing parameters with respect to performance characteristics. Therefore, a linear regression analysis was carried out to further investigate the correlation of fr with the MRR and TWR.

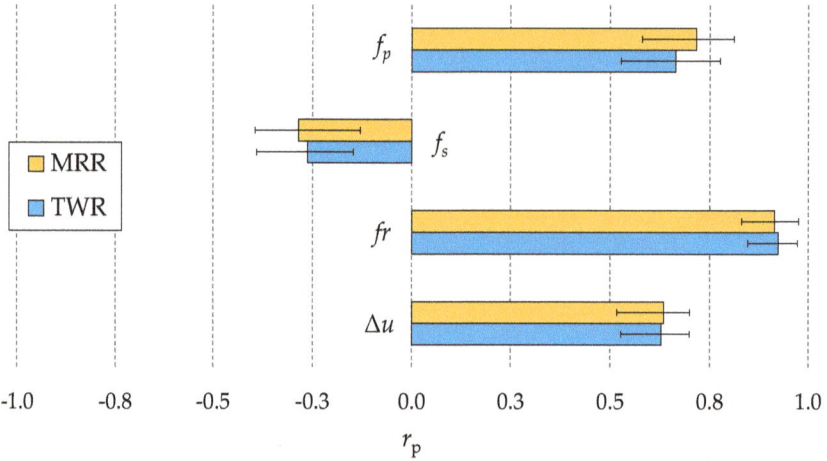

Figure 6. Pearson's correlation coefficients (r_p) of the four potential indicators for the process fingerprint with respect to the material removal rate (MRR) and tool wear rate (TWR). Mean values and ranges of the four experimental repetitions are shown.

Figures 7 and 8 show the correlation plots of fr against the MRR and TWR. A limited dispersion of the data points around the linear regression lines can be observed in both cases. This means strong positive linear correlations existed between fr and the MRR and TWR. The strength of the correlations is highlighted by the R^2 coefficients above 0.8. Despite the better sensitivity property of f_p, the result of this correlation analysis can be used to conclude that fr is the best quantity to be considered as process fingerprint for the purpose of optimizing the processing parameters.

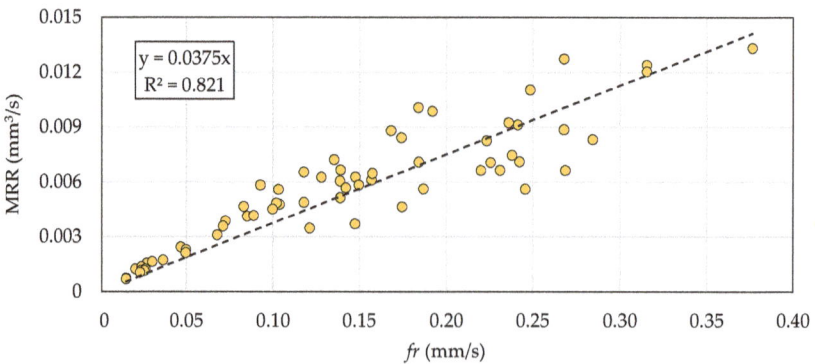

Figure 7. Correlation plot of the feed rate (fr) against the MRR. Each data point corresponds to one of the 64 experimental runs. The dashed line is the linear regression line.

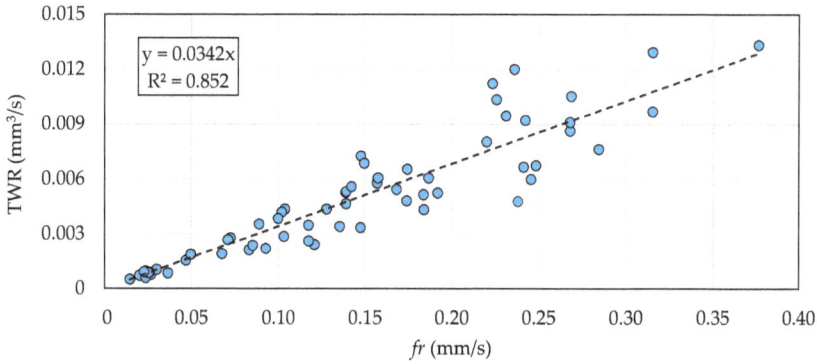

Figure 8. Correlation plot of *fr* against the TWR. Each data point corresponds to one of the 64 experimental runs. The dashed line is the linear regression line.

A possible algorithm to optimize the processing parameters could be based on a stochastic optimization technique [36]. However, in comparison with a previous application of this technique [6], the processing parameters could be varied during the drilling process of a single hole when following an approach based on in-line monitoring of *fr*. This would reduce the amount of time spent in each optimization iteration. As shown in Figure 9, after the touch-in stage, *fr* reached a stable value rather quickly when varying the processing parameters during the drilling process. This suggests that different settings of the processing parameters could be tried at regular steps during the drilling process (e.g., steps of 0.5 mm in depth). The measurement cycles to estimate the tool wear at the end of each drilling process could also be avoided. In this way, the time to reach the optimal setting of the processing parameters would be significantly reduced.

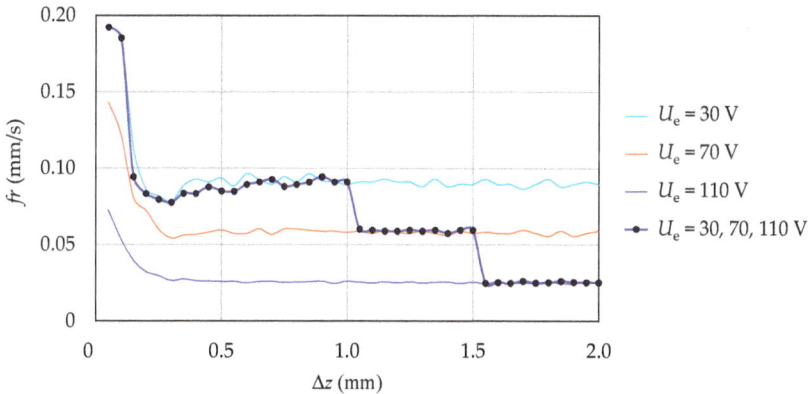

Figure 9. A comparison of the evolution of *fr* during four different drilling experiments. In three experiments the reference voltage (U_e) is unvaried, while in the other experiment U_e is varied at steps of 0.5 mm in depth after touch-in (first 0.5 mm in depth). The data points correspond to the in-line monitored values of *fr* during the latter experiment. The other parameters are unvaried during the four experiments. In particular, the levels of the factors are: A = +1, B = −1, C = −1.

3.3. Correlation with the Hole Quality

Figure 10 shows the values of the correlation coefficients of f_p, f_s, *fr*, and Δu to the DOC. Since short circuits and arcs do not contribute to material removal in a predictable manner, no correlation exists between f_s and the diameter of the hole inlet. On the contrary, the other three process fingerprint candidates showed a relatively good correlation with DOC. A possible reason can be that the DOC depends on the amount of discharge pulses occurring on the side of the tool and on the size of the discharge gap, which is dependent on the discharge energy. The main effect plots in Figure 5 confirm that f_p, fr and Δu tended to increase when the discharge energy increased or when the settings of the servo control system were more aggressive, a condition which might favour the ignition of discharges on the tool side. The fact that the r_p coefficient of f_p was lower than the ones of fr and Δu could imply that a higher frequency of discharges does not correspond to a higher probability of discharges on the side of the tool. Both fr and Δu could be used as indicators for the process fingerprint for in-line control of the DOC, since the mean values of r_p are above 0.8 for both quantities. Nevertheless, fr can be considered most suitable for controlling the DOC, considering that it displays a more limited variation of the r_p coefficient among the four experimental runs.

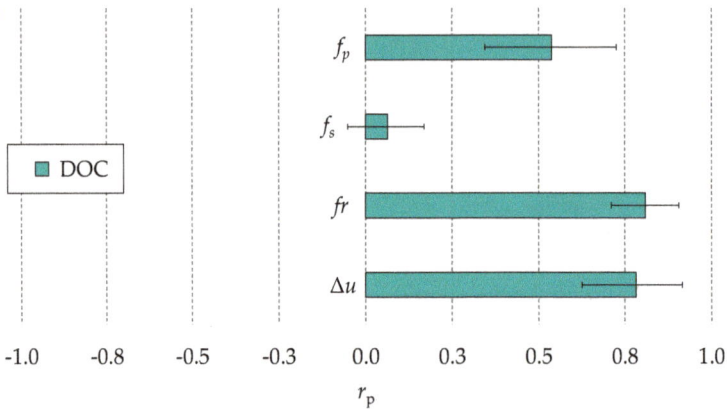

Figure 10. Pearson's correlation coefficients (r_p) of the potential indicators for the process fingerprint with respect to the diameter overcut (DOC). Mean values and ranges of the four experimental repetitions are shown.

Figure 11 depicts the evolution of fr during the two drilling experiments resulting in the minimum and maximum DOC. The trends of the two sets of in-line monitored data points are not overlapping each other, and a relevant difference in the average values of fr can be observed. This confirms the suitability of using the average value of fr as the output of a real-time algorithm for controlling the DOC during the drilling process.

Figure 12 shows the correlation coefficients of the potential indicators for the process fingerprint to the TR. It can be clearly noticed that the mean values or the r_p coefficients of f_p and f_s were extremely low (less than 0.25). Moreover, the ranges of these coefficients considering the four experimental runs were relatively wide and around the zero point. Hence the correlations of f_p and f_s to the TR were weak. The r_p coefficient of Δu was also low, while fr displayed a relatively good correlation to the TR. This means that fr is the only quantity that can be monitored in-line for controlling the taper of the micro holes among the four quantities considered in this research. The hole taper was mainly determined by the shape modifications of the tool tip due to the occurrence of discharge pulses on the side of the tool electrode. Therefore, the value of the correlation coefficients analysis suggest that the discharge pulses were more likely to occur on the side of the tool when the average feed rate of the tool electrode increased rather than when the total number of discharges was higher. This also explains why the correlation of fr and Δu to DOC were higher than the one of f_p.

It should be highlighted that a constant hole aspect ratio was considered in this research. However, the experimental results of Ali et al. [33] showed that the DOC and TR of micro-EDMed holes increase with the increase of the hole aspect ratio at almost the same rate. Therefore, similar correlation trends as the ones identified here can be expected when drilling micro holes of different aspect ratios. Overall, the process fingerprint approach is universal. Although this research has focused on a specific combination of tool and workpiece material, the approach can be applied in different process conditions.

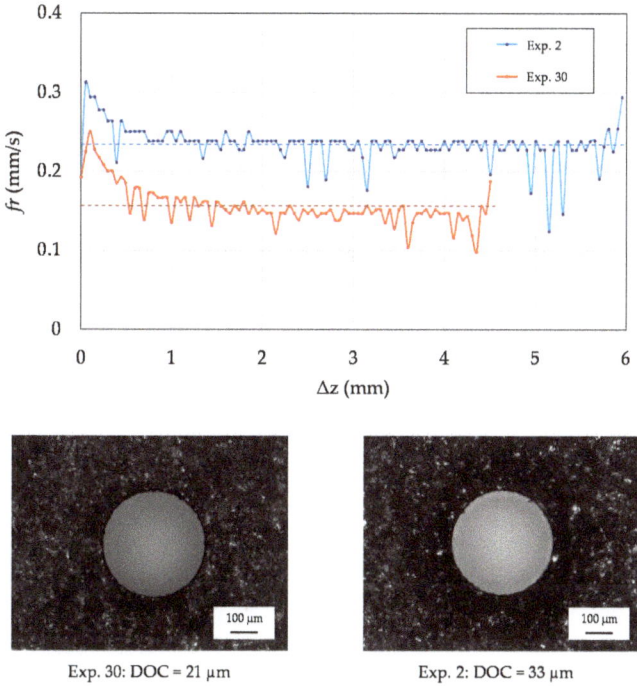

Exp. 30: DOC = 21 μm   Exp. 2: DOC = 33 μm

Figure 11. Evolution of *fr* during drilling. The data points represent the in-line monitored values of *fr*, while the dashed lines correspond to the average values. The evolution of *fr* and the hole inlets relative to two experimental runs are shown. Levels of the factors in Experiment 2: A = −1, B = +1, C = −1, D = +1. Levels of the factors in Experiment 30: A = −1, B = −1, C = +1, D = −1.

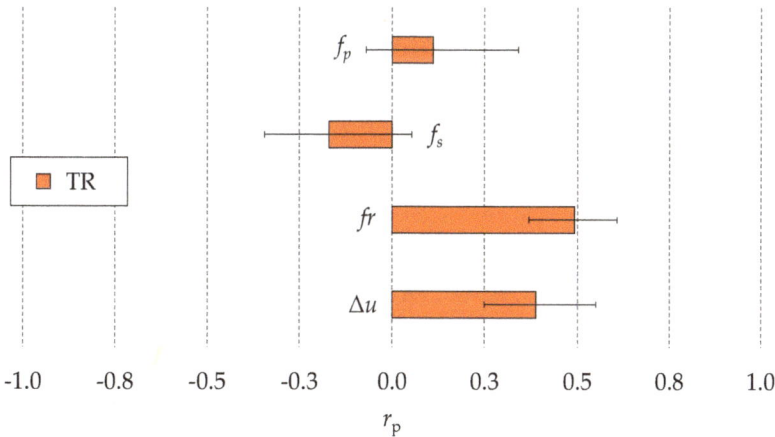

Figure 12. Pearson's correlation coefficients (r_p) of the potential indicators for the process fingerprint to the taper ratio (TR). Mean values and ranges of the four experimental repetitions are shown.

4. Conclusions

This research evaluated the applicability of the process fingerprint approach for reducing post-process metrology and shortening process optimization in micro-EDM drilling. In contrast with traditional approaches that focuses on establishing direct correlations between the processing parameters and the hole quality characteristics, this approach is based on in-line monitoring of a few selected physical quantities. The correlation of four different physical quantities to some of the main performance and hole quality characteristics (i.e., the material removal rate, tool wear rate, diameter overcut, and taper ratio) has been investigated within a wide process window.

The experimental results have shown that the average feed rate of the tool electrode displays a good correlation to the considered characteristics. Thus, it can be used as an indicator of the process fingerprint of micro-EDM drilling. This means that a real-time decision-making system could be implemented for optimizing the processing parameters or meeting a geometrical tolerance by keeping the average feed rate within a desired range during the drilling process.

The results of this research enable new solutions for automatic optimization of the processing parameters and in-line quality control using machine learning. Future work should focus on determining suitable control algorithms for these purposes and extending the approach to other relevant hole quality characteristics, such as the surface roughness and recast layer thickness.

Author Contributions: M.B. designed and carried out the experiments, M.B., J.Q. and D.R. critically analyzed the experimental results; M.B. wrote the paper; J.Q. and D.R. revised the paper.

Acknowledgments: This research work was undertaken in the context of MICROMAN project ("Process Fingerprint for Zero-defect Net-shape MICROMANufacturing"). MICROMAN is a European Training Network supported by Horizon 2020, the EU Framework Programme for Research and Innovation (Project ID: 674801). The authors would also like to acknowledge the partial support by Flanders Make vzw and the European Union's Horizon 2020 FoF project ProSurf (grant agreement number 767589).

Conflicts of Interest: The authors declare no conflict of interest.

References

1. Zhang, Y.; Xu, Z.; Zhu, D.; Qu, N.; Zhu, Y. Drilling of film cooling holes by a EDM/ECM in situ combined process using internal and side flushing of tubular electrode. *Int. J. Adv. Manuf. Technol.* **2016**, *83*, 505–517. [CrossRef]
2. Tong, H.; Li, Y.; Zhang, L.; Li, B. Mechanism design and process control of micro EDM for drilling spray holes of diesel injector nozzles. *Precis. Eng.* **2013**, *37*, 213–221. [CrossRef]
3. Yilmaz, O.; Okka, M.A. Effect of single and multi-channel electrodes application on EDM fast hole drilling performance. *Int. J. Adv. Manuf. Technol.* **2010**, *51*, 185–194. [CrossRef]
4. Moses, M.D.; Jahan, M.P. Micro-EDM machinability of difficult-to-cut Ti-6Al-4V against soft brass. *Int. J. Adv. Manuf. Technol.* **2015**, *81*, 1345–1361.
5. Sharp, M.; Ak, R.; Hedberg, T. A survey of the advancing use and development of machine learning in smart manufacturing. *J. Manuf. Syst.* **2018**, *48*, 170–179. [CrossRef]
6. Maradia, U.; Benavoli, A.; Boccadoro, M.; Bonesana, C.; Klyuev, M.; Zaffalon, M.; Gambardella, L.; Wegener, K. EDM Drilling optimisation using stochastic techniques. *Procedia CIRP* **2018**, *67*, 350–355. [CrossRef]
7. Lin, Y.C.; Tsao, C.C.; Hsu, C.Y.; Hung, S.K.; Wen, D.C. Evaluation of the characteristics of the microelectrical discharge machining process using response surface methodology based on the central composite design. *Int. J. Adv. Manuf. Technol.* **2012**, *62*, 1013–1021. [CrossRef]
8. Jung, J.H.; Kwon, W.T. Optimization of EDM process for multiple performance characteristics using Taguchi method and Grey relational analysis. *J. Mech. Sci. Technol.* **2010**, *24*, 1083–1090. [CrossRef]
9. Ay, M.; Caydas, U.; Haskalik, A. Optimization of micro-EDM drilling of inconel 718 superalloy. *Int. J. Adv. Manuf. Technol.* **2013**, *66*, 1015–1023. [CrossRef]
10. D'Urso, G.; Maccarini, G.; Ravasio, C. Process performance of micro-EDM drilling of stainless steel. *Int. J. Adv. Manuf. Technol.* **2014**, *72*, 1287–1298.

11. Jahan, M.P.; Wong, Y.S.; Rahman, M. A study on the quality micro-hole machining of tungsten carbide by micro-EDM process using transistor and RC-type pulse generator. *J. Mater. Process. Technol.* **2009**, *209*, 1706–1716. [CrossRef]

12. Suganthi, X.H.; Natarajan, U.; Sathiyamurthy, S.; Chidambaram, K. Prediction of quality responses in micro-EDM process using an adaptive neuro-fuzzy inference system (ANFIS) model. *Int. J. Adv. Manuf. Technol.* **2013**, *68*, 339–347. [CrossRef]

13. Brinksmeier, E.; Klocke, F.; Lucca, D.A.; Sölter, J.; Meyer, D. Process signatures—A new approach to solve the inverse surface integrity problem in machining processes. *Procedia CIRP* **2014**, *13*, 429–434. [CrossRef]

14. Eppinger, S.D.; Huber, C.D.; Pham, V.H. A methodology for manufacturing process signature analysis. *J. Manuf. Syst.* **1995**, *14*, 20–34. [CrossRef]

15. Sealy, M.P.; Liu, Z.Y.; Guo, Y.B.; Liu, Z.Q. Energy based process signature for surface integrity in hard milling. *J. Mater. Process. Technol.* **2016**, *238*, 284–289. [CrossRef]

16. Klink, A. Process signatures of EDM and ECM processes—Overview from part functionality and surface modification point of view. *Procedia CIRP* **2016**, *42*, 240–245. [CrossRef]

17. Annoni, M.; Rebaioli, L.; Semeraro, Q. Thin wall geometrical quality improvement in micromilling. *Int. J. Adv. Manuf. Technol.* **2015**, *79*, 881–895. [CrossRef]

18. Baruffi, F.; Calaon, M.; Tosello, G. Micro-injection moulding in-line quality assurance based on product and process fingerprints. *Micromachines* **2018**, *9*, 293. [CrossRef]

19. Yilmaz, O.; Bozdana, A.T.; Okka, M.A. An intelligent and automated system for electrical discharge drilling of aerospace alloys: Inconel 718 and Ti-6Al-4V. *Int. J. Adv. Manuf. Technol.* **2014**, *74*, 1323–1336. [CrossRef]

20. Li, M.S.; Chi, G.X.; Wang, Z.L.; Wang, Y.K.; Dai, L. Micro electrical discharge machining of small hole in TC4 alloy. *Trans. Nonferrous Met. Soc. China (Engl. Ed.)* **2009**, *19*, s434–s439. [CrossRef]

21. Qudeiri, J.E.A.; Mourad, A.I.; Ziout, A.; Abidi, M.H.; Elkaseer, A. Electric discharge machining of titanium and its alloys: Review. *Int. J. Adv. Manuf. Technol.* **2018**, *96*, 1319–1339. [CrossRef]

22. Plaza, S.; Sanchez, J.A.; Perez, E.; Gil, R.; Izquierdo, B.; Ortega, N.; Pombo, I. Experimental study on micro EDM-drilling of Ti6Al4V using helical electrode. *Precis. Eng.* **2014**, *38*, 821–827. [CrossRef]

23. Nirala, C.K.; Saha, P. Precise μEDM-drilling using real-time indirect tool wear compensation. *J. Mater. Process. Technol.* **2017**, *240*, 176–189. [CrossRef]

24. Wang, J.; Qian, J.; Ferraris, E.; Reynaerts, D. In-situ process monitoring and adaptive control for precision micro-EDM cavity milling. *Precis. Eng.* **2017**, *47*, 261–275. [CrossRef]

25. Bellotti, M.; Qian, J.; Reynaerts, D. Enhancement of the micro-EDM process for drilling through-holes. *Procedia CIRP* **2018**, *68*, 610–615. [CrossRef]

26. Xia, W.; Wang, J.; Zhao, W. Break-out detection for high-speed small hole drilling EDM based on machine learning. *Procedia CIRP* **2018**, *68*, 569–574.

27. Yang, F.; Bellotti, M.; Hua, H.; Yang, J.; Qian, J.; Reynaerts, D. Experimental analysis of normal spark discharge voltage and current with a RC-type generator in micro-EDM. *Int. J. Adv. Manuf. Technol.* **2018**, *96*, 2963–2972. [CrossRef]

28. Kendall, M.; Stuart, A. *The Advanced Theory of Statistics, Volume 2: Inference and Relationship*; Charles Griffin and Company Limited: London, UK, 1973; ISBN 0-85264-215-6.

29. Wang, D.; Ye, L.; Su, Z.; Lu, Y.; Li, F.; Meng, G. Probabilistic damage identification based on correlation analysis using guided wave signals in aluminum plates. *Struct. Health Monit.* **2010**, *9*, 133–144. [CrossRef]

30. Chang, Y.; Yang, D.; Guo, Y. Laser ultrasonic damage detection in coating-substrate structure via Pearson correlation coefficient. *Surf. Coat. Technol.* **2018**, *353*, 339–345. [CrossRef]

31. Natarajan, N.; Suresh, P. Experimental investigations on the microhole machining of 304 stainless steel by micro-EDM process using RC-type pulse generator. *Int. J. Adv. Manuf. Technol.* **2015**, *77*, 1741–1750. [CrossRef]

32. Saxena, K.K.; Bellotti, M.; Qian, J.; Reynaerts, D. Characterization of circumferential surface roughness of micro-EDMed holes using replica technology. *Procedia CIRP* **2018**, *68*, 582–587. [CrossRef]

33. Ali, M.Y.; Hamad, M.H.; Karim, A.I. Form characterization of microhole produced by microelectrical discharge drilling. *Mater. Manuf. Process.* **2009**, *24*, 683–687. [CrossRef]

34. Zhang, Y.; Xu, Z.; Zhu, Y.; Zhu, D. Effect of tube-electrode inner structure on machining performance in tube-electrode high-speed electrochemical discharge drilling. *J. Mater. Process. Technol.* **2016**, *231*, 38–49. [CrossRef]

35. Kao, C.-C.; Shih, A.J. Design and tuning of a fuzzy logic controller for micro-hole electrical discharge machining. *J. Manuf. Process.* **2009**, *10*, 61–73. [CrossRef]

36. Spall, J.C. *Introduction to Stochastic Search and Optimization*; Wiley: Hoboken, NY, USA, 2003.

micromachines

MDPI

Article

Process Control in Jet Electrochemical Machining of Stainless Steel through Inline Metrology of Current Density

Matin Yahyavi Zanjani [1,*], Matthias Hackert-Oschätzchen [1], André Martin [1], Gunnar Meichsner [2], Jan Edelmann [2] and Andreas Schubert [1,2]

[1] Professorship Micromanufacturing Technology, Faculty of Mechanical Engineering, Chemnitz University of Technology, 09107 Chemnitz, Germany; matthias.hackert@mb.tu-chemnitz.de (M.H.-O.); andre.martin@mb.tu-chemnitz.de (A.M.); andreas.schubert@mb.tu-chemnitz.de (A.S.)

[2] Fraunhofer Institute for Machine Tools and Forming Technology, 09126 Chemnitz, Germany; Gunnar.Meichsner@iwu.fraunhofer.de (G.M.); Jan.Edelmann@iwu.fraunhofer.de (J.E.)

* Correspondence: matin.yahyavi-zanjani@mb.tu-chemnitz.de; Tel.: +49-(0)-371-531-30522

Received: 28 February 2019; Accepted: 15 April 2019; Published: 18 April 2019

Abstract: Jet electrochemical machining (Jet-ECM) is a flexible method for machining complex microstructures in high-strength and hard-to-machine materials. Contrary to mechanical machining, in Jet-ECM there is no mechanical contact between tool and workpiece. This enables Jet-ECM, like other electrochemical machining processes, to realize surface layers free of mechanical residual stresses, cracks, and thermal distortions. Besides, it causes no burrs and offers long tool life. This paper presents selected features of Jet-ECM, with special focus on the analysis of the current density during the machining of single grooves in stainless steel EN 1.4301. Especially, the development of the current density resulting from machining grooves intersecting previous machining steps was monitored in order to derive systematic influences. The resulting removal geometry is analyzed by measuring the depth and the roughness of the machined grooves. The correlation between the measured product features and the monitored current density is investigated. This correlation shows that grooves with the desired depth and surface roughness can be machined by controlling current density through the adjustment of process parameters. On the other hand, current density is sensitive to the changes of working gap. As a consequence of the changes of workpiece form and size for the grooves intersecting premachined grooves as well as the grooves with a lateral gap, working gap, and current density change. By analyzing monitoring data and removal geometry results, the suitability of current density inline monitoring to enable process control is shown, especially with regards to manufacture products that should comply with tight predefined specifications.

Keywords: electrochemical machining (ECM); process control; current monitoring; current density; surface roughness; inline metrology

1. Introduction

Jet-ECM, like other electrochemical machining processes, is based on anodic dissolution of workpiece material. In electrochemical machining, the machined material is influenced neither thermally nor mechanically by the removal process. Hence, complex microstructures can be machined with high precision regardless of the mechanical properties like hardness and ductility of workpiece material. Due to the mentioned characteristics ECM has become an alternative process to conventional and other nonconventional machining processes. In most of the established electrochemical machining processes the flexibility is restricted, since the shape of removal geometry is defined by the shape of cathode. In contrast, Jet-ECM is a shape generating technique, where the motion strategy defines the shape of machined part. Hence, no complicated cathode geometries are required for machining complex structures [1–3].

One of the special characteristics of Jet-ECM is the application of an electrolyte jet ejected from a micronozzle. The electrolyte is ejected perpendicularly towards the workpiece surface surrounded by atmospheric air that forms a closed free jet. Microstructures can be machined by controlling the multidimensional motion of the nozzle [1]. The creation of microstructures can be controlled by switching the applied electric potential [4] or by controlling the gap between nozzle and workpiece surface [5]. In a recent research study, the effects which occur as a result of the variation of the incident jet angle were investigated [6]. The basic principle of Jet-ECM is shown in Figure 1.

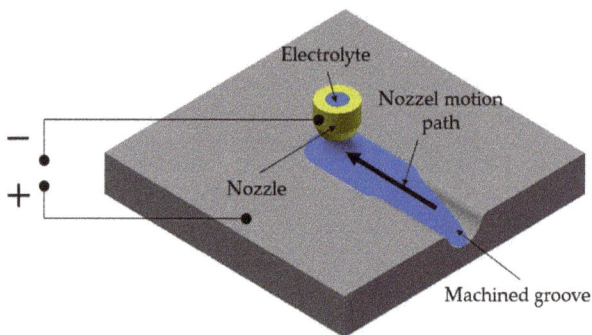

Figure 1. Scheme of Jet electrochemical machining (Jet-ECM).

Because of the specific technical arrangement, different machining tasks can be realized by Jet-ECM, such as microdrilling [3,7,8], cutting [3], or turning [9,10]. In microdrilling, bores with vertical walls can be generated by applying Jet-ECM with a trepanning movement of the nozzle [3]. Jet-ECM is also capable of realizing sharp edges [2]. Figure 2 shows five coaxial circles with varying radii from 1 mm to 2.2 mm and different numbers of crossings from 10 to 50, which has led to a minimal edge radius of ~1 μm. This makes Jet-ECM a potential machining process to produce sharp edges.

Figure 2. Jet-ECM of sharp edges in steel 1.4541, nozzle motion path from top right to left.

Furthermore, the suitability of Jet-ECM as a finishing process was investigated. Kawanaka et al. investigated the influences of current conditions and nozzle movement speed on surface roughness on stainless steel (SUS304) and realized a mirror-like surface with *Rz* surface values less than 0.2 μm. They have shown that surface roughness of grooves decreases with the increase of nozzle translating speed, and for a constant speed, the surface roughness of the grooves decreases and, after reaching a minimum value, increases again [11]. Moreover by help of inverse Jet-ECM, finish-machining of micro bores was demonstrated in simulations and experiments [12,13].

In Jet-ECM processes, the working gap is one important process parameter, which affects other process parameters among which current and, consequently, current density are of high importance. Schubert et al. have shown that by increasing the working gap, current and current density decrease and result in lower removal rates as well as different surface roughness, while the depth of cavities changes significantly with the working gap and no remarkable changes of width were seen [4]. They have shown that by applying 35 V, a 30% $NaNO_3$ solution as electrolyte, nozzle diameter $d_n = 100$ μm, and nozzle speed of 500 μm/s, mean current density decreases from 2100 A/cm^2 for working gap $a = 5$ μm to 400 A/cm^2 for $a = 100$ μm for machining of EN 1.4541 [4]. In order to control the working gap, different strategies based on electrostatic probing before machining have already been studied. When the normal vector of the surface is calculated based on touching three points on the surface of workpiece to adjust working gap, the strategy is called "adjusting by normal vector". In the "adjusting by grid" strategy, which is used for more complicated shape deviations, multiple points can be detected and the normal vectors of the corresponding areas can be calculated from the determined values. "Adjusting by reference points" is another strategy where an individual number of points along the removal geometry are detected. Besides, the working gap can also be controlled dynamically during the process where it is determined and adjusted when differing from a defined tolerance. This strategy is called "control dynamical" [5,14]. The precision of gap measurement increases when more points are detected. However, the measurement time becomes longer as well.

The applied voltage is considered as another important process parameter where higher voltages result in considerable higher currents and current densities consequently, especially with smaller working gaps less than 50 μm and d_n of 100 μm [4]. Several other process parameters, such as nozzle motion speed, nozzle diameter, electrolyte flow velocity, and electrolyte concentration, have significant influences on the machining result. The electrolyte flow is very important to ensure the complete removal of heat and gases produced by the reactions at either electrode and to let current flow to enable charge transport [15]. The nozzle diameter influences the distribution of current density and resulted removal geometry consequently. According to Schubert et al., it is shown that the depth of cut increases with nozzle diameter from 29 μm for $d_n = 60$ μm to 77 μm for $d_n = 200$ μm [16]. The type and concentration of salt in the electrolyte are chosen depending on the material and the need to provide sufficient conductivity to assure the dissolution of the workpiece material with adequate removal rate [17]. Table 1 shows the values of applied process parameters for different materials and the achieved results. Electrolyte type is usually selected based on workpiece material. According to the table, NaCl solutions are mostly used in Jet-ECM as a nonpassive electrolyte and $NaNO_3$ with the mass concentration of 20% to 30% as a passive electrolyte. Besides, nozzles diameters range from 100 μm to 510 μm; working gaps are also in the same range. Common potential values for Jet-ECM is up to 60 V, and nozzle speeds amounting to 1000 μm/s have been studied. These parameters result in up to 250 μm of depth and the Sa roughness of less than 1.5 μm. In this study, the used parameters were chosen with regards to Table 1.

Although several researches have already been done in the field of inline metrology, additional research efforts especially in Jet-ECM are still required to control this process. Thus, the objective of the present study is to focus on inline metrology of current density as a powerful method to control process. For this purpose, systematic experiments have been carried out to monitor the development of electric current during the machining of single and intersecting grooves. The measured data were used to calculate the average current density. The influence of different voltage levels as well as different working gaps on the resulting current densities and the removal geometries in machining singles grooves are shown. Furthermore, the influence of intersecting grooves on the resulting current density was analyzed. It is shown that changes in current density affect geometrical features such as the removal depth and surface roughness of the machined grooves. In order to highlight the importance of current density monitoring, the changes of the mentioned features in dependence of changes in current density were analyzed.

Table 1. Applied process parameters and achieved results in previous studies for different materials.

Parameter	Value									
Workpiece Material	Co [18]	WC [18]	WC-6% Co [18]	Nimonic 80A [19]	Ti-6Al-4V [20]	EN 1.2379 [21]	EN 1.4301 [21]	EN 1.4541 [21]	EN 1.5920 [21]	Brass, Cu39Zn2Pb [22]
Nozzle inner diameter (μm)	100	100	100	100	250	100				510
Electrolyte	20% NaCl	20% NaCl	20% NaCl	20% NaCl	2–4 M NaNO₃	30% NaNO₃				2.3 M NaNO₃
Working gap (μm)	100	100	100	100	500	100				500
Voltage (V)	50	50	10–55	1–56	-	56				-
Nozzle speed (μm/s)	200	200	200	150	0 Machining time: 10s	200–1000				500
Depth of removal (μm)	40	< 1	4–5	300	50–250	75–240	60–230	60–220	100–250	150 μm/C
Surface roughness (μm)	-	-	Ra < 0.65	-	-	0.35 < Ra < 0.45	0.1 < Ra < 0.15	0.15 < Ra < 0.33	0.3 < Ra < 0.45	0.3 < Sa < 1.5

2. Materials and Methods

Working gap and machining voltage are considered as main process parameters in Jet-ECM, since these parameters mainly influence the machining current and the current density consequently. The changes of these parameters lead to the variations of resulting removal geometries.

The mathematical basis for the calculation of the anodic material dissolution is described by Faraday's law of electrolysis. As a result of its reversal, the material removal in the form of mass m is determined quantitatively according to Equation (1) [23].

$$m = \frac{M}{z \cdot F} \cdot Q \tag{1}$$

Here, M is the molar mass of the dissolved material and Q is the electric charge that has been exchanged during the ECM process. The removal mass at the anode is therefore proportional to the molar mass and the exchanged electric charge [23]. F is the Faraday constant (F = 9.64853×10^4 C/mol) and z the electrochemical valence of an ion of the ablated material. The calculation of the electric charge Q results from the integration of the time-dependent electric current $I(t)$ over the processing time t. Mathematically, this relationship is described according to Equation (2) [24] where t_1 and t_2 are the times correspond to the start and the end of machining process.

$$Q = \int_{t_1}^{t_2} I(t) \, dt \tag{2}$$

Taking into account the density ρ of the removed material, the dissolved material volume V is calculated by extending Equation (1) according to Equation (3) [3].

$$V = \frac{M}{\rho \cdot z \cdot F} \cdot Q \tag{3}$$

As can be seen, the machinability of a material does not depend on its mechanical properties, like hardness or toughness, and is only characterized by its electrochemical properties. This makes EC machining an alternative technique, especially for hard-to-machine materials [23]. By a simplification assuming that $z = const.$, V_{sp} represents a material constant, which is calculated according to Equation (4) [21].

$$V_{sp} = \frac{M}{\rho \cdot z \cdot F} \tag{4}$$

Faraday's law presupposes that all the electrical charge Q exchanged during the process is consumed for material removal. However, this can only be achieved under idealized conditions and cannot be attained in practice. A part of the electrical energy is consumed at the electrodes for ablation-ineffective reactions, such as the formation of hydrogen or oxygen. In addition, the formation of an oxide layer on the workpiece surface, which is depending on the interaction of the electrolyte with the workpiece material, influences the removal process and significantly reduces the efficiency of the anodic metal dissolution [25].

In order to take this into account, the current efficiency η is used for determining the efficiency of the removal process. The current efficiency η is defined as the quotient of the effective dissolution volume $V_{eff} = V/Q$ and the specific dissolution volume V_{sp}, according to Equation (5) [3].

$$\eta = \frac{V_{eff}}{V_{sp}} \tag{5}$$

The current efficiency is not limited to a maximum of 100%, since it is mainly influenced by the determination of the electrochemical valence z, thus the current efficiency can be reasonably discussed only under consideration of z used for its calculation [26]. A variety of research topics includes the quantitative determination of z using micro flow cells or precision weighing [27–31]. Consequently, the electrochemical valence cannot be taken as an integer constant, but rather needs to be considered as a real number depending on the local electrical current density J.

The electric current density is thus a decisive process variable, which is calculated according to Equation (6) as a function of the local electric current I (consequently applied voltage and the electrical resistance according to Ohm's law) and the electrode surface A_E [23].

$$J = \frac{U}{R \cdot A_E} \tag{6}$$

The electrical resistance results from the working gap a between the tool and the workpiece and from the specific electrical conductivity κ of the electrolyte solution [23] according to Equation (7).

$$R = \frac{a}{\kappa \cdot A_E} \tag{7}$$

The electrical resistance changes during the process due to the anodic removal of workpiece material and the associated enlargement of the working gap as well as the pollution and temperature-related change in electrical conductivity of the electrolyte. Hence, the working gap and the machining voltage influence the current density and thus the EC removal. This highlights the importance of current monitoring for the development of an adequate process control. In the following paragraphs, the experimental analysis of the effect of these parameters on the product features will be discussed.

3. Experimental Setup

Figure 3 shows the applied in-house built Jet-ECM prototype system, which is composed of a table and a portal made of granite to guarantee the required mechanical and thermal stiffness. The relative movement between nozzle and workpiece is carried out by a linear three-axis positioning system. A pulsation-free pump transports the electrolyte to the nozzle. The electrolyte is ejected in Z direction towards the workpiece. The spent electrolyte is collected in a disposal tank. A process energy source supplies the electric voltage between the nozzle and the workpiece providing the required process current. A personal computer serves as a control system for all electrical and kinematic operations of the system [32].

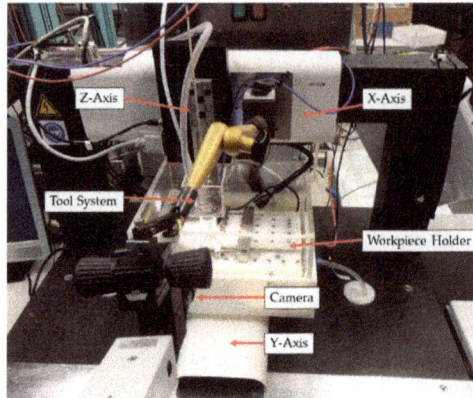

Figure 3. Photo of the applied Jet-ECM prototype system.

The current measurement was realized using a Keysight 34465A (Keysight Technologies, Santa Rosa, CA, USA) digital multimeter. For communication with the multimeter a custom control software was developed, based on National Instruments LabVIEW (14.0), in order to measure the required data and to visualize the time-dependent development during the process. The measured data are saved to text files in order to be used for analyses after the machining processes.

Design of Experiments

For the evaluation of the current measurement three sets of experiments were designed: (1) a set of single grooves were machined with different machining voltages and working gaps to analyze the influence of these parameters on the resulting current density, (2) a grid of single grooves to evaluate intersection characteristics, and (3) a set of parallel grooves with varying lateral distance between the grooves. The second and the third sets were executed to investigate the effects of previously machined grooves on the current density and resulting removal geometry of the postmachined grooves.

The process parameters of the above mentioned experiments are charted in Table 2.

Table 2. Process parameters for machining single, intersecting, and parallel grooves.

Parameter		Value
Workpiece material		EN 1.4301
Nozzle inner diameter		100 μm
Electrolyte		30% NaNO$_3$
Electrolyte supply rate		10 mL/min
Working gap	single grooves	100, 200, 300, 400, and 500 μm
	intersecting grooves	100 μm
	parallel grooves	100 μm
Voltage	single grooves	30, 40, 50, 60, 70, 80, and 90 V
	intersecting grooves	30, 40, 50, 60, and 70 V
	parallel grooves	60 V
Nozzle speed		200 μm/s

Various values of working gap and machining voltage were selected based on past experience to reach a wide range of current density which, as expected, resulted in the variations of product features of the machined single grooves. For the evaluation of product fingerprints, the depth *d* and the surface roughness *Sa* of the grooves were measured by a Keyence VK-9700 (Keyence Corporation, Osaka, Japan) confocal microscope. For single grooves, three or more areas of the groove were selected (as shown in Figure 4) and the roughness in the grooves bottom were measured. The average depth and roughness were calculated and used to evaluate the relation between the mean current density and the product features.

Figure 4. 3D image of a single groove with roughness measurement areas.

For investigations on the influence of premachined grooves on the subsequently machined grooves crossing each other rectangularly, a grid of grooves in the direction of X and Y was machined. The electric current was measured during machining the second grooves with special focus on deviations when the nozzle crossed the premachined groove. In order to investigate the influence of parallel grooves on each other, a set of parallel grooves with the machining voltages of 60 V (for both grooves) and the previously mentioned process parameters were machined. The lateral distance between the two parallel grooves was increased from 10 to 200 μm. The resulted data from current monitoring and depth measurement was analyzed to characterize the influence of the degree of superposition on the resulted mean current and removal depth of the subsequent grooves. The schematics of machining directions for intersecting and parallel grooves are shown in are shown in Figure 5A,B, respectively.

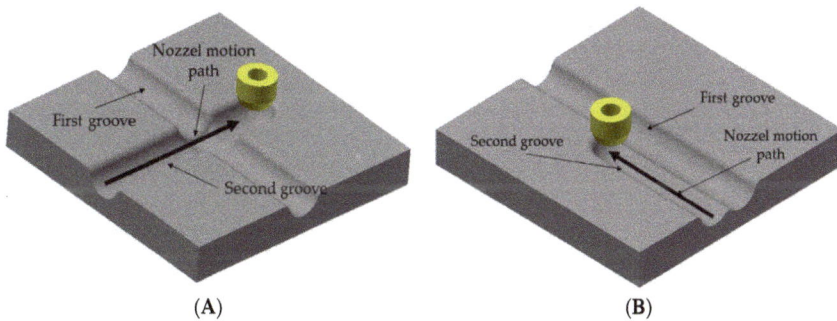

(A) **(B)**

Figure 5. Schematics of machining directions for (**A**) intersecting and (**B**) parallel grooves, the electrolyte is not shown.

4. Results

4.1. Single Grooves

After machining the measured current data was used to analyze the current density development for machining single grooves. The current values were divided by the inner area of the nozzle to calculate the plotted mean current density J_m [11]. As an example, the current density developed during the machining of single grooves with the machining voltage of 60 V as function of the nozzle displacement for the analyzed working gaps is shown as point diagram in Figure 6A. It can be seen in Figure 6A that no major changes in the current density were seen during machining single grooves over a plane workpiece surface.

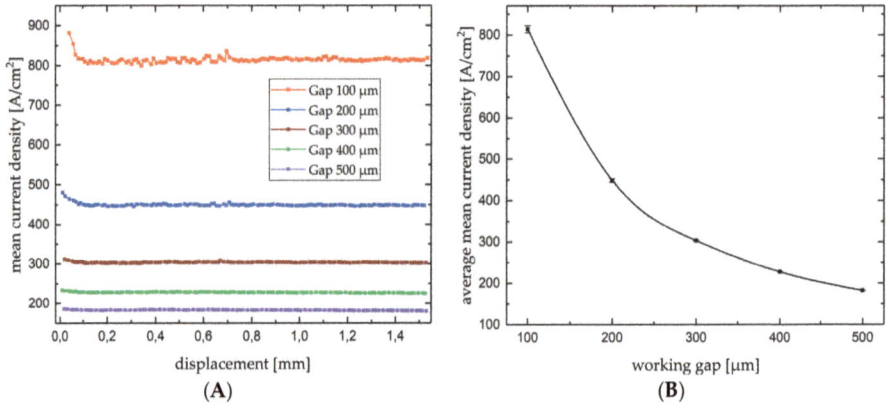

Figure 6. Mean current density as a function of the nozzle displacement (**A**) and average mean current density as a function of working gap for U = 60 V (**B**).

Figure 6B shows the average mean current density as a function of the working gap for 60 V. The error bars indicate the standard deviations calculated from the single measurement values of the Figure 6A. As can be realized, the standard deviations are comparatively low (<1%) for all working gaps. Hence, machining of single grooves on a plane workpiece surface only leads to slight deviations in mean current density, but significant changes in mean current density were detected due to changes in working gap while the voltage was kept constant. This indicates that mean current density is a proper process parameter for the control of the working gap during Jet-EC milling of single grooves. Besides, as shown in Table 2, apart from different working gap sizes, varying machining voltages were also applied. The variation of working gap as well as machining voltage lead to different values of current density.

Figure 7 shows the removal depth as a function of the current density in machining single grooves. The solid line shows the linear fit of the points. As can be seen in the graph, the depth of the groove increases linearly with increasing mean current density. The linear function between the mean current density and the depth of the single grooves underlines that controlling the mean current density is a useful tool for targeted machining of single grooves with predefined removal depth. This correlation has been stated in mathematical form in Equation (8).

$$d \ [\mu m] = 1.29 \ \mu m + 0.06 \times J_m \ [A/cm^2] \tag{8}$$

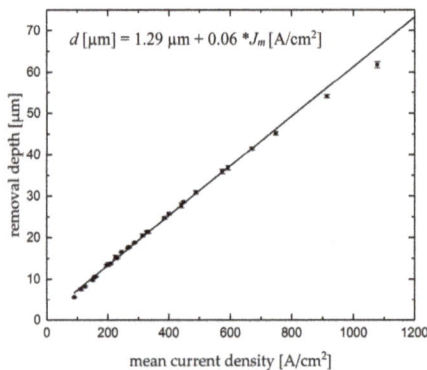

Figure 7. Removal depth of single grooves as function of the mean current density.

As an important product feature, the aerial roughness value *Sa* of the single grooves was measured. The result of the roughness measurements as a function of the mean current density is shown in Figure 8. The point diagram shows a decrease in roughness with an increase in mean current density up to a value of approximately 400 A/cm^2, while a further increase in mean current density results in an increasing roughness. An adequate control of the mean current density offers the possibility for finish-machining in order to achieve a predefined surface roughness *Sa*.

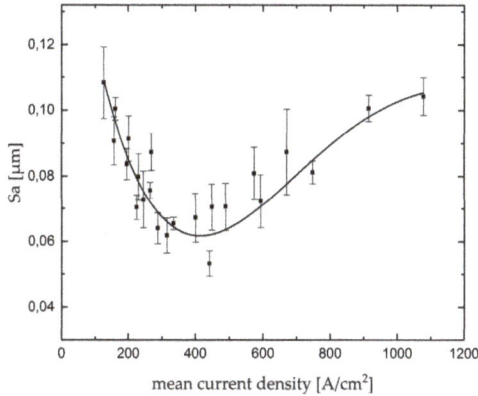

Figure 8. Aerial roughness Sa as function of the mean current density in Jet-EC milling of single grooves.

4.2. Intersecting Grooves

Figure 9 shows a 3D image of the intersection between two Jet-EC milled grooves, where both grooves were machined with 60 V.

Figure 9. 3D image of intersecting grooves.

The measured depths of the premachined grooves are shown in Table 3.

Table 3. Machining voltage and depth of first grooves of intersecting groove.

Process and Geometry Parameter	Value				
Machining voltage (V)	70	60	50	40	30
Depth of groove (µm)	59	52	44	36	28

Figure 10 shows a point diagram of the mean current density as function of the nozzle's displacement during machining a subsequent groove and crossing these five premachined grooves rectangularly. The premachined grooves were machined with a lateral distance of 1 mm from each other. As can be seen, the mean current density drops significantly when the nozzle crosses the premachined grooves, which indicates the sensitivity of the electric current to local changes of the working gap due to the deviations of workpiece surface.

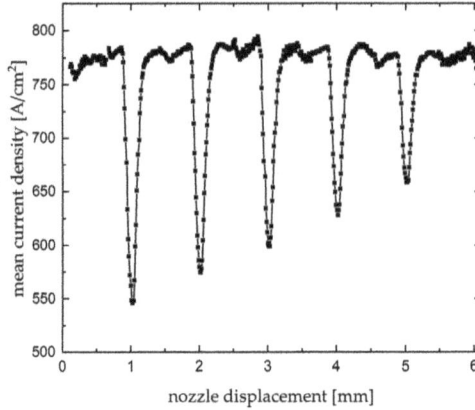

Figure 10. Mean current density as function of the nozzle displacement during Jet-EC milling of subsequent grooves with a voltage of 60 V crossing premachined grooves with different depths.

In Figure 11, the measured minimum mean current density J_{min} of the five intersecting positions are displayed as a function of the depth of the premachined groove for all the analyzed voltages. The point diagram shows that the minimum mean current density decreases linearly with increasing depth of the premachined groove. Hence, the minimum mean current density can be considered as an indicator for the value of surface deviations depending on the removal depth of the premachined grooves. As can be seen in Figure 11, the linearity of changes of the minimum current density with depth of premachined grooves is independent of machining voltage and therefore, with low or high machining voltages, the amount of surface deviations can be characterized.

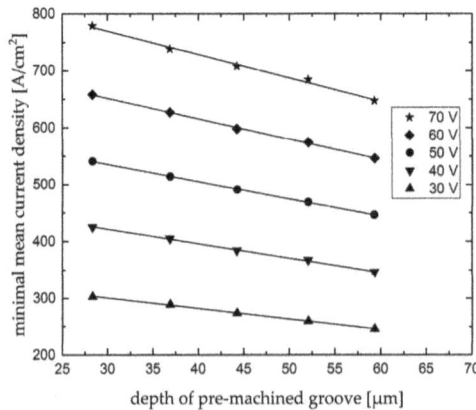

Figure 11. Minimum mean current densities of intersections as a function of the depth of premachined grooves for differing voltages.

For the intersecting grooves the measurements of the surface roughness and the removal depth of the intersections were carried out with the mentioned confocal measurement system. After preliminary investigations, including the evaluation of minimum and average current density over the intersections, it was found that the minimum current density occurs when the nozzle crosses the center of the intersections, which can be characterized as a specific product features. As another specific product feature, the relative depth d_r was calculated from the difference between the maximum depth in the intersection and the depth of the premachined groove. Figure 12 shows the calculated relative depths as function of the minimum current density.

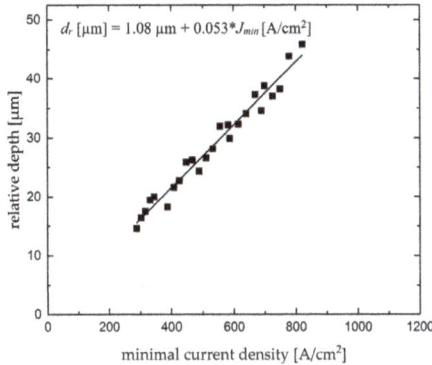

Figure 12. Changes of intersection relative depth with minimum current density.

Similar to the analyzed removal depth as a function of the mean current density in machining single grooves, the relative depth of the intersecting grooves increases linearly with an increase in minimum current density, as stated in Equation (9). In order to ease the comparison with the single grooves, the same scale was used for this graph. The data of this graph together with the results of Figure 11 can enhance the current monitoring with online control of the first groove depth as well as the intersection depth while the minimum value of current density over an intersection is proportional to the first groove depth, and this value can be used to estimate the relative depth of the intersection.

$$d_r \ [\mu m] = 1.08 \ \mu m + 0.053 \times J_{min} \ [A/cm^2] \tag{9}$$

Similar to the results in machining single grooves, the surface roughness of intersecting grooves decreases with increasing minimum current density up to a value of approximately 400 A/cm² and increases again at further increase in minimum current density, although the slope of changes is less significant than the slope determined for single grooves as can be seen in Figure 13.

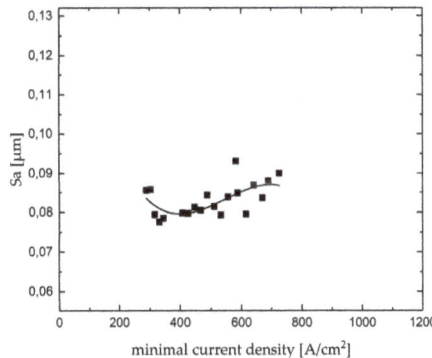

Figure 13. Aerial roughness (Sa) as function of the minimal current density in Jet-EC milling of intersecting grooves measured in the center of the intersections.

4.3. Parallel Grooves

Figure 14 shows a 3D image of parallel grooves with the lateral distance of 150 μm.

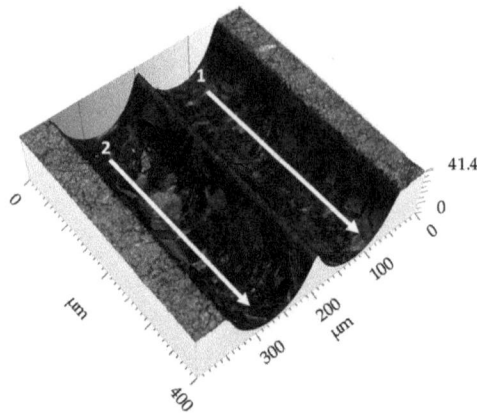

Figure 14. 3D image of parallel grooves.

The average means current density measured during machining subsequent groove as function of the lateral gap from the premachined parallel groove is shown in Figure 15. As can be seen, the average mean current density increases linearly with increasing lateral gap between the two parallel grooves in a range from 5 μm to 120 μm. Between 120 and 160 μm, the average mean current density rises slightly, and the influences of lateral gap are hardly detected for the gaps wider than 160 μm.

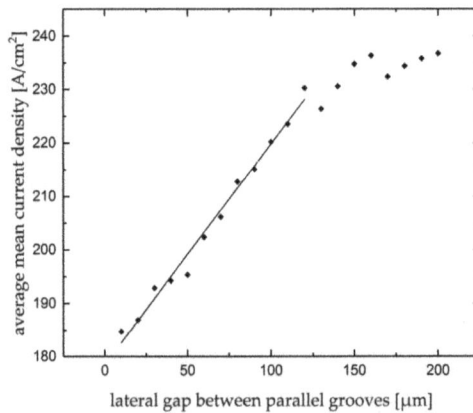

Figure 15. Average mean current density of the subsequent grooves as function of the lateral gap from the premachined parallel groove.

As a specific product feature in this case the depth of the subsequent groove was measured and characterized according to influences of the premachined groove. In Figure 16A, the changes of the depth of the subsequent groove with average mean current density is plotted. As can be seen, the depth of the subsequent groove decreases linearly at increasing average mean current density up to a value of 225 A/cm^2. This can be explained by the changes of current density distribution where by the increase of lateral gap, the actual working gap decreases and more material is ablated from the side of the groove rather than the bottom. According to Figure 15 this corresponds to the value at a lateral gap of approximately 120 μm, up to which the premachined groove affects the average mean current density when machining the subsequent groove. Hence, at a further increase in average mean current density only slight changes were detected.

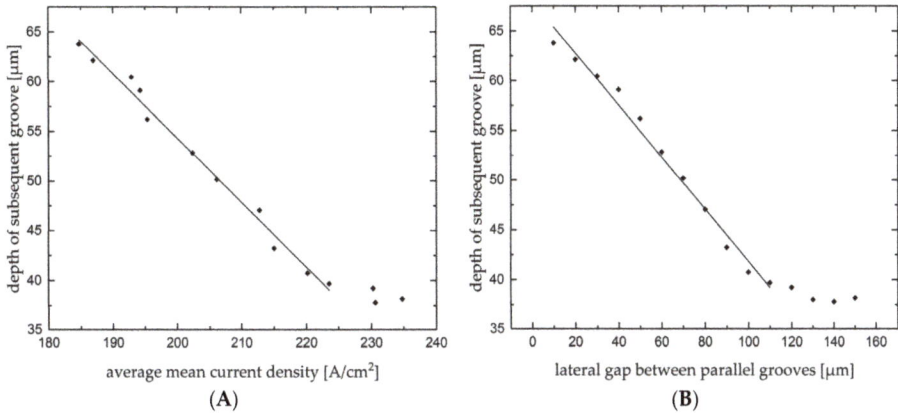

Figure 16. Depth of second parallel groove as function of (**A**) mean current density and (**B**) lateral gap.

Figure 16B shows the depth of the subsequent groove as function of the lateral gap between the grooves. The depth of the subsequent groove decreases linearly with increasing lateral gap up to a value of approximately 120 µm, which corresponds to the results asserted in Figures 15 and 16A, where little influence of the premachined groove on the subsequent groove was detected at wider lateral gaps. The results indicate that the control of the current density is a useful tool for targeted machining of parallel grooves with predefined removal depth and in specific ranges, when the lateral gap is smaller than the nozzle diameter, can be used to measure the actual lateral gap of the parallel grooves.

As another feature of the product, the roundness of the edges of the walls between grooves was investigated. Figure 17 shows the variation of this feature as a function of lateral gap. For lateral gaps smaller than 90 µm, the edge cannot be detected. As can be realized, the roundness of the edge decreases significantly up by increasing the gap up to 120 µm. The roundness of the edge changes slightly with the lateral gaps between 120 and 160 µm. As discussed before, the mean current density of the lateral groove increases slightly in this range of lateral gap. For lateral gaps bigger than 160 µm, no influence of the first groove was seen.

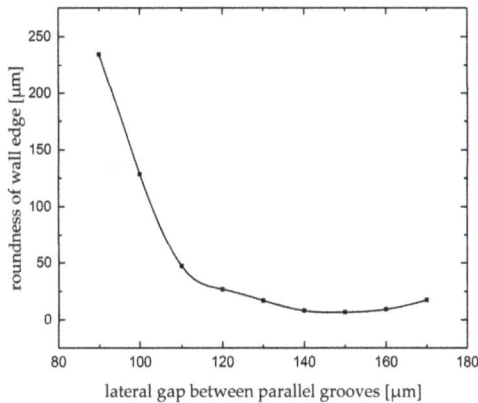

Figure 17. Roundness of the edge of thin wall between parallel grooves.

5. Conclusions

This paper has highlighted the importance of current density measurements in Jet-ECM process control. The results indicate that current density is very sensitive to the changes of working gap. For single grooves, provided there is a constant machining voltage, current density changes significantly by changing the preset working gap. On the other hand, product features of complex microstructures can be monitored during the process by measurement of current density.

The results of the analyses can be summarized as below.

For single grooves

- depth changes linearly with current density and
- surface roughness decreases with the increase in current density and then increases again.

Considering the above, in order to reach desired depth and roughness, a combination of process parameters which lead to specific current density should be selected. On the other hand, current monitoring during machining can be used to predict the product features before further measurements.

For intersecting and parallel grooves

- minimum current density over intersections changes proportionally to the depth of premachined grooves for each machining voltage level, which can be used as a monitoring tool for the first groove depth;
- the relative depth of intersections showed linear changes with the minimum current density over the intersection. Therefore, minimum depth over intersection can be applied for the prediction of the relative depth; and
- the depth and the mean current density of subsequent parallel grooves changes linearly with the lateral gap. This enhance the process monitoring with useful data of the actual lateral gap as well as the depth by monitoring the mean current density.

These results in an initial step toward Jet-ECM process control. In further research, experiments will be done to analyze the current density over different micro features on the workpiece.

Author Contributions: Conceptualization, M.Y.Z., M.H.-O., A.M., G.M., J.E. and A.S.; formal analysis, M.Y.Z., M.H.-O., A.M., G.M., J.E. and A.S.; writing—original draft preparation, M.Y.Z., M.H.-O., A.M. and G.M.; writing—review and editing J.E. and A.S.; supervision, M.H.-O., A.M., G.M., J.E. and A.S.; funding acquisition, M.H.-O., A.M. and A.S.

Funding: This research was funded by Horizon 2020, the EU Framework Programme for Research and Innovation (Project ID: 674801).

Acknowledgments: This research work was undertaken in the context of the MICROMAN Project ("Process Fingerprint for Zero-defect Netshape MICROMANufacturing", http://www.microman.mek.dtu.dk/). MICROMAN is a European Training Network supported by Horizon 2020, the EU Framework Programme for Research and Innovation (Project ID: 674801).

Conflicts of Interest: The authors declare no conflicts of interest.

References

1. Schubert, A.; Hackert-Oschätzchen, M.; Martin, A.; Winkler, S.; Kuhn, D.; Meichsner, G.; Zeidler, H.; Edelmann, J. Generation of Complex Surfaces by Superimposed Multi-dimensional Motion in Electrochemical Machining. *Procedia CIRP* **2016**, *42*, 384–389. [CrossRef]
2. Schubert, A.; Hackert-Oschätzchen, M.; Meichsner, G.; Zinecker, M.; Edelmann, J. Precision and Micro ECM with Localized Anodic Dissolution. In Proceedings of the 8th International Conference on Industrial Tools and Material Processing Technologies, Ljubljana, Slovenia, 2–5 October 2011; pp. 193–196.
3. Hackert, M. *Entwicklung und Simulation eines Verfahrens zum elektrochemischen Abtragen von Mikrogeometrien mit geschlossenem elektrolytischen Freistrahl*; Scripts Precision and Microproduction Engineering, Band 2; Schubert, A., Ed.; Verlag Wissenschaftliche Scripten: Auerbach, Germany, 2010; ISBN 9783937524955.

4. Schubert, A.; Hackert-Oschätzchen, M.; Meichsner, G.; Zinecker, M.; Martin, A. Evaluation of the Influence of the Electric Potential in Jet Electrochemical Machining. In Proceedings of the 7th International Symposium on Electrochemical Machining Technology, Wien, Austria, 3–4 November 2011; pp. 47–54.

5. Hackert-Oschätzchen, M.; Martin, A.; Meichsner, G.; Schubert, A. Evaluation of Gap Control Strategies in Jet Electrochemical Machining on Defined Shape Deviations. In Proceedings of the 10th International Symposium on Electrochemical Machining Technology, Saarbrücken, Germany, 13–14 November 2014; pp. 23–25.

6. Mitchell-Smith, J.; Speidel, A.; Clare, A.T. Advancing electrochemical jet methods through manipulation of the angle of address. *J. Mater. Process. Technol.* **2018**, *255*, 364–372. [CrossRef]

7. Goel, H.; Pandey, P.M.; Delhi, I.I.T.; Delhi, I.I.T. Experimental investigations into micro-drilling using air assisted jet electrochemical machining. In Proceedings of the 5th International & 26th All India Manufacturing Technology, Design and Research Conference, AIMTDR, Guwahati, India, 12–14 December 2014; Volume 1, pp. 1–7.

8. Hackert-Oschätzchen, M.; Meichsner, G.; Martin, A.; Zeidler, H.; Schubert, A. Fast Micromilling with Jet Electrochemical Machining. In Proceedings of the Danubia-Adria Symposium on Advances in Experimental Mechanics 2012, Belgrade, Serbia, 26–29 September 2012; pp. 54–55.

9. Martin, A.; Hackert-Oschätzchen, M.; Lehnert, N.; Schubert, A. Analysis of the fundamental removal geometry in electrochemical profile turning with continuous electrolytic free jet. *Procedia CIRP* **2018**, *68*, 466–470. [CrossRef]

10. Martin, A.; Hackert-Oschätzchen, M.; Lehnert, N.; Schubert, A. Analysis of the removal geometry in electrochemical straight turning with continuous electrolytic free jet. In Proceedings of the 13th International Symposium on ElectroChemical Machining Technology, Dresden, Germany, 30 November–1 December 2017; pp. 81–87.

11. Kawanaka, T.; Kunieda, M. Mirror-like finishing by electrolyte jet machining. *CIRP Ann. Manuf. Technol.* **2015**, *64*, 237–240. [CrossRef]

12. Hackert-Oschätzchen, M.; Kowalick, M.; Meichsner, G.; Schubert, A.; Hommel, B.; Jähn, F.; Scharrnbeck, M.; Garn, R.; Lenk, A. 2D Axisymmetric Simulation of the Electrochemical Finishing of Micro Bores by Inverse Jet Electrochemical Machining. In Proceedings of the European COMSOL Conference, Rotterdam, The Netherlands, 23–25 October 2013.

13. Oschätzchen, M.H.; Martin, A.; Meichsner, G.; Kowalick, M.; Zeidler, H.; Schubert, A. Inverse jet electrochemical machining for functional edge shaping of micro bores. *Procedia CIRP* **2013**, *6*, 378–383. [CrossRef]

14. Hackert-Oschätzchen, M.; Martin, A.; Meichsner, G.; Schubert, A. Analysis of strategies for gap control in jet electrochemical machining. In Proceedings of the 14th EUSPEN International Conference, Dubrovnik, Croatia, 2–6 June 2014; pp. 439–442.

15. Leese, R.J. Electrochemical Machining—New Machining Targets and Adaptations with Suitability for Micromanufacturing. Ph.D. Thesis, Brunel University, London, UK, 2016.

16. Schubert, A.; Hackert, M.; Meichsner, G. Simulating the Influence of the Nozzle Diameter on the Shape of Micro Geometries generated with Jet Electrochemical Machining. In Proceedings of the COMSOL Conference, Boston, MA, USA, 8–10 October 2009; pp. 2–5.

17. Leese, R.J.; Ivanov, A. Electrochemical micromachining: An introduction. *Adv. Mech. Eng.* **2016**, *8*, 1–13. [CrossRef]

18. Hackert-Oschätzchen, M.; Martin, A.; Meichsner, G.; Zinecker, M.; Schubert, A. Microstructuring of carbide metals applying Jet Electrochemical Machining. *Precis. Eng.* **2013**, *37*, 621–634. [CrossRef]

19. Hackert, M.; Meichsner, G.; Schubert, A. Generating Micro Geometries with Air Assisted Jet Electrochemical Machining. In Proceedings of the 10th Anniversary International Conference of the European Society for Precision Engineering and Nanotechnology, EUSPEN 2008, Zürich, Switzerland, 18–22 May 2008; pp. 420–424.

20. Speidel, A.; Mitchell-Smith, J.; Walsh, D.A.; Hirsch, M.; Clare, A. Electrolyte Jet Machining of Titanium Alloys Using Novel Electrolyte Solutions. *Procedia CIRP* **2016**, *42*, 367–372. [CrossRef]

21. Hackert-Oschätzchen, M.; Meichsner, G.; Zeidler, H.; Zinecker, M.; Schubert, A. Micro machining of different steels with closed electrolytic free jet. *AIP Conf. Proc.* **2011**, *1353*, 1337–1343. [CrossRef]

22. Speidel, A.; Mitchell-Smith, J.; Bisterov, I.; Clare, A.T. The importance of microstructure in electrochemical jet processing. *J. Mater. Process. Technol.* **2018**, *262*, 459–470. [CrossRef]

23. König, W. Elektrochemisches Abtragen (ECM). In *Fertigungsverfahren 3*; Springer: Berlin/Heidelberg, Germany, 2007; pp. 133–185.

24. Heinemann, H.; Krämer, H.; Zimmer, H. *Kleine Formelsammlung Physik*; Fachbuchverlag Leipzig: Leipzig, Germany, 1997.

25. Lindenlauf, H.-P. Werkstoff- und Elektrolytspezifische Einflüsse auf die Elektrochemische Senkbarkeit ausgewählter Stähle und Nickellegierungen. Ph.D. Thesis, RWTH Aachen, Aachen, Germany, 1977.

26. Rosenkranz, C. Elektrochemische Prozesse an Eisenoberflächen bei Extremen Anodischen Stromdichten. Ph.D. Thesis, Heinrich-Heine-Universität Düsseldorf, Düsseldorf, Germany, 2005.

27. Andreatta, F.; Lohrengel, M.M.; Terryn, H.; de Wit, J.H.W. Electrochemical characterisation of aluminium AA7075-T6 and solution heat treated AA7075 using a micro-capillary cell. *Electrochim. Acta* **2003**, *48*, 3239–3247. [CrossRef]

28. Haisch, T.; Mittemeijer, E.; Schultze, J.W. Electrochemical machining of the steel 100Cr6 in aqueous NaCl and NaNO3 solutions: Microstructure of surface films formed by carbides. *Electrochim. Acta* **2001**, *47*, 235–241. [CrossRef]

29. Lohrengel, M.M.; Klüppel, I.; Rosenkranz, C.; Bettermann, H. Microscopic investigations of electrochemical machining of Fe in NaNO3. *Electrochim. Acta* **2003**, *48*, 3203–3211. [CrossRef]

30. Rosenkranz, C.; Lohrengel, M.M.; Schultze, J.W. The surface structure during pulsed ECM of iron in NaNO3. *Electrochim. Acta* **2005**, *50*, 2009–2016. [CrossRef]

31. Schreiber, A.; Schultze, J.W.; Lohrengel, M.M.; Kármán, F.; Kálmán, E. Grain dependent electrochemical investigations on pure iron in acetate buffer pH 6.0. *Electrochim. Acta* **2006**, *51*, 2625–2630. [CrossRef]

32. Hackert-Oschätzchen, M.; Meichsner, G.; Zinecker, M.; Martin, A.; Schubert, A. Micro machining with continuous electrolytic free jet. *Precis. Eng.* **2012**, *36*, 612–619. [CrossRef]

micromachines

MDPI

Article

Structuring of Bioceramics by Micro-Grinding for Dental Implant Applications

Pablo Fook *, Daniel Berger, Oltmann Riemer and Bernhard Karpuschewski

Laboratory for Precision Machining (LFM), Leibniz Institute for Materials Engineering (IWT), MAPEX Center for Materials and Processes, University of Bremen, 28359 Bremen, Germany; berger@iwt.uni-bremen.de (D.B.); riemer@iwt.uni-bremen.de (O.R.); karpu@iwt.uni-bremen.de (B.K.)
* Correspondence: fook@iwt.uni-bremen.de; Tel.: +49-421-2185-1170

Received: 13 April 2019; Accepted: 1 May 2019; Published: 9 May 2019

Abstract: Metallic implants were the only option for both medical and dental applications for decades. However, it has been reported that patients with metal implants can show allergic reactions. Consequently, technical ceramics have become an accessible material alternative due to their combination of biocompatibility and mechanical properties. Despite the recent developments in ductile mode machining, the micro-grinding of bioceramics can cause insufficient surface and subsurface integrity due to the inherent hardness and brittleness of these materials. This work aims to determine the influence on the surface and subsurface damage (SSD) of zirconia-based ceramics ground with diamond wheels of 10 mm diameter with a diamond grain size (d_g) of 75 μm within eight grinding operations using a variation of the machining parameters, i.e., peripheral speed (v_c), feed speed (v_f), and depth of cut (a_e). In this regard, dental thread structures were machined on fully sintered zirconia (ZrO_2), alumina toughened zirconia (ATZ), and zirconia toughened alumina (ZTA) bioceramics. The ground workpieces were analysed through a scanning electron microscope (SEM), X-ray diffraction (XRD), and white light interferometry (WLI) to evaluate the microstructure, residual stresses, and surface roughness, respectively. Moreover, the grinding processes were monitored through forces measurement. Based on the machining parameters tested, the results showed that low peripheral speed (v_c) and low depth of cut (a_e) were the main conditions investigated to achieve the optimum surface integrity and the desired low grinding forces. Finally, the methodology proposed to investigate the surface integrity of the ground workpieces was helpful to understand the zirconia-based ceramics response under micro-grinding processes, as well as to set further machining parameters for dental implant threads.

Keywords: micro-grinding; bioceramics; materials characterisation; dental implant

1. Introduction

Dental implants aim to replace a partially or totally, damaged or diseased tooth structure, i.e., restoring the function and also the aesthetics [1,2]. In the last decades, this market has experienced growth and one of the solutions for dental treatment became the replacement of conventional metal-based dentures with ceramic materials [3]. The use of bioceramics is an alternate option to the toxic and allergic effects that might be caused by diffused metal ions due to corrosion and deterioration without wear of metal materials [4]. This alternative, however, is only possible because of the new developments in the field of biomaterials and computer-aided design/computer-aided manufacturing (CAD/CAM) technologies [5–7]. In this regard, the optimisation of CAD/CAM systems has enabled more efficient and cost-effective grinding processes in the scientific, industrial, and technological fields in a variety of sectors, such as aeronautics and biomedical [8].

Among the bioceramics in the market, aluminium oxide (alumina—Al_2O_3) and zirconium dioxide (zirconia—ZrO_2) have become the better alternatives due to their combination of biocompatibility, mechanical properties, like high flexural strength and wear resistance, as well as minimum thermal and electrical conductivity [5,9,10]. In the specific case of zirconia, its phase transformation toughening phenomenon is known to improve the properties of the material. This phenomenon stops crack propagation, resulting from the transformation of zirconia from the tetragonal phase into the monoclinic phase, as well as the consequential 3% to 5% volume expansion and induction of compressive stresses. The interest in the toughening mechanics of zirconia allowed for the development of further zirconia-based ceramics, such as alumina-toughened zirconia (ATZ) and zirconia-toughened alumina (ZTA) [1,2,5,11].

The structuring of bioceramics using micro-grinding is still an area under investigation, and the surface integrity characterisation of ground ceramics is considered to be a key aspect of their further applications as dental implants on the market [1–3]. In this study, three types of fully sintered zirconia-based ceramics machined by micro-grinding were characterised by monitoring the process forces and measuring the surface integrity of the ground workpieces. The grinding strategy suggested that replicating the square thread profiles of dental implants using diamond galvanic-bonded wheels and optimising the machining parameters, i.e., peripheral speed (v_c), feed rate (v_f), and depth of cut (a_e), using the design of experiments (DOE) method as the statistical approach for planning, conducting, analyzing and interpreting data from grinding experiments. The results evaluate the bioceramics with regard to their machinability and, thus, their suitability as materials for dental ceramics.

2. Materials and Methods

2.1. Case Study

The success of ceramic dental implants is correlated with effective osseointegration, such as the formation of direct contact between the implant and the surrounding bone [11,12]. According to the literature, to establish reliable osseointegration, six main factors should be considered: material selection, implant design, an optimum range of surface roughness, bone status, surgical technique, and loading conditions [13,14]. The last three points are correlated with the dentist's expertise and biological factors concerning surgical planning and tooth restoration. Therefore, the first three factors are directly influenced by the grinding conditions [9,13,14].

In order to avoid defect parts during manufacturing and failures during its use, enhanced micro-grinding processes are still necessary and, consequently, have been subject to several research investigations [3,6,7,9,13,15–19]. Crucial requirements for high surface integrity and mechanical reliability of dental ceramic implants are knowledge and control of the critical machining parameters that are based on the materials, the implant overall design, and custom-designed requirement, i.e., the patients' needs. Therefore, based on the current trends in dental implants, a wide variety of factors must be considered in threads design and component manufacturing. Figure 1 lists some of these characteristics according to the literature, as well as the relevant features and the ranges selected as optimal for the successful performance of the dental implants, namely shape, dimensions, surface roughness, and thread pattern. Specifically, implant threads are designed to maximize initial contact, provide primary stability, enhance the surface area, cause compression of bone, facilitate dissipation of loads at the bone-implant interface, and minimize the micro-movement to hazen osseointegration [12,13,20,21]. Multiple investigations have concluded that the square thread profile may provide the best primary stability and the most effective stress distribution in an immediate loading situation [20–22].

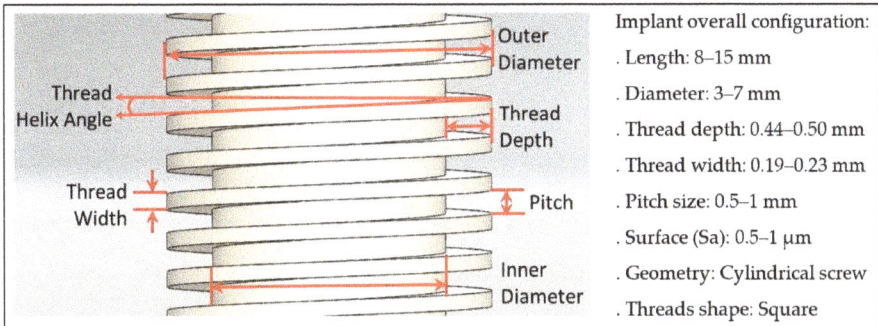

Figure 1. The overall design and characteristics of commercial dental implant threads [1,13,14,20–22].

2.2. Material Selection

The ceramic workpieces (5.0 mm × 7.0 mm × 33.0 mm) are fully sintered and commercially-available tetragonal polycrystalline zirconia (ZrO_2-TZP),—also known as zirconium dioxide (ZrO_2) and commonly called "zirconia"—alumina toughened zirconia (ATZ), and zirconia toughened alumina (ZTA). These materials are zirconium-based ceramics commonly used for dental applications and have intrinsic toughening mechanisms, but differ in their mechanical properties [5,23], as shown in Table 1.

Table 1. Description of the material properties [23].

Material	Density (g/cm^3)	Fracture Toughness K_{IC} (MPa m$^{1/2}$)	Young's Modulus (GPa)	Hardness HV_{10} (GPa)	Flexural Strength (MPa)	Fraction of ZrO_2 (%)
ZrO_2-TZP	6.03	4.8	200	11	1000	> 95
ATZ	5.50	7	220	14	820	76
ZTA	4.10	8	380	16	440	14

2.3. Process Kinematics and Experimental Conditions

In order to machine dental threads, as illustrated in Figure 1, the micro-grinding process kinematics were carried out on a DMG Sauer 20 linear machine tool (DMG Sauer GmbH, Bielefeld, Germany) under a water-based lubricant that also provided cooling, lubrication, and chip removal. The machining of the ceramic workpieces was performed by a tool feed (v_f) along the x-direction, while the tool spindle rotated (n_p). In this case, diamond galvanic-bonded wheels of 10 mm in diameter, commercialised by the company SCHOTT Diamantwerkzeuge GmbH, with a specific width (b_w) of 0.9 mm, and an average diamond grain size of 75 µm (D75) were used [24,25]. Moreover, the machining strategy and the diamond wheels used to grind the bioceramics were designed to follow the characteristic dental implant thread width (t_w) of 0.2 mm, as well as the overall design of a square thread profile of a dental implant. Figure 2 illustrates the process kinematics.

The machining conditions performed are summarised in Table 2, i.e., peripheral speed (v_c), feed speed (v_f), and depth of cut (a_e), which were selected after a screening campaign and a literature review concerning the critical depth of cut, as well as the equivalent chip thickness, to achieve ductile grinding mode machining [6,7,10,16–19,26].

In the present work, the Taguchi method was used as the design of experiment (DOE) approach to examine the influence of the grinding process parameters on the surface integrity and grinding forces of the three bioceramics materials [27–30]. As a result, eight process conditions (Table 3) were designed and every experiment (P1 to P8) was performed three times for statistical purposes.

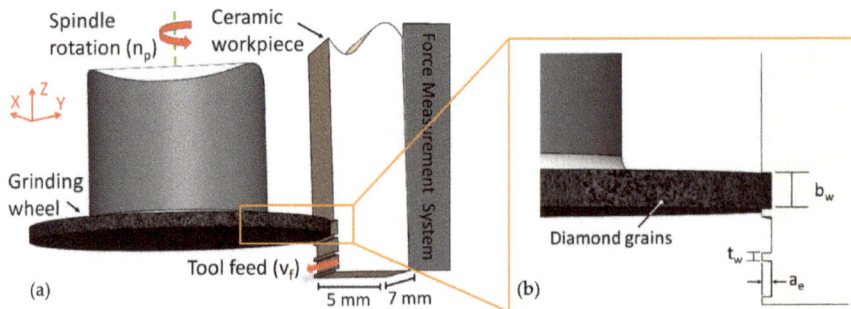

Figure 2. Illustration of the micro-grinding strategy performed in this work (**a**) in which the tool engagement and the workpiece features are highlighted (**b**).

Table 2. Grinding conditions and material selection.

Peripheral Speed, v_c (m/s)	Feed Speed, v_f (mm/min)	Depth of Cut, a_e (µm)	Material
10.00, 18.33	100, 300	50, 250	ZrO_2, ATZ, ZTA

Table 3. Design of the process conditions in this study.

Process Condition	Peripheral Speed, v_c (m/s)	Feed Speed, v_f (mm/min)	Depth of Cut, a_e (µm)
P1	18.33	300	250
P2	18.33	100	250
P3	10.00	300	250
P4	10.00	100	250
P5	18.33	300	50
P6	18.33	100	50
P7	10.00	300	50
P8	10.00	100	50

2.4. Workpiece Characterisation and Process Monitoring

In order to evaluate the surface and subsurface damage (SSD), the microstructure and surface topography were studied by means of a scanning electron tabletop microscope TM3030 (Hitachi Ltd., Hitachi, Japan) under magnifications of 60× and 2500×. An X-ray diffraction (XRD) machine (Bruker Co., Billerica, USA) was used to measure the residual stresses along the diagonal stress axis (Cu Kα-source, 30 kV, and 40 mA radiation). Finally, the surface roughness was measured by using a white light interferometer Talysurf CCI HD (WLI Taylor Hobson, Leicester, UK) within an air-conditioned laboratory. Herein, the roughness data was acquired using a Gaussian filter with a specified cut off λ_c of 0.08 mm and an objective of 50×.

Figure 3 depicts the force measurement system available at the DMG Sauer grinding machine. Herein, a three-component force dynamometer unit Kistler 9256-C2 (Kistler Holding AG, Winterthur, Switzerland) was used for the measurement of the grinding forces. A data acquisition and analysis software MesUSoft 2.5.23 (IWT, Bremen, Germany) was used for data collection and display. This study focused on the forces applied to the y-direction (Fy) as a methodology to monitor the machining process. The mean force (Fy) was estimated according to the average values of each grinding step in the y-axis.

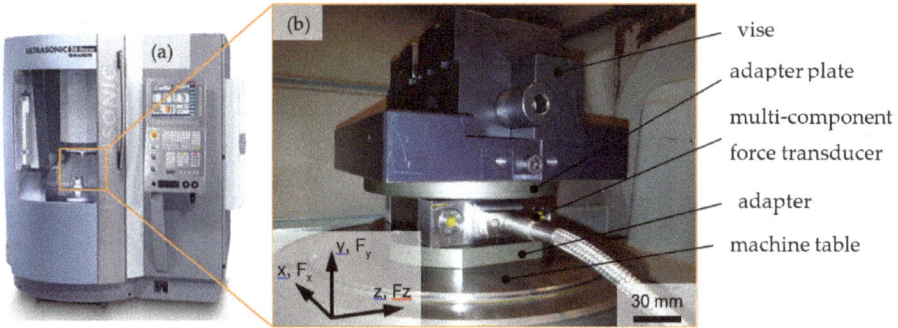

Figure 3. (**a**) DMG Sauer 20 linear machine tool and (**b**) setup for force measurement system.

3. Results

3.1. Surface and Subsurface Damage (SSD) Evaluation

3.1.1. Microstructure

Figures 4 and 5 show the microstructure modification of the ground threads as a result of the most representative process conditions investigated. Figure 4 corresponds to ground bioceramics machined with the highest possible parameters selected for this study (P1), while Figure 5 corresponds to the less demanding configuration (P8).

The surface of the ground ZrO_2 dental threads, with a 75-grit diamond wheel, at peripherical speed (v_c) 10.00 m/s, feed rate (v_f) 100 mm/min, and a depth of cut (a_e) 50 μm, i.e., process condition P8, showed ductile streaks and a smooth surface as indicated in Figure 5. The same bioceramic ground at v_c 18.33 m/s, v_f 300 mm/min, and an a_e 250 μm, i.e., process condition P1, as shown in Figure 4, had a ductile area with micro-ploughing deformation. In both images, ZrO_2 ground workpieces showed a greater amount of ductile areas than the ATZ and ZTA specimens, where the material was removed in a more partial ductile grinding and brittle mode.

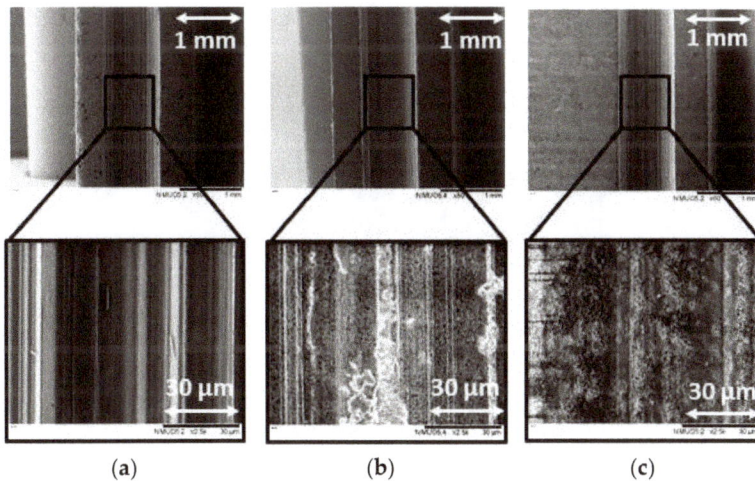

Figure 4. SEM analysis of the ground bioceramics under magnifications 60× and 2500×, i.e., for (**a**) ZrO_2, (**b**) ATZ and (**c**) ZTA specimens. The bioceramics were machined with process condition P1, i.e., $v_c = 18.33$ m/s, $v_f = 300$ mm/min, and $a_e = 250$ μm.

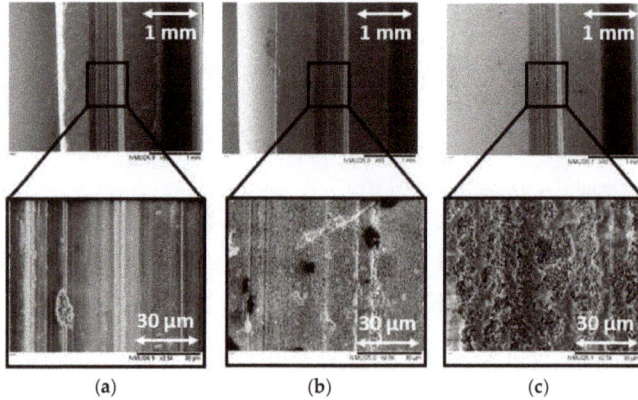

Figure 5. SEM analysis of the ground bioceramics under magnifications 60× and 2500×, i.e., for (**a**) ZrO_2, (**b**) ATZ, and (**c**) ZTA specimens. The bioceramics were machined with process condition P8, i.e., $v_c = 10.00$ m/s, $v_f = 100$ mm/min, and $a_e = 50$ μm.

3.1.2. Residual Stress

According to the XRD analysis, shown in Figure 6, compressive and tensile residual stresses of the ground samples were observed for P1 and P8. Grinding of the ZrO_2 workpieces predominantly increased the compressive stresses, i.e., −141 MPa for P1 and −52 MPa for P8. Both machined ATZ samples exhibited tensile stresses after grinding, for instance, 178 MPa and 133 MPa for P1 and P8, respectively. For ZTA specimens, the P1 tended to generate a slightly tensile stress of 6 MPa and substantial compressive stress for P8, herein −92 MPa. This phenomenon was due to the toughening mechanism, which also involves the phase transformation already mentioned. In general, compressive values are, likewise, desired in the specimen surface for biomedical application. For example, the compressive residual stress tends to increase the fatigue strength and the fatigue life of ceramic dental implants [1,5,10,19].

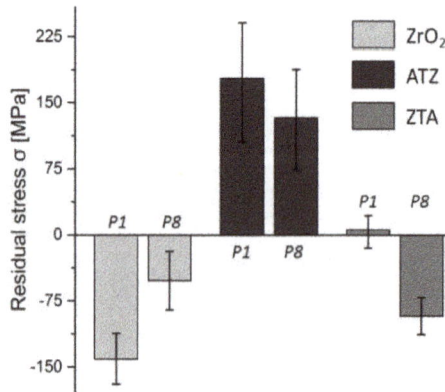

Figure 6. Residual stresses measured with X-ray diffraction technique of the ground ZrO_2, ATZ, and ZTA ceramics machined with parameters P1 ($v_c = 18.33$ m/s; $v_f = 300$ mm/min; $a_e = 250$ μm) and P8 ($v_c = 10.00$ m/s; $v_f = 100$ mm/min; $a_e = 50$ μm).

3.1.3. Surface Roughness

Figure 7 shows the surface roughness values, Sa (arithmetical mean height), of the as-received and ground ZrO_2, ATZ, and ZTA specimens. Although the same process conditions were applied for all the three bioceramics, different surface roughness were achieved. The values for the ATZ dental threads were all considered to be the optimum results for the successful osseointegration of biomedical implants. For example, dental implants are suggested to exhibit a surface roughness, Sa, of between 500 and 1000 nm [1,20,22]. Therefore, for further use of ZTA materials as an implant, processes P2, P3, P4, P7, and P8 had an optimum range for the dental application; for ground ZrO_2 ceramics, this was only observed for P3, P4, P7, and P8.

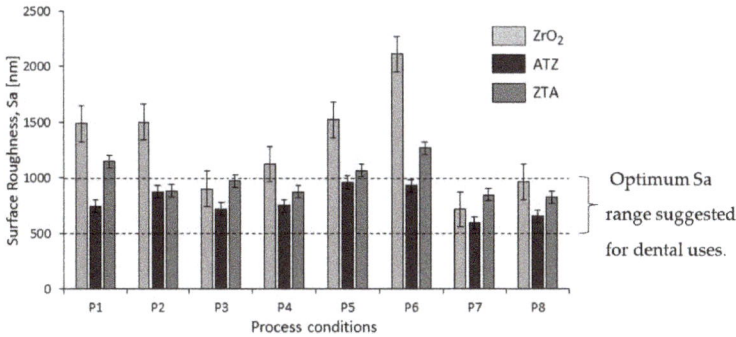

Figure 7. The surface roughness, Sa (nm), per material of the ground dental threads. (Table A1).

In general, the highest variation in Sa values, with respect to the eight machining conditions, was observed on ground ZrO_2 workpieces, i.e., from ca. 715 nm up to 2100 nm. In comparison, ATZ and ZTA ceramics ranged between ca. 510 nm and 960 nm, and between ca. 820 nm and 1200 nm, respectively.

3.2. Process Monitoring

Grinding Forces

Figure 8 shows the average values of each group of the bioceramic specimens and machining conditions, according to the methodology mentioned in Section 2.2. Due to the higher depths of cut in the first 4 processes (i.e., P1 to P4 and an a_e of 250 μm), higher mean forces (Fy) were also measured in comparison to the machining conditions that followed (i.e., P5 to P8 and an a_e of 50 μm). Specifically, for the machining of ATZ, higher forces were reported in regard to the grinding of ZrO_2 and ZTA, which obtained similar values. In general, lower forces are beneficial for the tool wear and tool life [6,31].

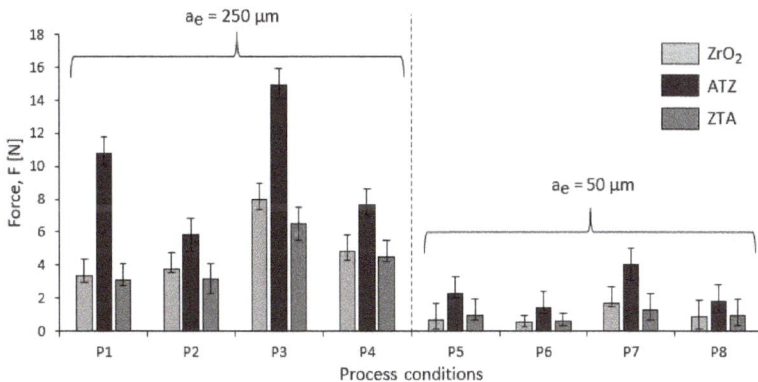

Figure 8. Mean forces Fy (N) for the grinding processes per material. (Table A2).

4. Discussion

In this section, further analysis of the surface and subsurface damage (SSD) of ground bioceramics threads are discussed. Moreover, the Taguchi method is used for understanding the mean Sa and F responses, based on the eight designed machining parameters [27–30].

4.1. Microstructure

In Figures 4 and 5, the surface of ATZ and ZTA workpieces show brittle intercrystalline breakouts, high roughness, and bulging at the scratch edges on the microstructure. The brittle outbreak marks are predominant in the ZTA specimens, which links to the higher hardness and lower flexural strength of the material in comparison to the ZrO_2 and ATZ ceramics. This showed that brittle materials led to different surface topographies although grinding conditions did not vary.

The grinding direction is clearly discernible in both figures. The surfaces consist mostly of a series of parallel grinding marks, and the width of these forms are better visible in Figure 4 than Figure 5 due to the higher machining conditions selected. For that reason, process condition P8 tended to generate less defects and flaws on the microstructure surface of the zirconia-based ceramics.

An important characteristic of a ceramic dental implant is the ability to create correct interaction between the ceramic implant and the bone tissue through the ground threads [32]. Since most of the implant surface is in direct contact with bone tissue, form and integrity of the microstructure surface have a great influence on successful osseointegration. Herein, the ZrO_2 ground ceramics have a tendency to have a higher osseointegration response once a better-machined surface is obtained than the ATZ and ZTA bioceramics [12,14,20–22,32].

4.2. Residual Stress

According to the XRD results, the highest machining parameters (P1, i.e., v_c = 18.33 m/s, v_f = 300 mm/min, and a_e = 250 µm) were more beneficial for ZrO_2 workpieces in comparison to the less demanding grinding conditions investigated (P8, i.e., v_c = 10.00 m/s, v_f = 100 mm/min, and a_e = 50 µm) once more compressive stress was measured. However, in the case of the ATZ specimens, condition P1 resulted in higher tensile stresses than P8. No significant stress differences were seen in ZTA ceramics ground with condition P1, but the machining parameter P8 generated considerable compressive stresses.

The different responses on the residual stresses of ZrO_2, ATZ, and ZTA ceramics are a function of their different intrinsic physical properties, material processing, machining history, and zirconia phase amount of each material [5,9–11]. The amount of zirconia phase was substantially distinct among the three bioceramics investigated (Table 1). Consequently, the machining effect on the crystallographic structure transformation from the zirconia tetragonal phase into the monoclinic phase to induce compressive stresses were also different.

The phase transformation in ceramics is a combination of the kinetics of diffusion-controlled as well as of diffusionless transformations at different strain rate and contact zone temperature by a martensitic transformation that occurs in the zirconia phase where the crystal structure changes from a tetragonal to a monoclinic structure and generates a 3% to 5% volume expansion [10,11,33,34].

Compressive stresses are ideally better accommodated by the complete implant–prosthesis system since the cortical bone is stronger in compression and weaker under shear and tensile forces [1,10,11,19]. In practical situations, the total contact area between the implant and bone may apply shear stresses that are transferred along with the interface, which can be harmful to the jaw and even destructive to the implant if a denture is wrongly chosen. Hence, the ground ZrO_2 ceramics were indicated as the best option due to the higher compressive residual stress observed in both grinding conditions analysed.

4.3. Influence of Processing Parameters on Surface Roughness and Grinding Forces

The Taguchi method was used as a statistical tool for the optimisation of the grinding process by analysing the machining parameters' influence on surface roughness and grinding forces. Therefore, an analysis of variance (ANOVA) was performed and evaluated using the statistical software Minitab 17 [35]. ANOVA results were carried out by separating the total variability of each machining parameter and its error. The main machining factors, peripheral speed (v_c), feed speed (v_f), and depth of cut (a_e), and their response on the surface roughness (Sa) and grinding forces (Fy) on the three bioceramics were analysed [27–30,35].

To examine the differences between the most and the least demanding machining parameters conditions (P1 and P8), the main effect plots were generated with Minitab 17 support [35]. Basically, there was a main effect response when the different levels of a grinding parameter affected the surface roughness as well as the mean forces in the y-axis differently. The main effect plot graphs were visualized by the response mean for each machining parameter connected by a line. When the line tended to be horizontal (parallel to the x-axis), there was no main effect. When the line was not horizontal, there was a main effect or influence between the two grinding parameters selected in relation to the response investigated—herein, the response to surface roughness (Sa) and forces (Fy) in the process. Therefore, different levels of the factor affected the response differently. The steeper the slope of the line, the larger the magnitude of the main effect.

4.3.1. Surface Roughness

The individual grinding parameters effects, i.e., peripheral speed (v_c), feed speed (v_f), and depth of cut (a_e), on the surface roughness of ZrO_2, ATZ and ZTA specimens are presented in Figure 9. All the bioceramic materials showed a similar trend regarding the peripheral speed (v_c), which had the highest influence on surface roughness. Feed speed (v_f) and depth of cut (a_e) had the lowest contribution factor, but different inclinations according to the tested material, i.e., the slope rose or fell with the increase of the respective machining parameter level tested.

For all three bioceramic materials, v_c had a very strong slope line, which indicated the highest influence on surface roughness (Sa). Therefore, increasing the rotation speed also led to an increase in surface roughness. The highest v_c response on surface roughness indicated that the setting of 10.00 m/s peripherical speed was beneficial to achieve an optimal Sa range, independent of the v_f and a_e designed in this study. This was the reason that the machining conditions P3, P4, P7, and P8 were indicated for grinding all three zirconia-based ceramics investigated once the Sa achieved between 500 and 1000 nm are an optimum range for further dental applications as also mentioned in Section 3.1.3. Surface Roughness [1,20,22].

Additionally, once the feed rate increased, the surface roughness decreased for the ZrO_2 and ATZ ceramics and increased for ZTA. Although the a_e main effect line tended to be horizontal for the machined bioceramics and, consequently, no significant Sa response was observed, the increase in the depth of cut for grinding ZrO_2 and ZTA seemed to have a slightly positive influence on the material surface, while the opposite was observed for ATZ specimens.

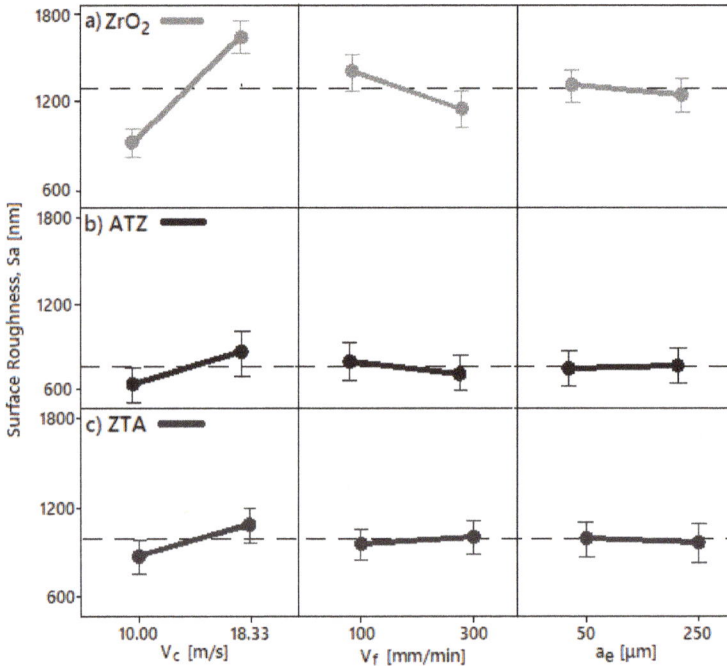

Figure 9. The main effect diagram for surface roughness (Sa) of (**a**) ZrO₂, (**b**) ATZ, and (**c**) ZTA in regard to peripheral speed (v_c), feed rate (v_f), and depth of cut (a_e).

4.3.2. Grinding Forces

The quality of the dental part produced by the micro-grinding process is influenced by the grinding tool and the conditions, which are linked in particular by the induced mechanical forces [2,6,10,29,33]. As a result, the normal forces (Fy) monitored for each bioceramic during the machining are based on the workpiece material properties and consequently chip formation and ploughing force [6,10,16,22,31].

Similar force values were monitored during the grinding of the ZrO₂ and ZTA ceramics. Both materials showed a significant lower process force response than the ATZ materials. Moreover, the highest forces measured in the process conditions with the higher depth of cut, i.e., a_e = 250 µm, were not desired because of expected higher tool wear during machining.

The forces measured while machining of the ATZ workpieces were essentially two- to three-times higher than the other two bioceramics. Basically, the long term machining of the ATZ ceramics tended to introduce more damage to the grinding tool life and to the surface integrity of the implant in comparison to the ZrO₂ and ZTA materials.

The parameter setting P3 (v_c = 10 m/s, v_f = 300 mm/min, and a_e = 250 µm) exhibited the highest forces monitored, while the lowest forces were seen with condition P6 (v_c = 18.33 m/s, v_f = 100 mm/min, and a_e = 50 µm). Therefore, in this study, high rotation, low feed speed, and low cutting depth tended to be beneficial to keep process forces low. This machining configuration is explained in Figure 10.

The Figure indicates the process forces response during the grinding. In general, the ZrO₂, ATZ, and ZTA materials showed similar tendencies regarding the machining factors influences. The depth of cut (a_e) had the highest contribution, and peripheral speed (v_c) and feed speed (v_f) had the lowest impact factor on the ceramics.

For all bioceramics, a_e had a very strong line inclination, which points out the highest influence on the grinding forces (Fy) response once the cutting depth was increased. Furthermore, the same tendency with a lower slope was seen for the v_f main effect, where the force response also increased when the feed speed set was higher. The opposite configuration was seen once the peripheral speed rose. Herein, the forces applied to the ceramics decreased.

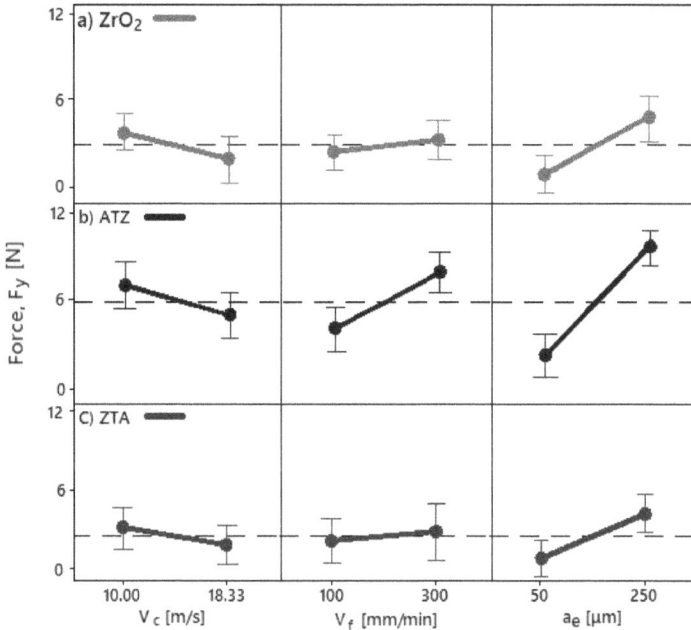

Figure 10. The main effect diagram for forces (Fy) of (**a**) ZrO$_2$, (**b**) ATZ, and (**c**) ZTA in regard to peripheral speed (v$_c$), feed rate (v$_f$), and depth of cut (a$_e$).

5. Conclusions

Three fully sintered types of zirconia-based ceramics, zirconium dioxide (ZrO$_2$), alumina-toughened zirconia (ATZ), and zirconia-toughened alumina (ZTA) were structured by micro-grinding process. In order to replicate dental threads with a square profile, a grinding wheel with a diameter of 10 mm and a specific width (b$_w$) of 0.9 (grain size: D75) was used. Eight machining conditions were designed and the process forces (F) were monitored. The microstructure of the ground bioceramics was analysed via SEM, the XRD technique accessed the residual stresses, and surface roughness (Sa) was measured with WLI.

The following conclusions were drawn from the investigation:

- The microstructures of the ground ATZ and ZTA workpieces showed brittle intercrystalline breakouts, high roughness, and bulging at the scratch edges. Although the ground ZrO$_2$ surfaces had parallel grinding marks with micro-ploughing deformation, their microstructure had a larger amount of ductile areas than the other specimens.
- For a successful implant and mechanical stability in the jaw, compressive residual stresses on the material surface are recommended. ZrO$_2$ ceramics had shown the best response concerning the residual stresses among the ceramics tested for dental application. Herein, higher compressive stresses after grinding were observed due to the toughening mechanics of the zirconia phase.
- The different surface roughness (Sa) and force (F) responses due to the different grinding parameters were directly correlated to the intrinsic physical properties and chemical composition of the bioceramics investigated. Basically, the machining conditions P3, P4, P7, and P8 generated the optimal surface roughness suggested for dental implants, i.e., between 500 and 1000 nm, on all machined bioceramic materials. This was due to the highest peripheral speed, v$_c$, response on the

surface roughness, which indicated that the low level tested (10.00 m/s) was beneficial to achieve an optimal Sa for dental uses. Regarding the process monitoring, the depth of cut (a_e) had the highest influence on the grinding forces (Fy) response when it was larger—herein, the machining process with an a_e of 50 μm are indicated for less tool wear and best implant integrity.

- ZrO_2 ceramics machined with the grinding conditions P7 (v_c = 10.00 m/s, v_f = 300 mm/min, and a_e = 50 μm) and P8 (v_c = 10.00 m/s, v_f = 100 mm/min, and a_e = 50 μm) are suggested for further dental applications due to their optimal Sa range, smoother microstructures, compressive residual stress, as well as low forces generated during machining.

Based on the results of this work, a future investigation should include a similar approach to machine aluminium oxide (alumina—Al_2O_3). This will extend the validity of this approach and will allow a fundamental material and process analysis in the micro-grinding of bioceramics. Finally, the next steps are the machining of ceramic dental parts based on the optimum grinding parameters investigated for further mechanical and biological evaluation.

Author Contributions: P.F. conceived and designed the experiments; P.F. performed experiments and measurements; P.F. and D.B. analysed the data; P.F. wrote the paper; D.B., O.R. and B.K. revised the paper.

Funding: This research work was undertaken in the context of the MICROMAN project ("Process Fingerprint for Zero-defect Net-shape MICROMANufacturing", http://www.microman.mek.dtu.dk/). MICROMAN is a European training network supported by Horizon 2020, the EU Framework Programme for Research and Innovation (Project ID: 674801).

Acknowledgments: The collaboration and technical advisement provided by Friedhelm Kleine (SCHOTT Diamantwerkzeuge GmbH), Darshan Jain (University of Applied Sciences Bremerhaven), Chimene Kenfack, and Devarsh Patel (University of Bremen) are also gratefully acknowledged.

Conflicts of Interest: The authors declare no conflict of interest.

Appendix A

Table A1. Surface roughness, Sa (nm), per material of the ground dental threads.

Process	ZrO_2	ATZ	ZTA
P1	1488.30 ± 152.46	744.74 ± 10.83	1148.46 ± 17.25
P2	1502.08 ± 41.03	875.75 ± 51.44	884.17 ± 45.23
P3	901.90 ± 57.05	720.76 ± 17.91	982.88 ± 10.78
P4	1125.77 ± 2.46	749.89 ± 51.89	874.31 ± 13.80
P5	1522.83 ± 89.90	961.05 ± 60.98	1065.36 ± 62.40
P6	2113.38 ± 54.18	932.95 ± 17.07	1272.16 ± 47.94
P7	716.62 ± 38.73	509.51 ± 27.51	850.40 ± 88.68
P8	965.73 ± 19.05	652.46 ± 33.69	826.89 ± 13.89

Table A2. Mean forces, F (N), in the grinding processes per material.

Process	ZrO_2	ATZ	ZTA
P1	3.37 ± 0.45	10.76 ± 0.72	3.07 ± 0.33
P2	3.77 ± 0.22	5.80 ± 1.01	3.12 ± 1.14
P3	7.95 ± 0.58	14.91 ± 1.12	6.50 ± 0.98
P4	4.85 ± 0.54	7.64 ± 0.59	4.50 ± 0.30
P5	0.65 ± 0.52	2.29 ± 0.25	0.94 ± 0.28
P6	0.56 ± 0.18	1.42 ± 0.33	0.58 ± 0.22
P7	1.67 ± 0.71	4.03 ± 0.94	1.31 ± 0.66
P8	0.86 ± 0.28	1.80 ± 0.67	0.95 ± 0.61

References

1. Gaviria, L.; Salcido, J.P.; Guda, T.; Ong, J.L. Current trends in dental implants. *J. Korean Assoc. Oral Maxillofac. Surg.* **2014**, *40*, 50–60. [CrossRef]

2. Shemtov-Yona, K.; Rittel, D. On the mechanical integrity of retrieved dental implants. *J. Mech. Behav. Biomed. Mater.* **2015**, *49*, 290–299. [CrossRef] [PubMed]

3. Özkurt, Z.; Kazazoğlu, E. Zirconia dental implants: a literature review. *J. Oral Implantol.* **2011**, *37*, 367–376. [CrossRef]

4. Muris, J.; Kleverlaan, C.J. Hypersensitivity to Dental Alloys. In *Metal Allergy: From Dermatitis to Implant and Device Failure*, 1st ed.; Chen, J.K., Thyssen, J.P., Eds.; Springer: Berlin, Germany, 2018; pp. 285–300.

5. Richerson, D.W.; Lee, W.E. *Modern Ceramic Engineering: Properties, Processing, and Use in Design*; CRC Press: Boca Raton, FL, USA, 2018.

6. Brinksmeier, E.; Mutlugünes, Y.; Klocke, F.; Aurich, J.C.; Shore, P.; Ohmori, H. Ultra-precision grinding. *CIRP Ann.* **2010**, *59*, 652–671. [CrossRef]

7. Bianchi, E.C.; de Aguiar, P.R.; Diniz, A.E.; Canarim, R.C. Optimization of ceramics grinding. In *Advances in Ceramics - Synthesis and Characterization, Processing and Specific Applications*, 1st ed.; IntechOpen: London, UK, 2011.

8. González, H.; Calleja, A.; Pereira, O.; Ortega, N.; López de Lacalle, L.; Barton, M. Super abrasive machining of integral rotary components using grinding flank tools. *Metals* **2018**, *8*, 24. [CrossRef]

9. Zhang, Y.; Lawn, B.R. Novel zirconia materials in dentistry. *J. Dent. Res.* **2018**, *97*, 140–147. [CrossRef] [PubMed]

10. Denkena, B.; Breidenstein, B.; Busemann, S.; Lehr, C.M. Impact of hard machining on zirconia based ceramics for dental applications. *Procedia CIRP* **2017**, *65*, 248–252. [CrossRef]

11. Allahkarami, M.; Hanan, J.C. Residual stress and phase transformation in Zirconia restoration ceramics. *Adv. Bioceram. Porous. Ceram. V Ceram. Eng. Sci. Proc.* **2012**, *574*, 37–47.

12. Branemark, P.I. Osseointegration and its experimental background. *J. Prosthet. Dent.* **1983**, *50*, 399–410. [CrossRef]

13. Manikyamba, Y.J.; Sajjan, S.; AV, R.R.; Rao, B.; Nair, C.K. Implant thread designs: An overview. *Trends Prosthodont. Dent. Implantol.* **2018**, *8*, 11–20.

14. Strickstrock, M.; Rothe, H.; Grohmann, S.; Hildebrand, G.; Zylla, I.M.; Liefeith, K. Influence of surface roughness of dental zirconia implants on their mechanical stability, cell behavior and osseointegration. *BioNanoMaterials* **2017**, *18*, 1–10. [CrossRef]

15. Secatto, F.B.S.; Elias, C.N.; Segundo, A.S.; Cosenza, H.B.; Cosenza, F.R.; Guerra, F.L.B. The morphology of collected dental implant prosthesis screws surface after six months to twenty years in chewing. *Dent. Oral Craniofac. Res.* **2017**, *3*, 7.

16. Zhong, Z.W. Ductile or partial ductile mode machining of brittle materials. *Int. J. Adv. Manuf. Technol.* **2013**, *21*, 579–585. [CrossRef]

17. Bifano, T.G.; Dow, T.A.; Scattergood, R.O. Ductile-regime grinding: A new technology for machining brittle materials. *J. Eng. Ind.* **1991**, *113*, 184–189. [CrossRef]

18. Yang, M.; Li, C.; Zhang, Y.; Jia, D.; Zhang, X.; Hou, Y.; Li, L.; Wang, J. Maximum undeformed equivalent chip thickness for ductile-brittle transition of zirconia ceramics under different lubrication conditions. *Int. J. Mach. Tools Manuf.* **2017**, *122*, 55–65. [CrossRef]

19. Fook, P.; Riemer, O. Characterisation of zirconia-based ceramics after micro-grinding. *ASME J. Micro. Nano-Manuf.* **2019**. [CrossRef]

20. Ryu, H.S.; Namgung, C.; Lee, J.H.; Lim, Y.J. The influence of thread geometry on implant osseointegration under immediate loading: a literature review. *J. Adv. Prosthodont.* **2014**, *6*, 547–554. [CrossRef]

21. Javed, F.; Ahmed, H.B.; Crespi, R.; Romanos, G.E. Role of primary stability for successful osseointegration of dental implants: Factors of influence and evaluation. *Interv. Med. Appl. Sci.* **2013**, *5*, 162–167. [CrossRef]

22. Dahiya, V.; Shukla, P.; Gupta, S. Surface topography of dental implants: A review. *J. Dent. Implants* **2014**, *4*, 66–71.

23. Werkstoffe TKC - Technische Keramik GmbH. Available online: https://tkc-keramik.de/ (accessed on 25 April 2019).

24. Egea, S.A.; Martynenko, V.; Krahmer, M.D.; López de Lacalle, L.; Benítez, A.; Genovese, G. On the cutting performance of segmented diamond blades when dry-cutting concrete. *Materials* **2018**, *11*, 264. [CrossRef]

25. Galvanic tools for hard and brittle material processing - Schott Diamantwerkzeuge GmbH. Available online: https://schott-diamantwerkzeuge.com (accessed on 25 April 2019).

26. İşerı, U.; Özkurt, Z.; Kazazoğlu, E.; Küçükoğlu, D. Influence of grinding procedures on the flexural strength of zirconia ceramics. *Braz. Dent. J.* **2010**, *21*, 528–532. [CrossRef]

27. Emami, M.; Sadeghi, M.H.; Sarhan, A.A.D.; Hasani, F. Investigating the Minimum Quantity Lubrication in grinding of Al_2O_3 engineering ceramic. *J. Clean. Prod.* **2014**, *66*, 632–643. [CrossRef]

28. Chang, C.W.; Kuo, C.P. Evaluation of surface roughness in laser-assisted machining of aluminum oxide ceramics with Taguchi method. *Int. J. Mach. Tools Manuf.* **2017**, *47*, 141–147. [CrossRef]

29. Singh, B.; Singh, G. Experimental investigation of cutting parameters influence on surface finish during turning of steel using Taguchi approach. *Int. J. Prod. Res.* **2018**, *13*, 49–67. [CrossRef]

30. Abdo, B.; Darwish, S.M.; El-Tamimi, A.M. Parameters optimization of rotary ultrasonic machining of zirconia ceramic for surface roughness using statistical Taguchi's experimental design. *Appl. Mech. Mater.* **2012**, *184*, 11–17. [CrossRef]

31. Zeng, W.M.; Li, Z.C.; Pei, Z.J.; Treadwell, C. Experimental observation of tool wear in rotary ultrasonic machining of advanced ceramics. *Int. J. Mach. Tools Manuf.* **2005**, *45*, 1468–1473. [CrossRef]

32. Lemons, J.E. Biomaterials, biomechanics, tissue healing, and immediate-function dental implants. *J. Oral Implantol.* **2004**, *30*, 318–324. [CrossRef]

33. Brinksmeier, E.; Cammett, J.T.; König, W.; Leskovar, P.; Peters, J.; Tönshoff, H.K. Residual stresses—measurement and causes in machining processes. *Cirp Ann.* **1982**, *31*, 491–510. [CrossRef]

34. Pradell, T.; Glaude, P.; Peteves, S.D.; Bullock, E. *Measurement of Residual Stress in Engineering Ceramics*; Physical Sciences; European Commission: Brussels, Belgium, 1990; p. 88, EUR 12635 EN.

35. Minitab, I. *MINITAB Release 17: Statistical Software for Windows*; Minitab Inc.: State College, PA, USA, 2014.

micromachines

MDPI

Article

Analysis of the Downscaling Effect and Definition of the Process Fingerprints in Micro Injection of Spiral Geometries

Antonio Luca * and Oltmann Riemer

Laboratory for Precision Machining (LFM), Leibniz Institute for Materials Engineering—IWT,
University of Bremen, Badgasteiner Straße 2, 28359 Bremen, Germany; riemer@iwt.uni-bremen.de
* Correspondence: aluca@iwt.uni-bremen.de; Tel.: +49-421-218-51169

Received: 4 April 2019; Accepted: 16 May 2019; Published: 22 May 2019

Abstract: Microinjection moulding has been developed to fulfil the needs of mass production of micro components in different fields. A challenge of this technology lies in the downscaling of micro components, which leads to faster solidification of the polymeric material and a narrower process window. Moreover, the small cavity dimensions represent a limit for process monitoring due to the inability to install in-cavity sensors. Therefore, new solutions must be found. In this study, the downscaling effect was investigated by means of three spiral geometries with different cross sections, considering the achievable flow length as a response variable. Process indicators, called "process fingerprints", were defined to monitor the process in-line. In the first stage, a relationship between the achievable flow length and the process parameters, as well as between the process fingerprints and the process parameters, was established. Subsequently, a correlation analysis was carried out to find the process indicators that are mostly related to the achievable flow length.

Keywords: microinjection moulding; process fingerprints; flow length; quality assurance

---

1. Introduction

The application of small components with dimensions in the micrometre and nanometre ranges has largely increased in various engineering fields over the recent decades [1]. Micro components are gaining importance in areas such as health care, optical products, automotive industry, communication, biotechnology and so forth. The demand for micro parts made of thermoplastic polymers is becoming increasingly widespread due to the need for reduced weight, high chemical resistance, low production costs and ease of fabrication—even in complex shapes. Therefore, advanced micro manufacturing technologies are fundamental to support their production. In this context, most of these products are nowadays manufactured by microinjection moulding. This process can be conceived of as a miniaturised variant of the conventional injection moulding process, with the intention of combining high productivity with the capability to manufacture micro components [2]. A substantial difference [3] between the conventional and the micro process is the filling of the cavity, which becomes much more challenging at the microscale. The downscaling of micro components leads to a relatively large surface-to-volume ratio and, consequently, to an increased heat flux at the mould–melt interface. Hence, solidification occurs quickly, hindering the complete filling of the cavity [4]. High levels of injection speed as well as melt and mould temperatures are typically required in order to favour the replication capability of the process [5–7]. However, the process window becomes narrower at the microscale, as the levels of the injection moulding process parameters are largely limited by the existence of polymer degradation. Therefore, process optimisation is a fundamental step for manufacturing products that comply with design specifications. When the geometric characteristics of the components are the response variables, experimental investigations based on off-line digital measurements are fundamental to tune the process, making process optimisation a time-consuming task. Additionally, high throughput rates do not allow for measurement of all the produced micro components in three dimensions [8]. A solution to this problem is the application of in-line monitoring techniques to the microinjection moulding process as a tool for quality assurance.

For example, in-line measurements can be carried out to monitor cavity pressure. This factor is in most studies identified [9–11] as the process variable that best outlines the evolution of the moulding cycle. However, in the micro process, the cavity dimensions can be comparable or even larger than that of typical sensors, therefore interfering with their use [12]. A solution to this issue is to monitor the hydraulic pressure provided by the injection plunger or screw, since this quantity can be extracted from the machine data for each moulding cycle without the need to use any further sensor. The limitation of this method is the difference between hydraulic and cavity pressures, which means that the pressure measured at the screw is not representative of the behaviour of the polymer melt inside the cavity.

The selection of the main injection moulding process parameters has a relevant impact on the recorded process indicators (e.g., cavity pressure [13], injection speed [14], demoulding forces [15], etc.). Thus, the process can be successfully monitored by controlling these variables. Nevertheless, a relationship between those indicators, called "process fingerprints", and the dimensional quality of the produced micro components has not been established so far. In this investigation, a correlation between the flow length (the response) and the monitored process fingerprints was achieved. Thus, quality assurance can be performed in-line by only controlling their values, enhancing its robustness and reducing the quality control time.

2. Materials and Methods

2.1. Spiral Geometry

In this work, micro components having the same spiral geometry but different cross-section dimensions were investigated. The aim was to evaluate the downscaling effect on the filling behaviour and the process conditions by measuring the manufactured parts and monitoring representative process variables. In this case, the geometric characteristic of the component considered as a response variable was represented by the achievable flow length of the polymeric material, which depended on the selected process parameters and the process conditions.

The three selected flow spiral geometries had rectangular cross sections of 1×1 mm^2, 0.5×0.5 mm^2 and 0.25×0.25 mm^2 (see Figure 1).

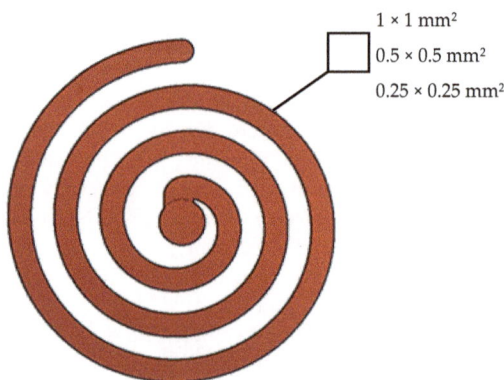

Figure 1. Spirals with different cross sections.

The cavities were designed in a changeable mould unit and integrated a central ejector (see Figure 2). The tooling process was carried out on a five-axis milling centre (DMG Sauer Ultrasonic 20 Linear).

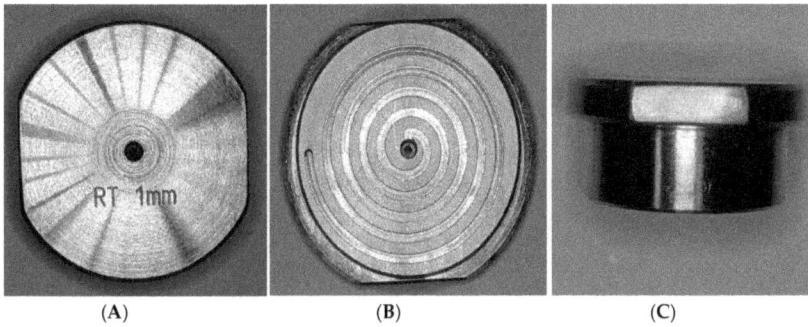

Figure 2. Changeable mould unit including a central ejector for spiral geometries with a cross section of 1×1 mm^2. (**A**) Bottom view; (**B**) top view; (**C**) side view.

2.2. Measurement Method

The flow length was determined by recording the solidified and demoulded part with a VHX-6000 digital microscope from the manufacturer Keyence. For this purpose, the flow spirals were divided into quadrants (see Figure 3).

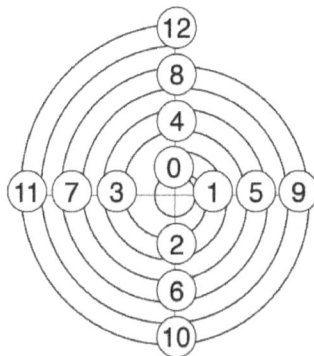

Figure 3. Quadrants of the flow spiral.

Each quadrant had a larger radius than the previous one. With the straight flow length from the sprue to the quadrant 0 and the different radii, the maximum flow length could be calculated up to the end of each quarter circle. For example, the achievable flow length of the spiral with a cross section of 1×1 mm^2 for different radii is illustrated in Table 1.

Table 1. Values of the flow length for different radii for the spiral with a 1×1 mm^2 cross section.

Quadrant	Radius (mm)	Max. Flow Length (mm)
0	1.5	1.5
1	2	4.63
2	2.5	8.55
3	3	13.25
4	3.5	18.75

If the last quadrant was not filled completely, a filling ratio of the incompletely filled quadrant was calculated according to the following equation:

$$l = r \cdot \pi \cdot \alpha / 180°. \tag{1}$$

As an example, in Figure 4, the measurements of the angle α for three different spiral parts (with cross sections of 1 × 1, 0.5 × 0.5 and 0.25 × 0.25 mm², respectively) are illustrated.

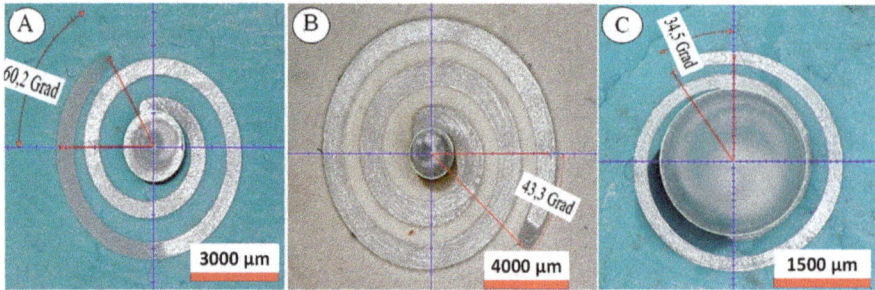

Figure 4. Example of measurements of the angle α for three different spiral parts with cross sections of (**A**) 0.5 × 0.5 mm², flow length: 28.83 mm; (**B**) 1 × 1 mm², flow length: 62.86 mm and (**C**) 0.25 × 0.25 mm², flow length: 11.21 mm.

2.3. Microinjection Moulding Machine

Injection moulding experiments were carried out using a microinjection moulding machine (Desma FormicaPlast 2K) consisting of a two-phase piston injection unit and a pneumatic injection drive (see Figure 5). The first phase refers to a heated plasticisation zone with a vertically positioned plasticising piston, while the second phase to a horizontally positioned piston for precision injection that has far more accurate control over the injected polymer melt than large diameter screws.

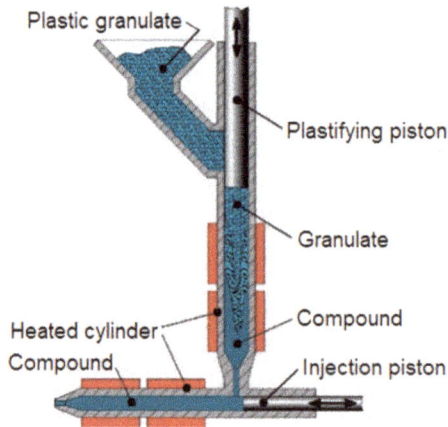

Figure 5. Working principle of FormicaPlast 2K [16].

2.4. Polymeric Material

The material chosen to perform the experiments was uncoloured POM N23200035, a thermoplastic polymer with an extremely low coefficient of friction and sliding wear when mated with smooth metal surfaces. Its main properties are listed in Table 2.

Table 2. Main properties of the material.

Property	Test Method	Units	Value
Density	ISO 1183	kg/m^3	1400
Melt volume rate (190 °C, 2.16 kg)	ISO 1133	cm^3/10 min	7.5
Melt temperature	ISO 11357-1/-3	°C	167

2.5. Design of Experiments (DoE)

DoE is a standardised approach to determine the relationship between factors affecting a process and its output. Particularly, it helps to identify the critical factors affecting the desired output, thus enabling optimisation of the entire process. The parameters selected for this investigation were: injection speed (Vinj), melt temperature (Tmelt), mould temperature (Tmould) and holding pressure (Phold). The values were based on the material recommendations and preliminary experiments. Two levels of each parameter were chosen. These values were chosen in such a way that they covered a wide range of variations for each of the selected process parameters (see Table 3).

Table 3. Process parameter settings for the design of experiments (DoE) plan.

Process Parameter	Low	High
Vinj (mm/s)	125	250
Tmelt (°C)	200	220
Tmould (°C)	80	100
Phold (bar)	200	400

In order to avoid excessive flash formation, the holding pressure was set at relatively low values. The velocity/pressure switch-over point was set at the packing pressure value. A four-factor full-factorial design consisting of 16 experiments was carried out (see Table 4). For statistical assurance, 10 moulded parts were moulded for every set of process conditions. In order to reach a steady state when changing from one experiment to another, the first seven test specimens were discarded and the following three were kept for evaluation.

Table 4. Process parameter settings for the DoE plan.

Experiment	Tmelt (°C)	Tmould (°C)	Vinj (mm/s)	Phold (bar)
1	200	80	125	400
2	200	80	125	200
3	200	80	250	400
4	200	80	250	200
5	200	100	125	400
6	200	100	125	200
7	200	100	250	400
8	200	100	250	200
9	220	80	125	400
10	220	80	125	200
11	220	80	250	400
12	220	80	250	200
13	220	100	125	400
14	220	100	125	200
15	220	100	250	400
16	220	100	250	200

2.6. Process Monitoring

Pressure and velocity of the injection plunger were recorded in-line during the injection moulding cycle. No external sensor was used, as these process variables were derived from the machine data.

These data are available to any machine user and easy to access. The injection pressure was recorded via a strain gauge transducer mounted on the back of the injection plunger, while the injection speed was acquired via the speed of the motor driving the plunger through the control unit of the machine. The recorded speed and pressure needed no alignment with respect to the timescale, as they were acquired synchronously.

The dependence of pressure and velocity on the injection moulding process parameters was investigated by identifying some variables referred to as process fingerprints. These process indicators well characterise the pressure and velocity curves and are defined as follows:

- Maximum injection pressure, Pmax: this value is defined as the maximum injection pressure recorded during each moulding cycle. This indicator is related to the filling behaviour of the cavity, since the pressure peak is increased by the small size of the channels.
- Mean injection pressure, Pmean: this quantity is calculated as the average of the pressure values recorded from the start to the end of the moulding cycle.
- Mean injection speed, Vmean: the mean injection speed is defined as the average velocity that characterises the filling phase. The speed values are recorded in the time interval between the start of the acceleration of the injection plunger and when it stops at the switch-over point.

3. Results and Discussion

3.1. Achievable Flow Length

Figure 6 shows the results of the achievable flow lengths for flow spirals with different cross sections (1×1, 0.5×0.5 and 0.25×0.25 mm^2) and different injection moulding process parameters. The error bars represent the standard deviations for the three parts taken into account for each experiment. Higher flow length values were achieved for higher levels of the process parameters. In the case of experiment 1, the achieved flow length for the spiral with a cross section of 0.5×0.5 mm^2 was slightly higher than the one with the cross section of 1×1 mm^2. This was an exception, since for all other experiments, the flow length increased with the size of the cross sections. For all other experiments, the different spirals showed similar behaviour. Comparing the results for cross sections of 1×1 and 0.5×0.5 mm^2, the difference in flow length could be estimated, in most cases, with a factor of two in favour of the larger cross section. Considering the results for cross sections of 0.5×0.5 and 0.25×0.25 mm^2, this difference was more evident. In this case, a factor of three in favour of the cross section of 0.5×0.5 mm^2 can be observed. Therefore, the downscaling of the cross section does not give flow lengths exactly proportional to the size of the cross section, but it is possible to find a factor representative of this behaviour.

The results from the experimental campaign were analysed considering the flow length as the response variable. An ANOVA with a main effect plot for identifying the factors that significantly affect the response was also employed. The statistical analysis was carried out using the statistical software Minitab 18. In Figure 7, the results for the three spirals are illustrated.

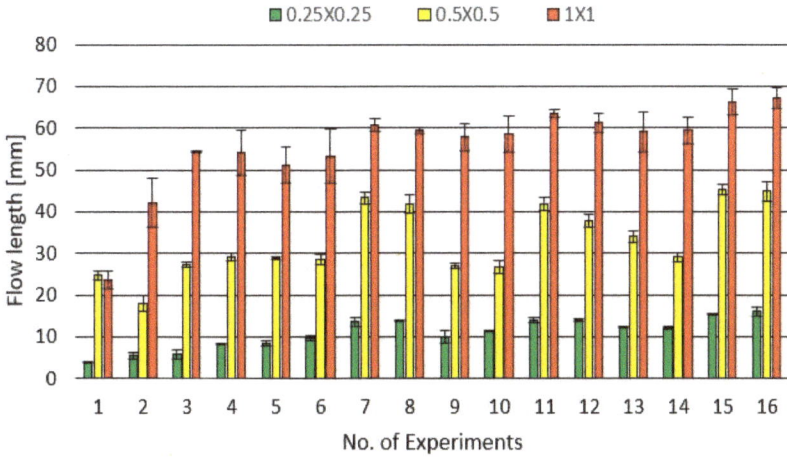

Figure 6. Achievable flow lengths for spiral geometries with different cross sections and for different process parameters.

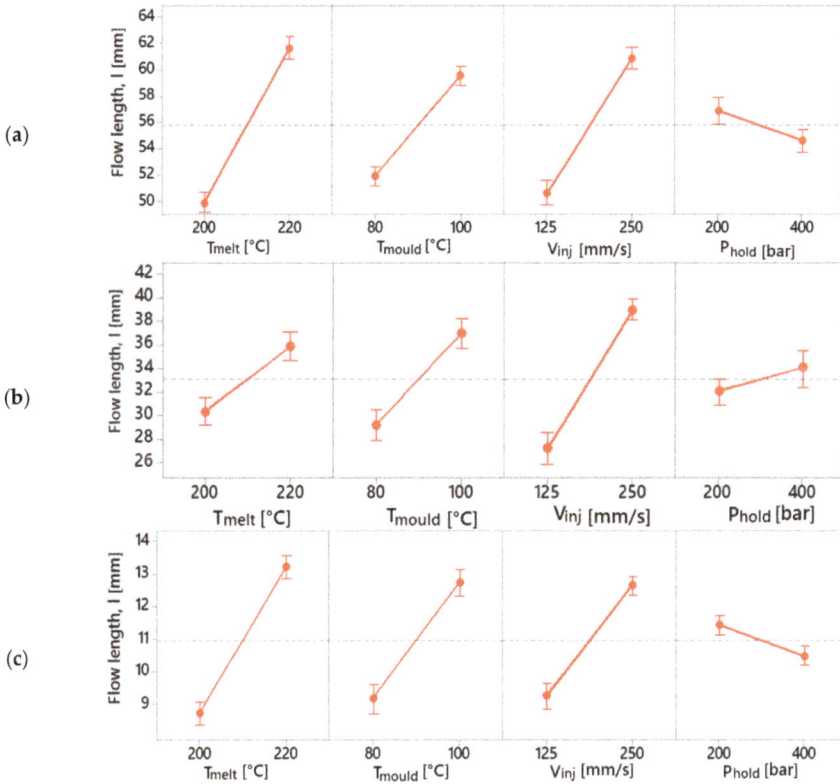

Figure 7. Main effect plot for spiral geometries with (**a**) 1×1 mm^2, (**b**) 0.5×0.5 mm^2 and (**c**) 0.25×0.25 mm^2 cross sections.

Figure 7a,c show a similar effect of the process parameters on the achievable flow length. In this case, the melt temperature was the most significant parameter. In the case of a 0.5×0.5 mm^2 cross section, the factors gave a substantially different response. The injection speed was the parameter that had the highest influence, followed by mould temperature, melt temperature and holding pressure. Furthermore, higher flow lengths were recorded for a holding pressure of 400 bar, unlike the other two cases, where higher flow lengths were obtained for a holding pressure of 200 bar. However, this difference can be neglected, since the holding pressure is the parameter that has less of an influence on the response.

3.2. Process Fingerprint Analysis

The process fingerprint analysis was carried out to identify the sensitivity of these process indicators with respect to the process parameters. This dependency is shown in the following diagrams (Figures 8–10). For each spiral geometry, the values of maximum injection pressure, mean injection pressure and mean injection speed were obtained during the experiments. Subsequently, an average of these values for the three spirals was carried out, considering experiments with the same process parameters.

Figure 8 shows the results for the maximum injection pressure. This indicator mostly depended on the selected injection speed and holding pressure values, as an increase in both parameters resulted in an increase of this process indicator. The influence of the holding pressure was due to the fact that the machine was set to switch from the filling to the holding phase at a given pressure value. In this case, the selection of a higher holding pressure implies that the injection pressure rises more before switching to the holding profile. The injection speed influenced the maximum injection pressure because a higher pressure requires a fluid flowing at a higher speed.

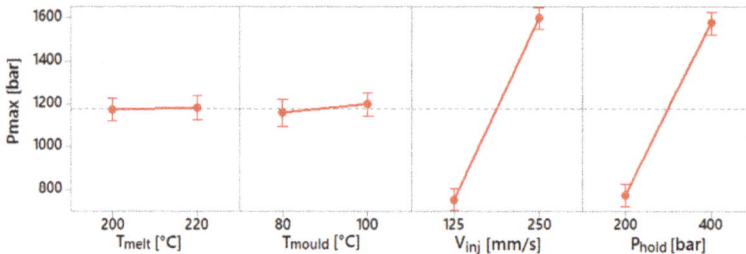

Figure 8. Main effect plot for maximum injection pressure.

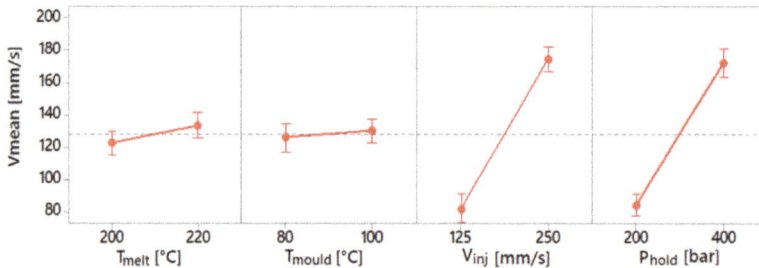

Figure 9. Main effect plot for mean injection speed.

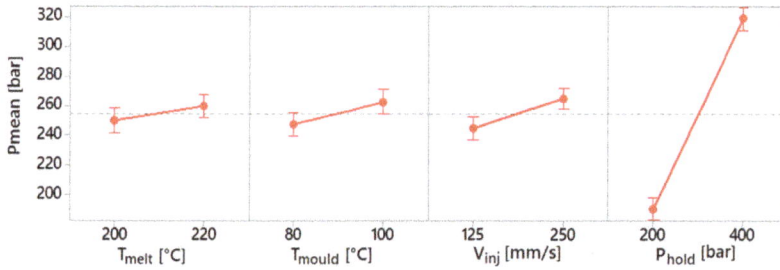

Figure 10. Main effect plot for mean injection pressure.

Figure 9 shows the results for the average injection speed. As can be observed, its value was predominantly influenced by the holding pressure and injection speed. The effect of the injection speed was obvious, due to the setting of a higher injection velocity. In the case of holding pressure, a higher level of this process parameter required a lower deceleration of the injection plunger. Therefore, a higher mean injection speed was observed during the moulding cycle. On the other side, melt and mould temperature had a slight effect.

Figure 10 shows the results for the mean injection pressure. This indicator was mostly influenced by the holding pressure. This was because the mean pressure was calculated considering the entire moulding cycle, which was dominated above all by the holding phase. The other parameters had a definitely lower influence on the mean injection pressure.

3.3. Correlation Analysis

A correlation analysis was carried out to establish an efficient quality control based only on the measurement of the process fingerprints. Thus, the achievable flow length was correlated with the process fingerprints (maximum injection pressure, mean injection pressure and mean injection speed). The coefficient of correlation was calculated with the following equation:

$$r(x, y) = \frac{\sum_i [(X_i - \overline{X})(Y_i - \overline{Y})]}{\sqrt{\sum_i [(X_i - \overline{X})^2 \cdot \sum_i [(Y_i - \overline{Y})^2}}} \tag{2}$$

where $\overline{X}$ and $\overline{Y}$ are the mean values of flow length and process fingerprints, respectively, and X_i and Y_i are their values obtained during the experiments. The coefficient of correlation can vary between +1 and −1. A value of 0 indicates that there is no correlation between the flow length and the process fingerprints, while a value of +1 or −1 indicates a perfect positive or negative correlation respectively. As the values of the process fingerprints for each experiment were an average of three values recorded for each of the three flow spirals, we also considered a single value of flow length for each experiment, making an average of the three values. The following Figure 11 shows the results of the correlation analysis:

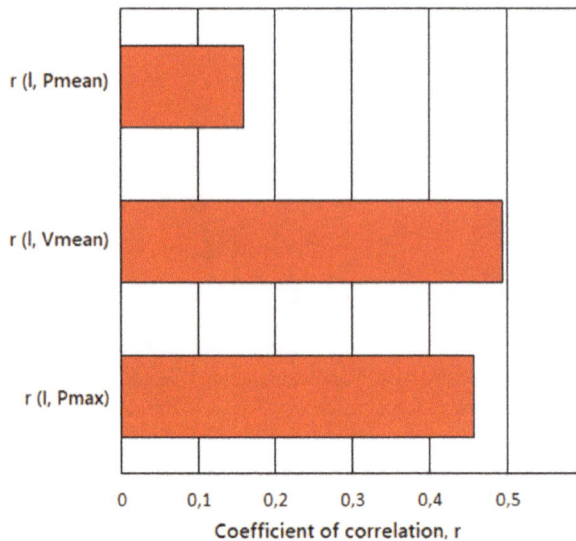

Figure 11. Values of the coefficient of correlation calculated between the flow length and the process fingerprints.

The results of the correlation analysis showed that the mean injection speed was the process fingerprint that had the strongest correlation with the achievable flow length, followed by maximum injection pressure and mean injection pressure. While the maximum injection pressure had a degree of correlation similar to the mean injection speed, the mean injection pressure showed a significantly lower correlation value, meaning that this process fingerprint is not useful to establish quality assurance based on this process indicator. On the other side, mean injection speed and maximum injection pressure act as a link between process monitoring and achievable flow length, representing useful process indicators for faster quality assurance of the injection-moulded micro spiral parts. By using this method, it is possible to have an estimation of the achievable flow length by analysing the process fingerprint values.

4. Conclusions

The aim of this research was dual: on one hand, the downscaling effect of spiral components with the same geometries but different cross sections was investigated, using the achievable flow length as the response variable; on the other hand, an optimisation method to reduce the off-line inspection effort of the moulded micro components was established. We demonstrated that the downscaling of the cross sections did not have a linear relationship with the achievable flow lengths. It was possible to identify a factor of two between the flow length achieved with cross sections 1×1 mm^2 and 0.5×0.5 mm^2 and a factor of three between the achieved flow length with cross sections of 0.5×0.5 mm^2 and 0.25×0.25 mm^2. A process in-line monitoring technique was implemented, finding a relationship between the process fingerprints and the microinjection moulding process parameters. Finally, a correlation analysis was carried out to relate the response (i.e., achievable flow length) with the process fingerprints. By monitoring these process indicators, it will be possible to predict the response, thus avoiding time-consuming off-line measurements and performing in-line quality assurance. This developed method has the potential to be applied to other micro components by defining their process fingerprints and responses (e.g., diameters, height of specific features, etc.) depending on the geometry of the micro components being investigated.

Author Contributions: A.L. conceived and designed the experiments; A.L. performed experiments and measurements; A.L. analysed the data; A.L. wrote the paper; O.R. revised the paper.

Funding: This research work was undertaken in the context of the MICROMAN project ("Process Fingerprint for Zero-defect Net-shape MICROMANufacturing", http://www.microman.mek.dtu.dk/). MICROMAN is a European Training Network supported by Horizon 2020, the EU Framework Programme for Research and Innovation (Project ID: 674801).

Acknowledgments: The support provided by Darshan Jain (University of Applied Sciences Bremerhaven) is gratefully acknowledged.

Conflicts of Interest: The authors declare no conflict of interests.

References

1. Tosello, G. *Micromanufacturing Engineering and Technology*; Qin, Y., Ed.; Elsevier: Amsterdam, The Netherlands, 2015; pp. 201–238.
2. Giboz, J.; Copponnex, T.; Mélé, P. Microinjection molding of thermoplastic polymers: A review. *J. Micromech. Microeng.* **2007**, *17*, R96–R109. [CrossRef]
3. Whiteside, B.; Martyn, M.T.; Coates, P.D.; Greenway, G.; Allen, P.S.; Hornsby, P. Micromoulding: Process measurements, product morphology and properties. *Plast. Rubber Compos.* **2004**, *33*, 11–17. [CrossRef]
4. Brousseau, E.B.; Dimov, S.; Pham, D.T. Some Recent Advances in Multi-Material Micro- And Nano-Manufacturing. *Int. J. Adv. Manuf. Technol.* **2010**, *47*, 161–180. [CrossRef]
5. Lucchetta, G.; Ferraris, E.; Tristo, G.; Reynaerts, D. Influence of mould thermal properties on the replication of micro parts via injection moulding. *Procedia CIRP* **2012**, *2*, 113–117. [CrossRef]
6. Meister, S.; Dietmar, D. Investigation on the Achievable Flow Length in Injection Moulding of Polymeric Materials with Dynamic Mould Tempering. *Sci. World J.* **2013**, *3*. [CrossRef]
7. Sha, B.; Dimov, S.; Griffiths, C.; Packianather, M.S. Micro-injection moulding: Factors affecting the achievable aspect ratios. *Int. J. Adv. Manuf. Technol.* **2007**, *33*, 147–156. [CrossRef]
8. Chen, Z.; Turng, L.S. A review of current developments in process and quality control for injection molding. *Adv. Polym. Technol.* **2005**, *24*, 165–182. [CrossRef]
9. Michaeli, W.; Schreiber, A. Online control of the injection molding process based on process variables. *Adv. Polym. Technol.* **2009**, *28*, 65–76. [CrossRef]
10. Speranza, V.; Vietri, U.; Pantani, R. Monitoring of injection moulding of thermoplastics: Adopting pressure transducers to estimate the solidification history and the shrinkage of moulded parts. *Strojniški Vestnik J. Mech. Eng.* **2013**, *59*, 677–682. [CrossRef]
11. Tsai, K.-M.; Lan, J.-K. Correlation between runner pressure and cavity pressure within injection mold. *Int. J. Adv. Manuf. Technol.* **2015**, *79*, 14–23. [CrossRef]
12. Mendibil, X.; Llanos, I.; Urreta, H.; Quintana, I. In process quality control on micro-injection moulding: The role of sensor location. *Int. J. Adv. Manuf. Technol.* **2017**, *89*, 3429–3438. [CrossRef]
13. Griffiths, C.A.; Dimov, S.; Scholz, S.G.; Hirshy, H.; Tosello, G. Process Factors Influence on Cavity Pressure Behavior in Microinjection Moulding. *J. Manuf. Sci. Eng.* **2011**, *133*. [CrossRef]
14. Griffiths, C.A.; Dimov, S.S.; Scholz, S.G.; Tosello, G.; Rees, A. Influence of Injection and Cavity Pressure on the Demoulding Force in Micro-Injection Moulding. *J. Manuf. Sci. Eng.* **2014**, *136*. [CrossRef]
15. Griffiths, C.A.; Dimov, S.S.; Brousseau, E.B. Microinjection moulding: The influence of runner systems on flow behaviour and melt fill of multiple microcavities. *Proc. Inst. Mech. Eng. Part. B J. Eng. Manuf.* **2008**, *222*, 1119–1130. [CrossRef]
16. Desma Tec. Formicaplast—Micro Injection for the Smallest Shot Weights (100—200 mg). Available online: http://www.formicaplast.de/en/index.php (accessed on 21 May 2019).

micromachines

MDPI

Article

One-Dimensional Control System for a Linear Motor of a Two-Dimensional Nanopositioning Stage Using Commercial Control Hardware

Lucía Candela Díaz Pérez [1,*], Marta Torralba Gracia [2], José Antonio Albajez García [1] and José Antonio Yagüe Fabra [1]

1   I3A, University of Zaragoza, C/María de Luna 3, 50018 Zaragoza, Spain; jalbajez@unizar.es (J.A.A.G.); jyague@unizar.es (J.A.Y.F.)
2   Centro Universitario de la Defensa, Ctra. Huesca s/n, 50090 Zaragoza, Spain; martatg@unizar.es
*   Correspondence: lcdiaz@unizar.es; Tel.: +34-97-676-2561

Received: 16 July 2018; Accepted: 21 August 2018; Published: 22 August 2018

Abstract: A two-dimensional (2D) nanopositioning platform stage (NanoPla) is in development at the University of Zaragoza. To provide a long travel range, the actuators of the NanoPla are four Halbach linear motors. These motors present many advantages in precision engineering, and they are custom made for this application. In this work, a one-dimensional (1D) control strategy for positioning a Halbach linear motor has been developed, implemented, and experimentally validated. The chosen control hardware is a commercial Digital Motor Control (DMC) Kit from Texas Instruments that has been designed to control the torque or the rotational speed of rotative motors. Using a commercial control hardware facilitates the applicability of the developed control system. Nevertheless, it constrains the design, which needs to be adapted to the hardware and optimized. Firstly, a dynamic characterization of the linear motor has been performed. By leveraging the dynamic properties of the motor, a sensorless controller is proposed. Then, a closed-loop control strategy is developed. Finally, this control strategy is implemented in the control hardware. It was verified that the control system achieves the working requirements of the NanoPla. It is able to work in a range of 50 mm and perform a minimum incremental motion of 1 μm.

Keywords: positioning platform; Halbach linear motor; commercial control hardware

---

1. Introduction

Positioning stages are becoming fundamental devices in nanotechnology and nanomanufacturing processes [1,2], where they act as a supplementary unit for measuring or manipulating samples [3,4]. Depending on the application, a certain combination of working range and metrological performance is required [5]. To obtain effective positioning, several metrological systems are currently available [6–8]. These systems have been designed for demanding and accurate operations. Nevertheless, their measuring and positioning range is often very limited [9,10]. Other applications such as measuring or manipulating solar cells or silicon wafers require working with larger areas in a planar part, where cutting of specific samples may be necessary. Therefore, the nanotechnology industry is demanding not only more accurate positioning systems but also larger working ranges [11]. Within this line of research, a nanopositioning platform stage (NanoPla) has been developed and manufactured at the University of Zaragoza [12,13]. It is expected to provide effective positioning at the nanometre scale inside a large working range of 50 mm × 50 mm. Its first application integrates an atomic force microscope (AFM) as a suitable technique for micro- and nanometrology [14], due to the high vertical as well as lateral resolution in the topographic characterization task of specimens.

Depending on their structure, nanopositioning stages can be classified into stages with stacked linear axes and plane stages. Stages with stacked linear axes are characterized for long kinematic chains with an unfavourable force transfer behaviour [15,16]. Whereas the absence of linear motion in plane stages minimizes geometrical errors and presents many other advantages in precision engineering [17]. For these reasons, it has already been implemented in multiple systems [18,19].

In the NanoPla design, the principles of precision engineering have been applied, including planar motion. However, planar motion conditions the actuator selection, since the motor design or its guiding system should not impede the displacement of the motor along the orthogonal direction of its driving axes. Halbach linear motors [20] suppose a solution to this issue, whose movement in the 2D plane is only limited by the size of its winding area. Other advantages of Halbach linear motors are that they provide non-contact motion and, in addition to the propulsion force, they generate a levitation force. Although one of the design criterions of the NanoPla is to implement as many commercial devices as possible, unguided Halbach linear motors are not commercialised yet. Therefore, they have been custom-made for this application due to the advantages of performing accurate and long travel range positioning.

The fact that the use of this kind of Halbach linear motors is not yet widespread means that there is no available commercial solution for the driving task. In other positioning stages described in the literature [6,21], the control hardware and software were specifically designed and built for this purpose. Nevertheless, as was mentioned, one of the targets of the NanoPla design is to develop it with commercial devices when possible, which will facilitate a future industrial applicability of the developed system. Thus, a commercial generic solution for the hardware has been chosen: a Digital Motor Control (DMC) Kit from Texas Instruments (Dallas, TX, USA). This control hardware has been designed for rotary permanent magnet synchronous motors (PMSM), where the aim is to control the rotation speed or the torque generated. According to the literature, the integration of completely generic control hardware with linear motor actuators is a novelty that is presented in this study. Such integration presents many limitations that need to be overcome by optimizing the control system design. Nevertheless, this has been done in this work always by using the available options of the control hardware modules. The hardware has not been modified and no additional electronic has been required. The use of only one commercial hardware and no custom-made electronics facilitates the applicability and replication of the developed control strategy, which is in line with the targets of the NanoPla design. This work can be very useful for other developers willing to implement commercial devices for the control of linear motors.

This article presents and experimentally validates a challenging one-dimensional (1D) control system for a custom-made Halbach linear motor that works as an actuator in the two-dimensional (2D) long working range NanoPla. The control system is characterized by the integration of a commercial solution hardware which is commonly implemented with rotary actuators. Thus, this paper first presents an overview of the NanoPla, which is necessary to define the working requirements of the control system. Secondly, the working principle of Halbach linear motors is described, and the materials used in this work are presented. Then, a dynamic characterization of these motors is performed, and a sensorless open-loop solution is proposed. Afterwards, the 1D control strategy is defined, and the proposed control strategy is implemented in the chosen commercial control hardware. Finally, the experimental results are shown and conclusions are withdrawn.

2. Two-Dimensional Nanopositioning Platform Stage (NanoPla) Overview

As shown in Figure 1a, the NanoPla consists of a three-layered architecture: an inferior and a superior base that are fixed, and a moving platform that is placed in the middle. Three air bearings lift the moving platform and levitate it. The planar motion is performed by four Halbach linear motors that are symmetrically assembled in an inverted position. In other words, the stators are fixed to the superior base, and the magnet arrays are assembled to the moving platform (see Figure 1a). The horizontal forces of each pair of parallel motors will move the platform in the X and Y direction, as Figure 1b shows. In addition, the vertical forces of the four motors will favour the levitation of the moving platform. A 2D laser interferometer system works as positioning sensor. The laser heads are fixed to the inferior base, and the mirrors are placed in the moving platform. In addition, in the NanoPla, a two-stage scheme has been applied. That means that the XY-long range positioning of the moving platform is complemented by an additional fine nanopositioning system for the more demanding scanning operations. This second stage is a commercial piezo-nanopositioning device with a working range of $100 \times 100 \times 10$ μm^3, which increases the number and variety of applications of the NanoPla.

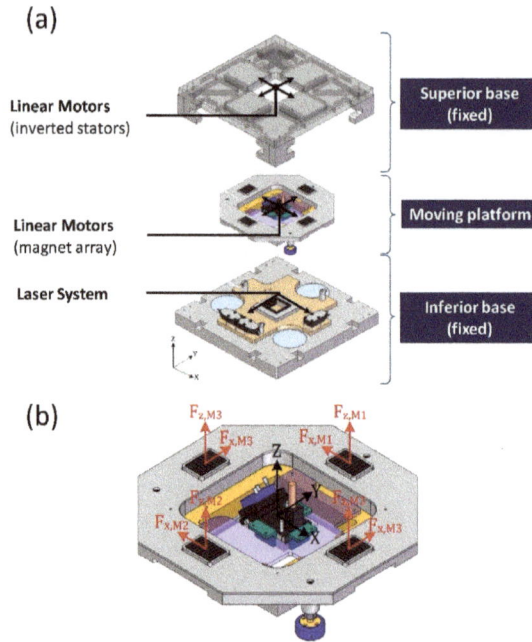

Figure 1. (**a**) Nanopositioning platform (NanoPla) prototype; (**b**) vertical and horizontal forces generated by the motors in the moving platform.

The first device that is going to be integrated in the NanoPla is an AFM, which will be placed on the moving platform. The NanoPla will position the AFM along the working range of 50 mm × 50 mm. The sample will be placed in the commercial piezo-nanopositioning stage fixed to the inferior base. During the scanning task, the moving platform and the AFM will be static (air bearings off) and the commercial nanopositioning stage will perform the fine motion of the sample.

A preliminary modelling of the 2D positioning control of the NanoPla was presented in [22] with a different approach. The input currents are controlled independently, which is not possible with the commercial control hardware solution proposed. Nevertheless, the control strategy requirements were initially defined. According to this, the 1D linear motion control strategy must be able to work in a range of 50 mm. In addition, the positioning error must be at least on order of magnitude smaller than the maximum XY range of the commercial piezo-nanopositioning stage, i.e., 10 μm. The settling time is not critical. Finally, other considerations are related to the transient behaviour of the positioning response. In addition, oscillation should be avoided.

3. Halbach Linear Motors

The linear motors used as actuators in the NanoPla were developed by Trumper et al. [20] and custom-made in the Center for Precision Metrology of the University of North Carolina at Charlotte (Charlotte, NC, USA). They consist of a Halbach permanent magnet array and three-phase ironless coils (stator). In this section, the motor law and the commutation law that define the working principle of the motor are described.

3.1. Motor Law

In a Halbach array of permanent magnets, the configuration of the magnets augments the magnetic field generated on one side and nullifies the magnetic field on the other side. That is, the rotating pattern of the permanent magnets forces the cancellation of magnetic components resulting in a one-sided flux. In a Halbach linear motor (Figure 2a), this flux is concentrated between the magnet array and the stator. When a DC current flows through the coils of the stator, these currents interact with the magnetic field of the magnet array. The electromagnetic interaction generates two orthogonal forces: one is horizontal (F_x) and the other is vertical (F_z), as can be seen in Figure 2b.

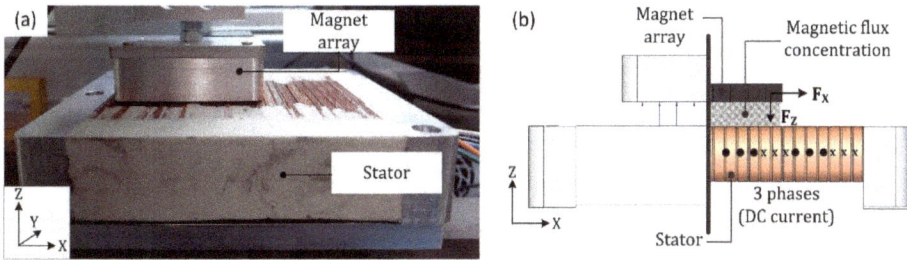

Figure 2. (**a**) Halbach motor (magnet array and stator); (**b**) graphical representation of the dual forces generated by the Halbach motor.

The direction and amplitude of the two forces depend on the relative position between the magnet array and the coils as well as on the magnitude of the DC phase currents flowing through the coils. The motor law (Equation (1)) represents the mathematical relationship between the generated forces (F_x and F_z), the input phase currents (I_a, I_b and I_c) and the relative position between the stator and the magnet array (x_0), along the axis of movement.

$$\begin{bmatrix} F_x \\ F_z \end{bmatrix} = A \begin{bmatrix} \cos(kx_0 + \varphi) & \cos\left(kx_0 - \frac{2\pi}{3} + \varphi\right) & \cos\left(kx_0 + \frac{2\pi}{3} + \varphi\right) \\ \sin(kx_0 + \varphi) & \sin\left(kx_0 - \frac{2\pi}{3} + \varphi\right) & \sin\left(kx_0 + \frac{2\pi}{3} + \varphi\right) \end{bmatrix} \begin{bmatrix} I_a \\ I_b \\ I_c \end{bmatrix} \quad (1)$$

In Equation (1), A and k are constant parameters of the motor. The constant of the motor, A, depends on design parameters of the motor, as represented in Equation (2). On the other hand, k is the fundamental wave number, which is calculated according to Equation (3), where l is the pitch or the spatial period of the array wavelength. In Table 1, a description of all these parameters is provided [23].

$$A = N_m \eta_0 \mu_0 M_0 G e^{-kz_0} \quad (2)$$

$$k = \frac{2\pi}{l} \quad (3)$$

Table 1. Description and theoretical values of the motor parameters.

Parameter	Description	Theoretical Value
N_m	Number of spatial periods of the magnet array	2
η_0	Winding density of the stator coil	832,400 turns/m^2
$\mu_0 M_0$	Remanence of the permanent magnets	0.4 T
G	Effects of the motor geometry	2.62×10^{-6}
k	Fundamental wave number	211.1285 rad/m
z_0	Separation gap between stator-magnets array	400 μm
l	Spatial period of the array wavelength	29.76 mm

Their values were calculated first theoretically and then experimentally in a previous work [24]. This was done by measuring with a load cell, along the travel range of the motor, the vertical and horizontal force generated by certain known phase currents. These results are shown in Table 2. The initial position $x_0 = 0$ can be adjusted by changing the phase difference φ in Equation (1). φ must have the same value in the F_x row as in the F_z row because the two forces are orthogonal. In this paper, for simplicity reasons, the value of φ will be considered null. φ acts as an initial offset and, thus, this assumption only affects the absolute initial position of the motor but not the results.

Table 2. Theoretical and experimental fitting parameters of the motor law.

Parameter	Theoretical Value	Experimental Value
A (N/A)	1.6	1.6067
k (rad/m)	211.1285	211.0001
l (mm)	29.760	29.778

3.2. Commutation Law

The commutation law is defined as the inverse of the motor law (Equation (1)), and it allows the calculation of the phase currents that are required in order to generate a certain F_x and F_z in a specific position. On the basis of Equation (1), the phase currents can be determined with one degree of freedom (three unknowns, two equations). As there are three input currents, one more equation needs to be considered to uniquely determine them. In [25], an additional constraint was proposed for power minimization, which is possible due to the fact that the control strategy acts independently on the input currents using linear transconductance power amplifiers built for that purpose. By contrast, this paper proposes the use of a generic DMC Kit from Texas Instruments. This control hardware imposes a star-connection on the phases of the motor, which adds an additional constraint (Equation (4)) that prevents the control of the three currents independently. Thus, it also impedes the implementation of the power minimization constraint.

$$I_a + I_b + I_c = 0 \tag{4}$$

Therefore, combining Equations (1) and (4) and considering $\varphi = 0$, the commutation law for the case of this study is defined as in Equation (5):

$$\begin{bmatrix} I_a \\ I_b \\ I_c \end{bmatrix} = \frac{2}{3A} \begin{pmatrix} \cos kx_0 & \sin kx_0 \\ \cos\left(kx_0 - \frac{2\pi}{3}\right) & \sin\left(kx_0 - \frac{2\pi}{3}\right) \\ \cos\left(kx_0 + \frac{2\pi}{3}\right) & \sin\left(kx_0 + \frac{2\pi}{3}\right) \end{pmatrix} \begin{bmatrix} F_x \\ F_z \end{bmatrix} \tag{5}$$

4. Experimental Setup and Hardware Description

Before the implementation of the control system designed in this work into the two-dimensional NanoPla, its validation is firstly performed in a separate experimental setup. In this manner, the experimental validation has been carried out in a metrology laboratory with standard conditions of temperature 20 ± 1 °C and humidity 50–70% controlled 24/7. The scheme of the experimental setup is shown in Figure 3. This setup installs one of the linear motors of the NanoPla and the same DMC Kit that will be implemented in the NanoPla. A pneumatic 1D-linear stage was used to imitate the frictionless motion of the NanoPla. The stator of the linear motor is mounted over the pneumatic linear guide, and the magnet array is fixed to the bridge part. The actuator is connected to the three-phase power stage of the control hardware, while the control card is connected to a computer by a USB port. As positioning sensor, a laser interferometer system has been used (i.e., laser head source and reflectors). The laser system is also connected to the computer. In addition, an oscilloscope has been used to monitor the signals of the control hardware.

As can be observed in Figure 3, in this preliminary setup, the stator is the moving part while the magnet array is static, in contrast to the design of the NanoPla. Nevertheless, the relative motion between parts is the same in both cases. Therefore, this does not affect the design of the control system nor the experimental validation.

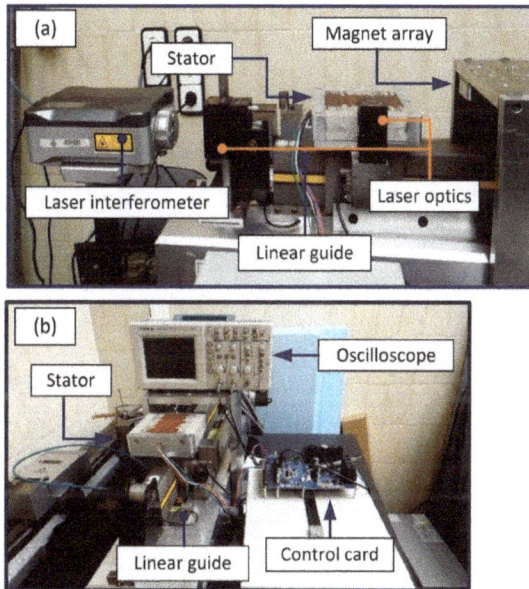

Figure 3. Lateral (**a**) and front (**b**) view of the experimental setup for the implementation of the control system of one linear motor.

As stated, this work proposes to facilitate the control issue by implementing a commercial solution for the control hardware. The selected device to perform the control is a DMC Kit (DRV8302-HC-C2-KIT) from Texas Instruments. This control hardware is designed to operate with generic rotary permanent magnet synchronous motors. It provides closed-loop digital control feedback, analogue integration and comprises a microcontroller unit (MCU) and the inverter stage that generates the phase voltages. The MCU is a C2000 microcontroller and is able to perform real time control by working with 32-bit data. The control hardware is able to generate three phase voltages. That means that in the NanoPla each motor will need one control DMC Kit.

A Renishaw XL-80 laser interferometer (Renishaw, Gloucestershire, UK) has been integrated to provide the position feedback. The readouts of the laser system are sent to the computer and then from the computer to the control card by a serial communication interface. The Renishaw XL-80 laser system has a resolution of 1 nm, and the measured noise under laboratory controlled conditions has a range of 400 nm. The purpose of this work is to develop, integrate and validate a 1D control strategy for one linear motor so that this control system can be implemented later in the four linear motors of the NanoPla for a 2D movement. Renishaw XL-80 laser system performance is similar to the laser system of the NanoPla, and it is perfectly suitable for this validation.

In contrast, the NanoPla includes a 2D laser system that belongs to the Renishaw RLE10 laser interferometer family. It consists of a laser unit (RLU), two sensor heads (RLD), two plane mirrors (one per axis), and an environmental control unit (RCU). In addition, an external interpolator improves the resolution to 1.58 nm. The measured noise of this system is 20 nm. In [26] an analysis of the performance of the NanoPla 2D laser system was presented and its suitability as positioning sensor was confirmed. This laser system will be used once the 2D positioning system is implemented in the NanoPla.

5. Dynamic Characterization

Now the driving actuators and the experimental setup have been described, in this section a dynamic characterisation of the system is performed. In other work that introduces the use of Halbach linear motors for metrology applications [20], electromechanical modelling was presented. In contrast, this section focuses on observing and understanding the dynamic behaviour of the motor under the electromagnetic forces that are generated when a DC current flows through the coils. This dynamic characterisation allows the definition of an open-loop control system, which will facilitate the design of the closed-loop control strategy described in the next section. Firstly, the conditions of the equilibrium of the system are studied. After defining the equilibrium state, a sensorless controller is developed. This controller moves the motor by varying the force distribution along the axis, and it does not require a positioning feedback sensor.

5.1. Equilibrium Position

In the system under study, the only forces that act on the motor are the orthogonal electromagnetic forces F_x and F_z, F_x being the only propulsion force that acts along the axis of movement. As mentioned in Section 3, when certain phase currents flow through the stator, F_x and F_z are generated, and their magnitude depends on the relative position between stator and magnet array. Figure 4 represents the sinusoidal shape of the horizontal force (F_x) and the vertical force (F_z) generated by certain phase current values along the axis of movement (x_s).

The motor will remain motionless once it arrives at a position x_0 where the propulsion force is null; that is, the equilibrium position. As can be observed in Figure 4, in each magnetic spatial period (pitch: l = 29.778 mm), there are two equilibrium positions. For instance, in the first pitch, F_x is equal to 0 N at the positions $x_s = 0$ mm and $x_s = 14.889$ mm. However, these two equilibrium positions have different characteristics. The second one, where the slope is negative, is a stable equilibrium position. As can be seen in Figure 4, if a perturbation displaces the motor from this stable equilibrium position, then the electromagnetic force pushes it forwards if the displacement is negative or backwards if the displacement is positive, always returning it to the stable equilibrium position. On the contrary, where the slope is positive, there is an unstable equilibrium position, where a small disturbance moves the motor away from its position to the nearest stable equilibrium position. According to Figure 4, at the stable equilibrium position, the value of the vertical force F_z is maximum and positive, while at the unstable equilibrium position, the value of F_z is minimum and negative.

In the NanoPla, once the motor arrives to the target or reference position ($x_s = x_{ref}$), it must remain motionless ($F_x = 0$). In addition, the magnet arrays are fixed to the moving part that is levitating by means of three air bearings. In order to leverage the vertical force generated by the motor, F_z, must be positive (Figure 2), favouring the levitation by lifting the magnet array. In other words, the target position must fulfil the conditions of a stable equilibrium position. For the experimental validation presented in this work, the target value of F_z has been defined as 1 N.

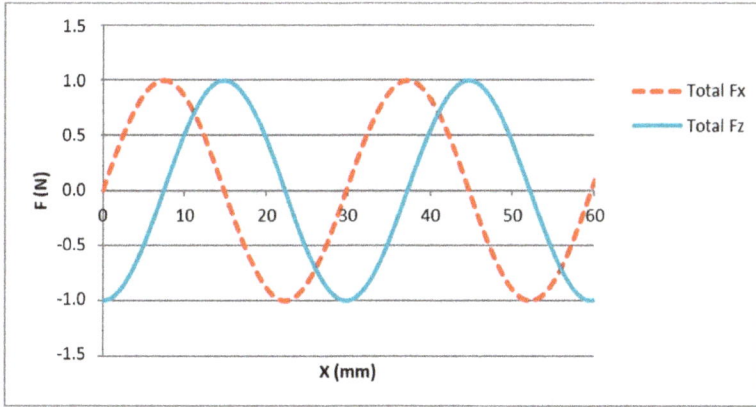

Figure 4. F_x and F_z along the axis of movement, for $I_a = 0$ A; $I_b = 0.3593$ A; $I_c = -0.3593$ A.

5.2. Electromagnetic Sensorless Controller

As stated in the previous subsection, according to the working conditions of the NanoPla, when the motor achieves the target position, it must be in a stable equilibrium state. By introducing the conditions of the stable equilibrium ($F_x = 0$ and $F_z = 1$ N) for a particular desired target position ($x_s = x_{ref}$) in the commutation law (Equation (5)), the required phase currents that create this state can be calculated. When these currents flow through the coils, the electromagnetic forces are generated. Therefore, by combining the phase currents, the equilibrium state can be created at any desired position. Then, the horizontal force moves the motor to the stable equilibrium position (x_{ref}) where it is maintained under small perturbations.

Thus, when the phase currents create a stable equilibrium state, the electromagnetic horizontal force acts as a controller, with the stable equilibrium position as the reference position. This system consists of the electromagnetic controller and the load elements of the plant, as represented in Figure 5. This electromagnetic controller does not require a positioning sensor.

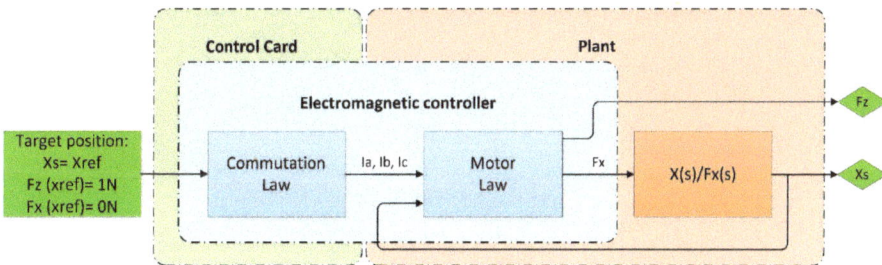

Figure 5. Scheme of the linear motor system in an open-loop system.

Nonetheless, this electromagnetic sensorless controller presents many limitations. The first one is the working range; it works only inside the range of 1 pitch (29.778 mm). That is because each combination of phase currents creates a sinusoidal distribution of the forces along the axis, with one stable equilibrium position in each pitch. Thus, the electromagnetic horizontal force takes the motor to the nearest stable equilibrium position, which may be in a maximum distance of ±14.889 mm. Another limitation of the electromagnetic controller is that it does not allow the tuning of the transient response. However, these two limitations can be overcome by introducing, as an input position (x_{ref}), a discrete ramp that moves the motor in small steps until it arrives at the target position. This allows control of the movement from the initial position to the target position, working in the full range of the linear motor.

The most significant disadvantage that cannot be overcome in this open-loop system is the positioning accuracy. The constant parameters k and A of the motor law (Equation (1)) have been determined theoretically and experimentally (Table 2). Nevertheless, the values of these parameters are an approximation. They may vary from point to point and from pitch to pitch as the motor is not ideal. Similarly, the generated phase currents may also present deviations. Hence, the electromagnetic controller will take the motor to the stable equilibrium position; however, due to these inaccuracies, the equilibrium position may not be exactly coincident with the target position.

6. One-Dimensional Control Strategy and Hardware Implementation

In the previous section, a dynamic characterisation was performed, and it was stated that the electromagnetic force could be defined to behave as a sensorless controller. Nevertheless, as mentioned, the electromagnetic sensorless controller presents many limitations: a working range of one pitch, positioning errors and uncontrolled transient response. Therefore, in order to detect the movement errors, a positioning sensor should be implemented. The readouts of this positioning sensor can be used as feedback for a proportional–integral–derivative (PID) position controller that compares them to the reference position and defines the action necessary to correct the error. Moreover, it will allow the full travel range of the motor to be used. Finally, by tuning the PID controller, it is also possible to adjust the transient response.

The resultant position control system has been represented in Figure 6. The reference position (x_{ref}) is the input to the PID controller, whose output is the required horizontal force (F_x^*). Knowing the desired vertical force (F_{zref}) and the required horizontal force (F_x^*), the commutation law calculates the required phase currents that are needed to generate those forces at the present position (x_s). The resultant phase currents are generated by the control hardware and, according to the motor law, the electromagnetic forces F_x and F_z are produced. The horizontal force F_x displaces the motor to the desired position while F_z favours the levitation. The real position of the motor is read by a positioning sensor and fed back to the PID controller where it is compared to the reference position and corrected. The positioning sensor readouts are also used as inputs to the commutation law.

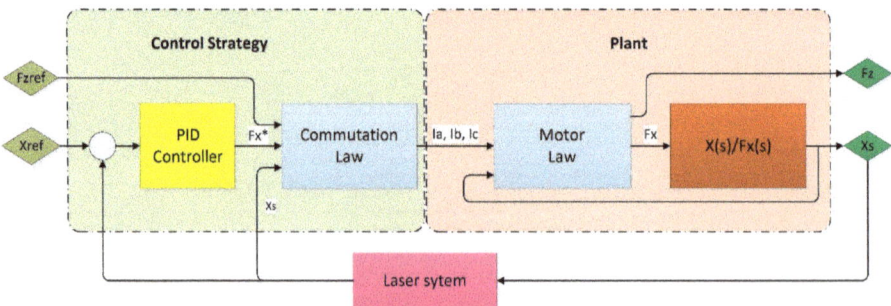

Figure 6. Position control scheme of the linear motor.

Once the 1D control strategy has been developed, it can be implemented in the control hardware and the positioning sensor that were presented in Section 4. The control hardware that has been chosen is a DMC Kit from Texas Instruments, and a Renishaw XL-80 laser system acts as positioning sensor. In this section, the implementation of the control system in the control hardware is presented and schematized.

The selected control hardware presents many advantages, such as being commercial, having a low cost and being suitable for this application. However, it is generic hardware for the control of rotary motors, where the main target is to control the rotation speed or the torque. Therefore, it has some limitations that need to be taken into account.

The first drawback is that the three phases must be wired, presenting a start connection. This constraint does not allow the introduction of an additional constraint of minimum power losses, as mentioned in Section 3.2, when the commutation law was defined.

In addition, the control hardware is not able to act on the phase currents directly as in [21,25], where hardware with current amplifiers was built for the control. Instead, it generates phase voltages by pulse wide modulation (PWM) [27]. The control program must calculate the duty cycles (DC) that are required to generate a desired phase voltage. Then, the transistor bridge generates the pulses of the PWM according to the DC, producing the corresponding voltages of each motor phase. The resulting phase currents flowing through the phases are constant, and their value depends on the phase voltages and the winding phase resistances that, in this case, are approximately 1 Ω for every phase. The DMC Kit from Texas Instruments includes high-resolution PWM (HRPWM) modules based on micro-edge positioner (MEP) technology which are able to extend the time resolution capabilities of the conventionally derived digital pulse [28]. The PWM working frequency must be higher than 10 kHz, according to the manufacturer. In order to get the best DC resolution, it was set to 14.64 kHz. The phase voltage resolution obtained when using the HRPWM modules at this frequency is 2.64×10^{-5} V.

Besides this, the control card is able to communicate in real-time with other peripherals, such as a computer, and transmit data through the serial communication interface (SCI).

Figure 7 presents the control system implementation in the control hardware, having the laser system as positioning sensor. The control strategy reference inputs are the desired position (x_{ref}) and the vertical force (F_{zref}). The outputs are the required phase currents (I_a*, I_b* and I_c*). As already mentioned, the control hardware does not act directly on the phase currents. Instead, it generates phase voltages by PWM. Hence, the PWM modules must generate the phase voltages that correspond to those phase currents. The required DCs must be calculated for the transistor bridge to generate these phase voltages. The voltage drop between the phase terminals and the neutral point of the motor creates the phase currents (I_a, I_b and I_c). Thus, by the motor law, two orthogonal forces (F_x and F_z) are generated, and their magnitude depends on the relative position between the stator and the magnet array (x_s). The motor position is measured by the laser system, and the readouts are extracted to the PC and directly sent to the control card through the SCI, together with the reference position command (x_{ref}). The control strategy is performed at the sampling speed of the positioning sensor. In this case, the fastest sampling speed of the laser system is 0.05 s.

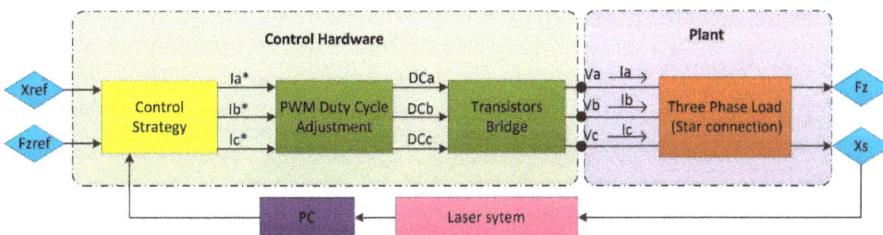

Figure 7. Implementation of the one-dimensional (1D) control strategy in the control hardware.

7. Experimental Results

This section presents and analyses the performance of the implementation of the developed electromagnetic sensorless controller and the 1D control system, which were previously described. The aim of the experiments is to confirm that the 1D control system fulfils the working conditions that the NanoPla design demands. For every experiment, the vertical force value defined as reference is 1 N, which defines the phase currents working range, that, in this case, is ±0.5 A.

7.1. Electromagnetic Sensorless Controller Results

Firstly, the electromagnetic sensorless controller that was presented in Section 5.2 was implemented and analysed. As expected, the electromagnetic controller is able to perform displacements inside a range of ±14.889 mm. The repeatability of the system when performing the same displacement of 5 mm inside the same pole 10 times is ±0.018 mm, and the average positioning error when reaching the position of 5 mm is 0.757 mm. The cause of this variation when performing the same displacement is that the electromagnetic controller displaces the motor to a certain position by creating a stable equilibrium state at this point. The stable equilibrium position is defined by the combinations of phase currents. Even though the command for these currents does not vary for the same target position, the real resultant phase currents may not be the same as they depend on other factors, such as the voltage generation noise and the winding resistor, which may vary with the temperature.

By introducing a stepped ramp as the input for the reference position, it was confirmed that the electromagnetic sensorless controller is able to work in the full range of 50 mm (Figure 8) as stated in Section 5.2. Nevertheless, as expected, this sensorless controller is unable to correct the positioning error, which increases the farther it moves away from the zero position. At the end of the travel range, when the reference position (blue line) is 50 mm, the real position (red line) of the motor is 48.920 mm, which supposes a positioning error equal to 1.080 mm. This positioning error is not acceptable for the NanoPla operation, and thus another control strategy approach is necessary, such as the one proposed below.

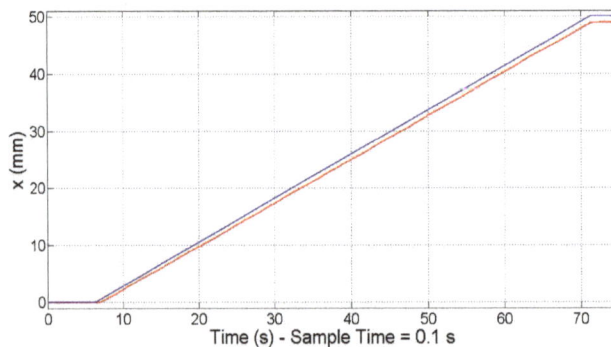

Figure 8. Electromagnetic sensorless controller: 50 mm travel range at constant speed.

7.2. One-Dimensional Control Strategy Results

The control system was implemented in the control hardware and its PID was experimentally tuned. As was done in the previous subsection for the electromagnetic sensorless controller, the repeatability of the system was measured by performing the same displacement of 5 mm 10 times. In this case, the system always reaches the target position; that is, its repeatability is equal to 0, and the average position error is 0 μm. Nevertheless, when the motor is at a stationary state, it slightly oscillates around the target position. This positioning noise has a root mean square (RMS) deviation of ±0.143 μm.

In order to confirm that the control system works along the travel range of 50 mm that is required in the NanoPla, the response to a 50 mm travel range input at a constant speed was recorded. As can be seen in Figure 9 the measured position (red line) follows the reference position (blue line) along the travel range, even doubling the sample time to 0.1 s.

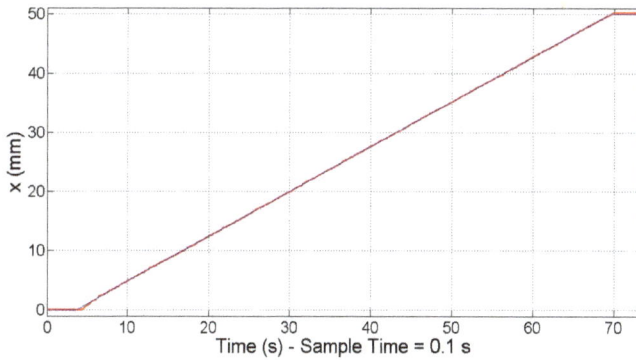

Figure 9. Closed-loop proportional–integral–derivative (PID) controller: 50 mm travel range at constant speed.

Moreover, it was verified that the motor is able to respond to the minimal required motion, that is, 10 μm, as stated in the Introduction. In Figure 10, the response (red line) to a 10 μm step (blue line) is represented.

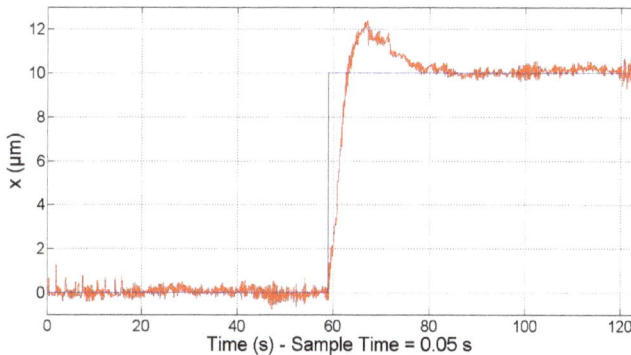

Figure 10. Closed-loop PID controller: 10 μm step response.

Besides this, the smallest step input that the motor can generate was also tested; that is, the minimum incremental motion. It must be noted that, in order to perform the motion, a change in the phase currents must occur. In turn, the variation of the phase currents is produced by a variation of the phase voltages that are controlled by the HRPWM module. As mentioned in the previous section, the minimum voltage variation that this module can perform is 2.62×10^{-5} V, which corresponds to a variation of approximately 2.62×10^{-5} A in the phase currents. Ideally, a change of this magnitude in the phase currents produces a displacement of approximately 600 nm. Nevertheless, the phase currents are also affected by the noise of the voltage source and the PWM signals. Therefore, the power stage is not able to work in the full range of the needed phase currents to perform a motion step in the submicrometre scale. However, the PID controller is fast enough to switch between two combinations of phase currents in order to reduce the positioning error, resulting in an improvement of the effective motion resolution. As shown in Figure 11, the system is able to respond to a staircase of 1 μm (reference position in blue and response in red). The magnitude of the positioning sensor measuring noise together with the resolution and noise of the voltage generation are the main contributors to the positioning noise of the control system. It must be taken into account that the laser system used for the experimental validation has a measuring noise in the range of 400 nm. In contrast, the laser system of the NanoPla has a measuring noise in the range of 20 nm. Therefore, better results are expected when the driving system is implemented in the NanoPla.

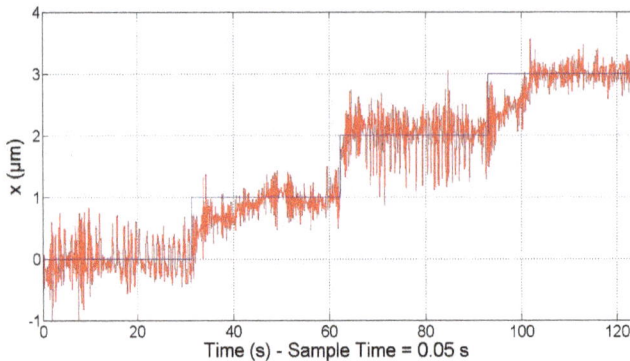

Figure 11. Closed-loop PID controller: 1 μm staircase response.

It should be noted that this results are for the motor under study, the design parameters of which are represented in Table 2. Experimental tests have shown that other motors having similar design parameters provide the same results. Nevertheless, for other values of k and A in the motor law (Equation (1)), the results could be different. The main aspect that should be taken into account when implementing this control system is the phase current's working range and the minimum variation in the phase current that the voltage generation module is able to perform. The phase current's working range is defined by the design parameters of the motor and the reference vertical force, while the minimum variation in the phase current is limited by the hardware.

In the control strategy, a reference value is also set for the generated vertical force (Figure 6). As mentioned, although the levitation of the moving platform of the NanoPla is performed by three air bearings, the design includes the use of the vertical force generated by the motors to favour the levitation. The inverted stators placed on the superior base of the NanoPla will attract the magnet arrays fixed to the moving platform. However, most of the load will be supported by the air bearings, and the linear motors will provide a levitation force of 1 N each. In the experimental setup, the vertical force generated by the motor was measured with a load cell. When the motor arrives to the target position, the vertical force is positive and constant, as required. However, when the motor moves from the initial position to the target position during the transient period, the vertical force varies slightly. It was observed that the transient response of the vertical force improves with the closed-loop control system compared to the open-loop system. During the transient response of the open-loop system, the value of F_z decreases by 18%. In contrast, in the closed-loop system, the value of F_z increases by 7%. According to the manufacturer of the air bearings, considering an air gap of 5 μm, they have a stiffness of 13 N/μm. Therefore, a change of 0.07 N in the load will compress the gap by 5 nm, which is acceptable for the application.

8. Conclusions

In this work, a control system for a Halbach linear motor in 1D has been designed, implemented and experimentally validated in commercial control hardware. The chosen hardware is a Digital Motor Control Kit from Texas Instruments. The usual application of this DMC Kit is the control of the rotation speed or the torque in rotatory motors. As a novelty, in this work it is used to control the position of a linear motor.

The developed control system will be implemented in the four Halbach linear motors that work as actuators in a 2D-nanopositioning stage (NanoPla). The NanoPla is currently in development at the University of Zaragoza. Halbach linear motors have been chosen as actuators because they allow movement along the moving axes and, also, in the orthogonal direction. Therefore, the 2D movement of the NanoPla is achieved in one plane. In addition, besides the propulsion force, Halbach linear motors generate a vertical force that favours the levitation of the moving part of the NanoPla. Developing the control system in a commercial hardware facilitates the future industrial applicability of the NanoPla.

Firstly, this work has proposed an open-loop control system that uses the electromagnetic horizontal force generated by the motor as a controller. Being a sensorless control system, it presents positioning errors that cannot be corrected. Consequently, a laser system has been chosen as a positioning sensor, and a closed-loop control system has been designed. Then, the developed control system has been implemented in the chosen hardware. The limitations of the commercial hardware have been overcome by optimizing the design.

Once the control system was designed and implemented, its performance was validated in the experimental setup. It has been verified that the system fulfils the working requirements of the NanoPla, which is a working range of 50 mm and a step response of 10 μm. It was also tested that the system is able to respond to steps of 1 μm. It must be noted that the laser system used in the experimental validation has a noise of 400 nm, while the laser system of the NanoPla has a noise of 20 nm. Therefore, the performance is expected to improve in the NanoPla. Additionally, the vertical force generated by the motor has also been measured, being constant at steady state and varying slightly (+7%) during the transient period; that is, when the motor is moving from the initial position to the target position. In the NanoPla, the main support of the levitation of the moving platform comprises three air bearings, while the motors favour the levitation with a force of 1 N each. Thus, a variation of 0.07 N does not affect the stability of the system.

The fact that the developed control strategy implemented in the chosen control hardware is able to operate according to the NanoPla design requirements, make unnecessary the use of more advance control devices. In future works, the control system that has been presented in this paper for one Halbach linear motor along one dimension will be implemented in the four NanoPla actuators to provide a two-dimensional travel range. One DMC Kit will be needed for each motor and the whole control strategy and experimental results will be obtained to assure the desired performance requirements of the NanoPla. At this point, different variants of the control strategy should be tested, in order to find the best performance. These variants could be for instance, operating at constant speed or applying a feedforward loop.

Author Contributions: M.T.G., J.A.A.G. and J.A.Y.F. conceived a preliminary design of the control strategy; L.C.D.P., M.T.G., J.A.A.G. and J.A.Y.F. optimized the control strategy for the implementation in the commercial hardware and performed the implementation; L.C.D.P. performed the experiments; L.C.D.P. wrote the manuscript. All authors contributed to the editing of the manuscript.

Funding: This project was funded by the Spanish government project DPI2015-69403-C3-1-R "MetroSurf" with the collaboration of the Diputación General de Aragón—Fondo Social Europeo. Appreciation is expressed to the FPU Program of the Ministerio de Educación, Cultura y Deporte of the Spanish government, which sponsored the first author.

Conflicts of Interest: The authors declare no conflict of interest. The founding sponsors had no role in the design of the study; in the collection, analyses, or interpretation of data; in the writing of the manuscript, and in the decision to publish the results.

Micromachines **2018**, *9*, 421

References

1. Kramar, J.A.; Dixson, R.; Orji, N.G. Scanning probe microscope dimensional metrology at NIST. *Meas. Sci. Technol.* **2011**, *22*, 24001–24011. [CrossRef]
2. Manske, E.; Jäger, G.; Hausotte, T.; Füßl, R. Recent developments and challenges of nanopositioning and nanomeasuring technology. *Meas. Sci. Technol.* **2012**, *23*, 74001–74010. [CrossRef]
3. Gao, W.; Kim, S.W.; Bosse, H.; Haitjema, H.; Chen, Y.L.; Lu, X.D.; Knapp, W.; Weckenmann, A.; Estler, W.T.; Kunzmann, H. Measurement technologies for precision positioning. *CIRP Ann.* **2015**, *64*, 773–796. [CrossRef]
4. Sinno, A.; Ruaux, P.; Chassagne, L.; Topu, S.; Alayli, Y.; Lerondel, G.; Blaize, S.; Bruyant, A.; Royer, P. Enlarged atomic force microscopy scanning scope: Novel sample-holder device with millimeter range. *Rev. Sci. Instrum.* **2007**, *78*, 095107. [CrossRef] [PubMed]
5. Sato, K. Trend of precision positioning technology. *ABCM Symp. Ser. Mechatron.* **2006**, *2*, 739–750.
6. Fesperman, R.; Ozturka, O.; Hockena, R.; Ruben, S.; Tsao, T.; Phipps, J.; Lemmons, T.; Brien, J.; Caskey, G. Multi-scale alignment and positioning system—MAPS. *Precis. Eng.* **2012**, *36*, 517–537. [CrossRef]
7. Werner, C. A 3D Translation Stage for Metrological AFM. Ph.D. Thesis, Eindhoven University of Technology, Eindhoven, The Netherlands, 2010.
8. Klapetek, P.; Valtr, M.; Matula, M. A long-range scanning probe microscope for automotive reflector optical quality inspection. *Meas. Sci. Technol.* **2011**, *22*, 094011. [CrossRef]
9. Balasubramanian, A.; Jun, M.B.G.; DeVor, R.E.; Kapoor, S.G. A submicron multiaxis positioning stage for micro- and nanoscale manufacturing processes. *J. Manuf. Sci. Eng.* **2008**, *130*, 031112. [CrossRef]
10. Ducourtieux, S.; Poyet, B. Development of a metrological atomic force microscope with minimized Abbe error and differential interferometer-based real-time position control. *Meas. Sci. Technol.* **2011**, *22*, 094010. [CrossRef]
11. Jäger, G.; Hausotte, T.; Manske, E.; Büchner, H.J.; Mastylo, R.; Dorozhovets, N.; Hofmann, N. Nanomeasuring and nanopositioning engineering. *Measurement* **2010**, *43*, 1099–1105. [CrossRef]
12. Torralba, M.; Yagüe-Fabra, J.A.; Albajez, J.A.; Aguilar, J.J. Design optimization for the measurement accuracy improvement of a large range nanopositioning stage. *Sensors* **2016**, *16*, 84. [CrossRef] [PubMed]
13. Torralba, M.; Valenzuela, M.; Yagüe-Fabra, J.A.; Albajez, J.A.; Aguilar, J.J. Large range nanopositioning stage design: A three-layer and two-stage platform. *Measurement* **2016**, *89*, 55–71. [CrossRef]
14. Hansen, H.N.; Carneiro, K.; Haitjema, H.; De Chiffre, L. Dimensional micro and nano metrology. *CIRP Ann.* **2006**, *55*, 721–743. [CrossRef]
15. Kim, J.A.; Kim, J.W.; Kang, C.S.; Eom, T.B. Metrological atomic force microscope using a large range scanning dual stage. *Int. J. Precis. Eng. Manuf.* **2009**, *10*, 11–17. [CrossRef]
16. Liu, C.H.; Jywe, W.Y.; Jeng, Y.R.; Hsu, T.H.; Li, Y. Design and control of a longtraveling nano-positioning stage. *Precis. Eng.* **2010**, *34*, 497–506. [CrossRef]
17. Lu, X.; Usman, I.-U.-R. 6D direct-drive technology for planar motion stages. *CIRP Ann.* **2012**, *61*, 359. [CrossRef]
18. Hesse, S.; Schäffel, C.; Mohr, H.U.; Katzschmann, M.; Büchner, H.J. Design and performance evaluation of an interferometric controlled planar nanopositioning system. *Meas. Sci. Technol.* **2012**, *23*, 074011. [CrossRef]
19. Holmes, M.; Hocken, R.; Trumper, D. Long-range scanning stage: a novel platform for scanned-probe microscopy. *Precis. Eng.* **2000**, *24*, 191–209. [CrossRef]
20. Trumper, D.; Kim, W.; Williams, M. Design and analysis framework for linear permanent-magnet machines. *IEEE Trans. Ind. Appl.* **1996**, *32*, 371–379. [CrossRef]
21. Yu, H. Design and Control of a Compact 6-Degree-of-Freedom Precision Positioner with Linux-Based Real-Time Control. Ph.D. Thesis, Texas A&M University, College Station, TX, USA, 2009.
22. Torralba, M.; Albajez, J.A.; Yagüe-Fabra, J.A.; Aguilar, J.J. Preliminary modelling and implementation of the 2D-control for a nanopositioning long range stage. *Procedia Eng.* **2015**, *132*, 824–831. [CrossRef]
23. Kim, W.J.; Trumper, D.L.; Lang, J.H. Modeling and vector control of planar magnetic levitator. *IEEE Trans. Ind. Appl.* **1998**, *34*, 1254–1262.
24. Torralba, M.; Yagüe-Fabra, J.A.; Albajez, J.A.; Aguilar, J.J. Caracterización de motores lineales tipo Halbach para aplicaciones de nanoposicionado. In Proceedings of the 2014 Congreso Nacional de Ingeniería Mecánica (CNIM), Málaga, Spain, 24–26 September 2014.

25. Ruben, S. Modeling, Control, and Real-Time Optimization for a Nano-Precision System. Ph.D. Thesis, University of California, Los Angeles, CA, USA, 2010.

26. Díaz-Pérez, L.; Torralba, M.; Albajez, J.A.; Yagüe-Fabra, J.A. Performance analysis of laser measuring system for an ultra-precision 2D-stage. In Proceedings of the 17th International Conference & Exhibition, Hannover, Germany, 29 May–2 June 2017.

27. Texas Instruments. TMS320x2802x, 2803x Piccolo Enhanced Pulse Width Modulator (ePWM) Module, Reference Guide. Available online: http://www.ti.com/lit/ug/spruge9e/spruge9e.pdf (accessed on 18 August 2018).

28. Texas Instruments. TMS320x2802x, 2803x Piccolo High Resolution Pulse Width Modulator (HRPWM), Reference Guide. Available online: http://www.ti.com/lit/ug/spruge8e/spruge8e.pdf (accessed on 18 August 2018).

micromachines

MDPI

Article

Generation of Color Images by Utilizing a Single Composite Diffractive Optical Element

Jiazhou Wang [1,2,†], Liwei Liu [1,2,†], Axiu Cao [1], Hui Pang [1], Chuntao Xu [1], Quanquan Mu [3], Jian Chen [4], Lifang Shi [1,*] and Qiling Deng [1]

[1] Institute of Optics and Electronics, Chinese Academy of Sciences, Chengdu 610209, China; wangjiazhou0401@163.com (J.W.); a1252909630@163.com (L.L.); longazure@163.com (A.C.); wuli041@126.com (H.P.); xuchuntao@live.com (C.X.); dengqiling@ioe.ac.cn (Q.D.)

[2] School of Optoelectronics, University of Chinese Academy of Sciences, Beijing 100049, China

[3] State Key Laboratory of Applied Optics, Changchun Institute of Optics, Fine Mechanics and Physics, Chinese Academy of Sciences, Changchun 130033, China; muquanquan@ciomp.ac.cn

[4] State Key Laboratory of Transducer Technology, Institute of Electronics, Chinese Academy of Sciences, Beijing 100190, China; chenjian@mail.ie.ac.cn

* Correspondence: shilifang@ioe.ac.cn; Tel.: +86-28-8510-1178

† These authors contributed equally to this work.

Received: 30 August 2018; Accepted: 4 October 2018; Published: 9 October 2018

Abstract: This paper presents an approach that is capable of producing a color image using a single composite diffractive optical element (CDOE). In this approach, the imaging function of a DOE and the spectral deflection characteristics of a grating were combined together to obtain a color image at a certain position. The DOE was designed specially to image the red, green, and blue lights at the same distance along an optical axis, and the grating was designed to overlay the images to an off-axis position. We report the details of the design process of the DOE and the grating, and the relationship between the various parameters of the CDOE. Following the design and numerical simulations, a CDOE was fabricated, and imaging experiments were carried out. Both the numerical simulations and the experimental verifications demonstrated a successful operation of this new approach. As a platform based on coaxial illumination and off-axis imaging, this system is featured with simple structures and no cross-talk of the light fields, which has huge potentials in applications such as holographic imaging.

Keywords: diffractive optics; gratings; microfabrication; computer holography

1. Introduction

A diffractive optical element (DOE) is a phase modulation element that has the potential to significantly promote the miniaturization, integration, and arrangement of conventional optical systems [1–6]. More specifically, DOEs play a key role in the field of holographic imaging [7,8]. However, the working wavelengths of traditional DOEs are monochromatic; thus, multiple DOEs are always needed in the holographic imaging of color pictures, leading to complex and bulky systems.

In order to address this issue, previous studies were conducted to generate color images using one single DOE, which relied on coaxial imaging and coaxial illumination [9–17] or coaxial imaging and off-axis illumination [18–20]. In coaxial imaging and coaxial illumination, the phase modulations of incident lights with different wavelengths are obtained by properly designing DOEs. More specifically, Bengtsson proposed an algorithm that can produce a specific image of two wavelengths at a certain distance [9–11]. Ogura et al. optimized the weighting factors of this algorithm to realize multi-wavelength imaging [12,13]. Jesacher et al. proposed the concept of equivalent phases to optimize the designs of DOEs, which mainly relied on the modification of the Gerchberg–Saxton (GS) algorithm [14,15]. However, these approaches suffered from the limited numbers of imaging points, close distances of output fields, and complex fabrications of DOEs.

Meanwhile, DOEs were also designed using imaging features of the Fresnel DOE, where the imaging position varies for multi-wavelength incident lights to obtain color holograms [16,17]. However, the proposed optical field was prone to cross-talk among the individual color components. In order to address this issue, coaxial imaging and off-axis illumination were proposed to generate color images using one single DOE, where the RGB (red, green, and blue) components of color images are encoded at different positions within the same plane, and thus the superposition of the light field can be realized by changing the irradiation angles of the incident lights [18–20].

However, since the incident angles of each wavelength's components need to meet specific design requirements, a complicated mechanical structure and a bulky system were required in this approach. In addition, slight deviations of the angles of the incident lights can affect the quality of the color images, and thus the stability of the proposed systems was under question.

In order to deal with the aforementioned problems, we report in this paper the design of a composite diffractive optical element (CDOE) for generating color images. This design offers significant improvements in comparison to the traditional approach based on coaxial incidences and off-axis imaging to generate color images, whereby (1) only a RGB mixed light was used as the incident light on the CDOE, significantly reducing the complexity of the optical system; and (2) the principle of spatial superposition was adopted, ensuring that the obtained target field was well recognizable. The structures of this paper are as follows: 1. Introduction, 2. Principle, 3. Simulation, 4. Experiments, and 5. Summary.

2. Principle

2.1. Schematics

The principle of this method is illustrated in Figure 1, where the CDOE is irradiated by a composite RGB light beam along an optical axis, generating a color image at off-axis positions. The CDOE proposed in this study contained the modulation phase to generate the target field composing RGB components of the color images, and the phase information of the spectral grating, which was used to modulate the three light fields to a specific position for the superposition. The design process and the parameter analysis will be discussed carefully in the following sections.

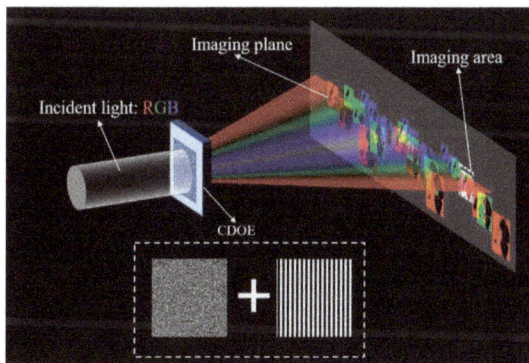

Figure 1. The schematic of an imaging method by using composite diffractive optical elements (CDOE) comprising a diffractive optical elements (DOE) and a grating.

2.2. The Process of Designing and Parameter Analysis

The amplitude distribution of the target color light was denoted by $A(x, y)$. Since all of the colors can be synthesized by RGB components, the target color field could be represented by RGB components, in which the individual amplitude distributions of the three color components were denoted by $A_R(x, y)$, $A_G(x, y)$, and $A_B(x, y)$, respectively.

In the design process, the chromatic aberrations of the RGB lights were addressed. The revised amplitude distributions $A'_R(x, y)$, $A_G(x, y)$, and $A'_B(x, y)$ were obtained by making compensations to the R and the B components, while the G component remained unchanged. The three distributions were arranged from left to right in the design plane, and the distances between each of the two components were represented as D_{RG} and D_{GB}, respectively, as shown in Figure 2a.

By combining $A'_R(x, y)$, $A_G(x, y)$, and $A'_B(x, y)$ together, the target amplitude distribution $E_{m0}(x, y)$ was obtained first. A DOE was then designed for the green light in order to obtain a phase distribution of $\varphi_G(x, y)$. When the RGB lights passed through the DOE, the light fields of $E_G(x, y)$, $E_R(x, y)$, and $E_B(x, y)$ were obtained, respectively. However, the three light fields were not imaged in the correct position to form a color image. So, a grating phase $\varphi_T(x, y)$ was designed and superimposed on $\varphi_G(x, y)$ to alter the positions of the three light fields, whose periods determined the values of the D_{RG} and D_{GB}. The accurate modulations of the RGB lights were realized by carefully choosing the interactions of φ_G and φ_T.

A superposition of $E_R(x, y)$, $E_G(x, y)$, and $E_B(x, y)$ created a color image at the first-order position of the green light. In the design process, the parameters determining D_{RG} and D_{GB}, the corrections of chromatic aberrations, and the interactions between φ_G and φ_T were analyzed in detail in order to obtain the final amplitude distribution $A(x, y)$. In addition, the impacts of the zero-level positions and its optimization measures were considered carefully.

2.2.1. Determination of the Parameters D_{RG} and D_{GB}

In the design process, the target field of the DOE was obtained by coding the three sets of amplitude distributions of the RGB components. Then, the issue was how to determine the parameters of D_{RG} and D_{GB}, which express the intervals between amplitudes. The intervals were cooperated with the wavelength-dependent spectral deflection abilities of the grating to realize the superposition of the light fields. Thus, the values of D_{RG} and D_{GB} were closely related to the grating periods. At the same time, three amplitude distributions were generated by the DOE. Therefore, the values were also closely related to the structural parameters of the DOE, with the detailed analysis discussed below.

Suppose that the period of grating was d, and the spectral deflection ability of grating was expressed by the grating equation:

$$\theta = \arcsin\left(\frac{m\lambda}{d}\right) \tag{1}$$

where λ is the wavelength of the incident light, and θ is the angle between the m-order of the light and optical axis in the imaging plane. The distance between the imaging plane and the DOE was z. The sampling interval δ of the imaging plane was closely related to the period D of the DOE, which was expressed by the following equation:

$$\delta = \lambda z / D \tag{2}$$

Both the sampling interval δ and the angle θ depend on light wavelength. Thus, the position of m-order light generated by the grating depends also on light wavelength. For convenience, a pixel number M was used to express the position of m-order light (relative to that of the zero-order light), which was expressed by the following equation:

$$M = \frac{\tan(\theta)z}{\delta} \tag{3}$$

The position of the m-order green light was taken as the image area. Since the angle θ between the image area and the axis was very small, Equation (3) was reduced to Equation (4) in case of paraxial:

$$M = \frac{mD}{d} \tag{4}$$

The pixel numbers for each of the three RGB wavelength components were expressed by Equation (5):

$$M_R = \frac{mD\lambda_G}{d\lambda_R}; M_G = \frac{mD}{d}; M_G = \frac{mD\lambda_G}{d\lambda_B} \tag{5}$$

The green light wavelength was regarded as the nominal working wavelength of the DOE, and the green component of the color image was coded in the middle of the target field. Thus, the intervals D_{RG} and D_{GB} were expressed as the difference of pixel numbers (M_G–M_R) and (M_B–M_G), respectively (see Figure 2a):

$$\begin{aligned} D_{RG} &= \frac{mD}{d} - \frac{mD\lambda_G}{d\lambda_R} \\ D_{GB} &= \frac{mD\lambda_G}{d\lambda_B} - \frac{mD}{d} \end{aligned} \tag{6}$$

2.2.2. Correction for the Chromatic Aberration

In order to avoid the influence of zero-order on the color image, the position of the color image was shifted vertically, and the shifting distance was represented as L, ensuring that the zero-level does not appear on the image. As can be seen from Section 2.2.1, the sampling interval δ of the imaging plane was related to the incident wavelength, and thus, there was a lateral chromatic aberration, which needed to be corrected (see Figure 2b). The amount of shift in pixel numbers for each wavelength component was calculated with the help of Equation (2), as expressed by Equation (7):

$$N_R = \frac{LD}{\lambda_R z}; N_G = \frac{LD}{\lambda_G z}; N_B = \frac{LD}{\lambda_B z} \tag{7}$$

There was also a magnification of the chromatic aberration due to the difference of the sampling interval. The imaging sizes of RGB lights were different from each other. In order to ensure the perfect superposition of each component, it was necessary to adjust the pixel number $X \cdot Y$ of the RGB components to ensure that the sizes of each image were the same. The scaling relation was related to δ, and in the case that the z and D were constants, they can be expressed by Equation (8):

$$\begin{aligned} X_R : X_G &= Y_R : Y_G = \lambda_G : \lambda_R; \\ X_B : X_G &= Y_B : Y_G = \lambda_G : \lambda_B; \end{aligned} \tag{8}$$

where $X_R \cdot Y_R$, $X_G \cdot Y_G$, and $X_B \cdot Y_B$ were the corrected pixel numbers of RGB components, respectively. According to the aforementioned analysis, the color image was firstly decomposed into RGB components. The RGB components were then encoded and scaled in the same plane according to the calculations. Finally, the desired target field of the DOE was obtained as shown in Figure 2c.

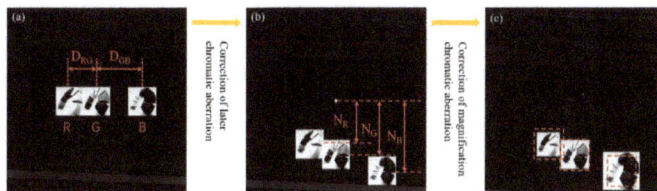

Figure 2. (**a**) The position shifts of red, green, and blue (RGB) components according to the parameters D_{RG} and D_{GB}; (**b**) the correction for lateral chromatic aberration; (**c**) the correction for magnification chromatic aberration.

2.2.3. Interaction Mechanism between Modulation Phase φ_G and Grating Phase φ_T

In this work, the distribution of the modulation phase φ_G was calculated by the GS algorithm, as shown in Figure 3a, and the corresponding grating phase φ_T (see Figure 3c) was superimposed on the φ_G to generate the final phase distribution of the DOE (see Figure 3e). When the RGB lights were regarded as incident lights, the light field of φ_G and φ_T in the imaging plane were expressed as E_1 and E_2, respectively, by Equation (9):

$$E_1 = \begin{pmatrix} E_R^1 \\ E_G^1 \\ E_B^1 \end{pmatrix} = \begin{pmatrix} FFT\left(A\exp\left(i\frac{\lambda_G}{\lambda_R}\varphi_G\right)\right) \\ FFT(A\exp(i\varphi_G)) \\ FFT\left(A\exp\left(i\frac{\lambda_G}{\lambda_B}\varphi_G\right)\right) \end{pmatrix}$$

$$E_2 = \begin{pmatrix} E_R^2 \\ E_G^2 \\ E_B^2 \end{pmatrix} = \begin{pmatrix} FFT\left(A\exp\left(i\frac{\lambda_G}{\lambda_R}\varphi_T\right)\right) \\ FFT(A\exp(i\varphi_T)) \\ FFT\left(A\exp\left(i\frac{\lambda_G}{\lambda_B}\varphi_T\right)\right) \end{pmatrix} \tag{9}$$

where A is the amplitude distribution of the incident light, which is a constant for the uniform parallel light. The resulting light field distributions were shown in Figure 3b,d. By superimposing the two phases to obtain the final phase distribution, the corresponding light field E was expressed by Equation (10):

$$E = \begin{pmatrix} FFT\left(A\exp\left(i\frac{\lambda_G}{\lambda_R}(\varphi_G+\varphi_T)\right)\right) \\ FFT(A\exp(i(\varphi_G+\varphi_T))) \\ FFT\left(A\exp\left(i\frac{\lambda_G}{\lambda_B}(\varphi_G+\varphi_T)\right)\right) \end{pmatrix} = \begin{pmatrix} FFT\left(A\exp\left(i\frac{\lambda_G}{\lambda_R}\varphi_G\right)\cdot\exp\left(i\frac{\lambda_G}{\lambda_R}\varphi_T\right)\right) \\ FFT(A\exp(i\varphi_G)\cdot\exp(i\varphi_T)) \\ FFT\left(A\exp\left(i\frac{\lambda_G}{\lambda_B}\varphi_G\cdot\exp\left(i\frac{\lambda_G}{\lambda_B}\varphi_T\right)\right)\right) \end{pmatrix}$$

$$= A\cdot \begin{pmatrix} FFT\left(\exp\left(i\frac{\lambda_G}{\lambda_R}\varphi_G\right)\right)*FFT\left(\exp\left(i\frac{\lambda_G}{\lambda_R}\varphi_T\right)\right) \\ FFT(\exp(i\varphi_G))*FFT(\exp(i\varphi_T)) \\ FFT\left(\exp\left(i\frac{\lambda_G}{\lambda_B}\varphi_G\right)\right)*FFT\left(\exp\left(i\frac{\lambda_G}{\lambda_B}\varphi_T\right)\right) \end{pmatrix} = A\cdot \begin{pmatrix} E_R^1*E_R^2 \\ E_G^1*E_G^2 \\ E_B^1*E_B^2 \end{pmatrix} \tag{10}$$

where $*$ means convolution. It was noticed from Equation (10) that the final light field E can be obtained by the convolution of E_1 and E_2. The light field E_1 was repeated in each order of the light field E_2 generated by the grating phase. It was observed that the order position of different incident lights was different from Figure 3d, and the superposition of RGB components was achieved by using this property to complete the generation of color images (see Figure 3f).

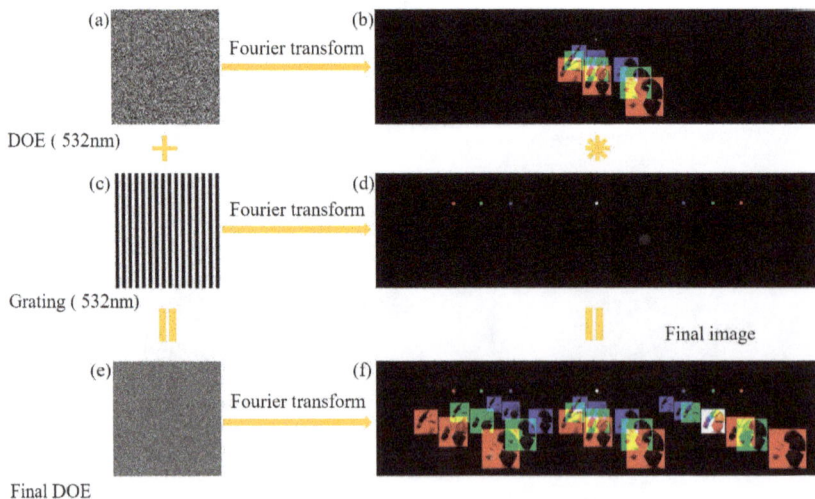

Figure 3. The schematic diagram of the interaction mechanism between modulation phase φ_G and grating phase φ_T. The phase distribution of (**a**) DOE, (**c**) grating, and (**e**) the final composite DOE. Their corresponding imaging effects are shown in images (**b**), (**d**), and (**f**), respectively.

3. Simulation

To validate the design methodologies described above, numerical simulations were conducted to obtain the target images (see Figure 4a). The simulation was carried out by using Matlab-based programs that we wrote. The design of the CDOE and the imaging effect can all be simulated by the programs. The wavelengths λ_R, λ_G, and λ_B that were used in simulation were 650 nm, 532 nm, and 406 nm, respectively. The pixel size of the color target image was 512×512, which was decomposed into RGB components, and arranged based on the aforementioned analysis. The imaging position was selected at the first-order position of the green light imaging. Then, the values of the D_{RG} and D_{GB} intervals were quantified as 492 pixels and 842 pixels, respectively, based on Equation (6). To avoid the effect of zero order, a longitudinal offset was conducted with a vertical offset of 200 pixels relative to the green component. The periods, duty ratios of the grating, and phase depth were quantified as 3 μm, 1:1, and π for the green light, respectively. The characteristic feature size of the DOE was 1.5 μm. Based on the above data, a synthetic target field with 5335×5335 pixels was obtained. In order to use a CCD to receive the image, a Fourier lens with the focal length of 37.5 cm was added into the system. The simulated imaging light field is shown in Figure 4b.

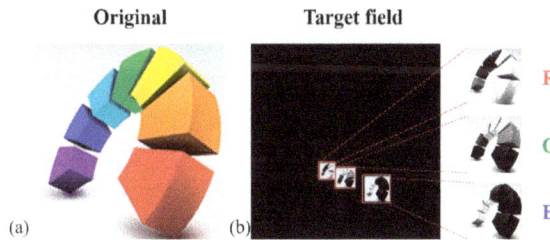

Figure 4. (**a**) The target color field; (**b**) the simulated imaging light field of RGB components.

Using the aforementioned parameters, where 532 nm was used as the working (green) wavelength, the calculated target field was translated to the phase distribution of the single-wavelength DOE. As shown in Figure 5a, the grating phase distribution critical feature size of 1.5 μm, and periods of grating of 3 μm as shown in Figure 5b were superimposed on the DOE phase to obtain the final composite phase distribution of the CDOE as shown in Figure 5c with a level height of 1.156 μm.

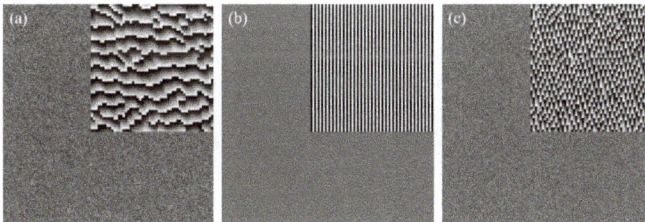

Figure 5. The simulation of the phase distribution: (**a**) the phase distribution of the DOE; (**b**) the phase distribution of the grating; and (**c**) the final composite phase distribution of CDOE.

The simulated imaging results are shown in Figure 6, with (a), (b), and (c) representing the imaging fields generated by red, green, and blue lights passing through the DOE, respectively. Each light field has an amplitude distribution of RGB components, and the central position of three images was used for imaging combination, producing the color images shown in Figure 7d. The generated color images were consistent with the target image, indicating a high degree of color reproduction, and validating the feasibility of this approach to produce full color images.

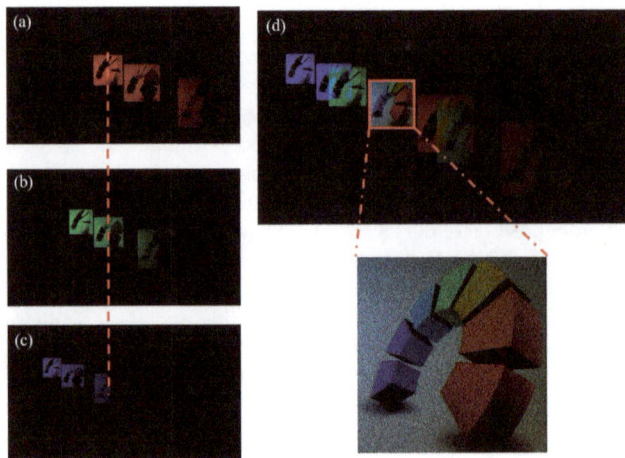

Figure 6. Simulation results: (**a**), (**b**), and (**c**) are the imaging light field of red, green, and blue incident lights, respectively; (**d**) the composite color imaging light field.

4. Experiments

To further validate the feasibility of the proposed approach, corresponding experiments were carried out. Firstly, the CDOE was fabricated by using the photolithography. Silica was chosen as the substrate, and AZ9260 was chosen as the photoresist. The photoresist was spin-coated on the substrate at a rotation speed of 5000 rad/min. The process parameters of coating time, prebake temperature, prebake time, and the resist thickness were 30 s, 100 °C, 5 min, and 3 µm, respectively. A photolithography system was employed to perform the exposure. The illumination light source was an Hg lamp with a central wavelength of 365 nm. The total exposure time was 20 s. The process parameters of development, after-bake temperature, and after-bake time were 3 min, 120 °C, and 30 min, respectively. Reactive ion etching (RIE) was carried out to transfer the structure into the substrate. The etching gases were SF_6 and CHF_3, and the etching time was 80 min. The pictures of the obtained elements are shown in Figure 7b. The light beam paths were constructed as shown in Figure 7a, employing three light sources operated at wavelengths of 460 nm, 532 nm, and 650 nm, respectively. The three parallel beams produced by the laser modules were combined through two beam splitters/combiners (in the form of dispersion prisms) to obtain a mixed light of RGB. The mixed light was irradiated onto the CDOE and imaged by a lens onto a CCD camera. An example image captured by the CCD is shown in Figure 7c, which was consistent with the simulation results, producing nice imaging without optical cross-talk. Furthermore, the incident light is an RGB mixed light functioning as an incident light on the DOE, confirming the coaxiality of the proposed method.

As a phase modulation element, the efficiency of the CDOE is related to the number of levels in the phase construction. The more levels there are, the higher the efficiency. In this method, since the phase level of the CDOE is two, the efficiency of the CDOE is not as high as when using eight or 16 levels. In fact, by using a CDOE with eight or 16 phase levels, the symmetric images can be eliminated, and the efficiency can be increased greatly. In the current research, we focused on the proposal and verification of this new method.

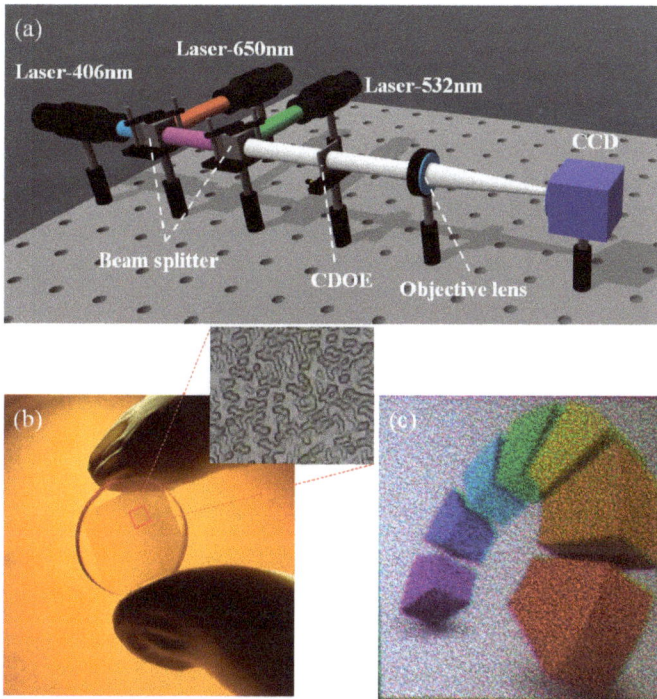

Figure 7. (**a**) Experimental setup of the coaxial illumination system to generate color images and experiment results by using the fabricated CDOE; (**b**) the photograph of the fabricated composite diffractive optical element; (**c**) the generated color image.

5. Summary

This paper presents a novel method to generate color images based on coaxial illumination and off-axis imaging by utilizing a single composite diffractive optical element (CDOE). The principle of this method was analyzed in detail, and the corresponding simulations and experimental implementations were conducted successfully. The simulation and experimental results validated the proposed approach, which featured a simplified system without a cross-talk of light fields. This provides a new perspective for generating color images based on a single CDOE.

Author Contributions: Conceptualization, J.W. and L.L.; Methodology, A.C.; Software, H.P.; Validation, C.X., Q.M. and J.C.; Formal Analysis, J.W.; Investigation, L.S.; Resources, Q.D.; Data Curation, J.C.; Writing—Original Draft Preparation, J.W.; Writing—Review & Editing, L.L. and L.S.; Visualization, A.C.; Supervision, Q.D.; Project Administration, L.S.; Funding Acquisition, L.S.

Funding: This research was supported by the National Key R&D Program of China (2017YFC0804900), National Natural Science Foundation of China (NSFC) (Nos. 61505214, 61605211), the Applied Basic Research Programs of Department of Science and Technology of Sichuan Province (Nos. 2016JY0175, 2016RZ0067, 2017JY0058), and the Youth Innovation Promotion Association of the Chinese Academy of Sciences (CAS).

Acknowledgments: The authors also thank Yanyun Qin and Xue Luo for their help in the fabrication process of the CDOE.

Conflicts of Interest: The authors declare no conflict of interest.

References

1. Amako, J.; Nagasaka, K.; Fujii, E. Direct laser writing of diffractive array illuminators operable at two wavelengths. *Proc. SPIE* **2001**, *4416*, 360–363.

2. Yaras, F.; Kang, H.; Onural, L. State of the art in holographic displays: A survey. *J. Display Technol.* **2010**, *6*, 443–454. [CrossRef]

3. Lapchuk, A.; Yurlov, V.; Kryuchyn, A.; Pashkevich, G.A.; Klymenko, V.; Bogdan, O. Impact of speed, direction, and accuracy of diffractive optical element shift on efficiency of speckle suppression. *Appl. Opt.* **2015**, *54*, 4070–4076. [CrossRef]

4. Lapchuk, A.; Pashkevich, G.A.; Prygun, O.V.; Yurlov, V.; Borodin, Y.; Kryuchyn, A. Experiment evaluation of speckle suppression efficiency of 2D quasi-spiral m-sequence-based diffractive optical element. *Appl. Opt.* **2015**, *54*, 47–54. [CrossRef] [PubMed]

5. Goncharsky, A.; Goncharsky, A.; Durlevich, S. Diffractive optical element for creating visual 3D images. *Opt. Express* **2016**, *24*, 9140. [CrossRef] [PubMed]

6. Piao, M.; Cui, Q.; Mao, S. Optimal design method on diffractive optical elements with antireflection coating. *Opt. Express* **2017**, *25*, 11673–11678.

7. Wu, L.; Cheng, S.B.; Ta, S.H. Simultaneous shaping of amplitude and phase of light in the entire output plane with a phase only hologram. *Sci. Rep.* **2015**, *5*, 15426. [CrossRef] [PubMed]

8. Pang, H.; Wang, J.; Cao, A.; Deng, Q. A high-accuracy method for holographic image projection with suppressed speckle noise. *Opt. Express* **2016**, *24*, 22766. [CrossRef] [PubMed]

9. Bengtsson, J. Design of fan-out kinoforms in the entire scalar diffraction regime with an optimal-rotation-angle method. *Appl. Opt.* **1997**, *36*, 8435–8444. [CrossRef] [PubMed]

10. Bengtsson, J. Kinoforms designed to produce different fan-out patterns for two wavelengths. *Appl. Opt.* **1998**, *37*, 2011–2020. [CrossRef] [PubMed]

11. Bengtsson, J.; Johansson, M. Fan-out diffractive optical elements designed for increased fabrication tolerances to linear relief depth errors. *Appl. Opt.* **2002**, *41*, 281–289. [CrossRef] [PubMed]

12. Ogura, Y.; Shirai, N.; Tanida, J.; Ichioka, Y. Wavelength-multiplexing diffractive phase elements: Design, fabrication, and performance evaluation. *J. Opt. Soc. Am. A* **2001**, *18*, 1082–1092. [CrossRef]

13. Ogura, Y.; Shirai, N.; Tanida, J.; Ichioka, Y. Wavelength-multiplexing diffractive phase element with quantized phase structure. *Opt. Rev.* **2001**, *8*, 245–248. [CrossRef]

14. Jesacher, A.; Bernet, S.; Ritsch-Marte, M. Colour hologram projection with an slm by exploiting its full phase modulation range. *Opt. Express* **2014**, *22*, 20530–20541. [CrossRef] [PubMed]

15. Wang, J.; Pang, H.; Zhang, M.; Shi, L.; Cao, A.; Deng, Q. Design method for multi-wavelength diffractive optical element. *Acta Opt. Sin.* **2015**, *35*, 1005002. [CrossRef]

16. Makowski, M.; Sypek, M.; Ducin, I.; Fajst, A.; Siemion, A.; Suszek, J. Experimental evaluation of a full-color compact lensless holographic display. *Opt. Express* **2009**, *17*, 20840–20846. [CrossRef] [PubMed]

17. Yue, W.; Song, Q.; Yu, C.; Yue, W.; Zhu, J.; Situ, G. A simple method reconstructing colorful holographic imaging with gpu acceleration based on one thin phase plate. *Optik* **2015**, *126*, 3457–3462. [CrossRef]

18. Ito, T.; Okano, K. Color electro holography by three colored reference lights simultaneously incident upon one hologram panel. *Opt. Express* **2004**, *12*, 4320–4325. [CrossRef] [PubMed]

19. Xing, J.; Zhou, H.; Wu, D.; Hou, J.; Gu, J. Color hologram reconstruction based on single DMD. *Proc. SPIE* **2016**, *9684*, 1–6.

20. Li, X.; Chen, L.; Li, Y.; Zhang, X.; Pu, M.; Zhao, Z.; Ma, X.; Wang, Y.; Hong, M.; Luo, X. Multicolor 3D meta-holography by broadband plasmonic modulation. *Sci. Adv.* **2016**, *2*, e1601102. [CrossRef] [PubMed]

micromachines

MDPI

Article

Manufacturing Signatures of Injection Molding and Injection Compression Molding for Micro-Structured Polymer Fresnel Lens Production

Dario Loaldi [1],*, Danilo Quagliotti [1], Matteo Calaon [1], Paolo Parenti [2], Massimiliano Annoni [2] and Guido Tosello [1]

[1]   Department of Mechanical Engineering, Technical University of Denmark (DTU),
     2800 Kgs. Lyngby, Denmark; danqua@mek.dtu.dk (D.Q.); mcal@mek.dtu.dk (M.C.); guto@mek.dtu.dk (G.T.)
[2]   Politecnico di Milano, Department of Mechanical Engineering, 20156 Milan, Italy;
     paolo.parenti@polimi.it (P.P.); massimiliano.annoni@polimi.it (M.A.)
*   Correspondence: darloa@mek.dtu.dk; Tel.: +45-4525-4847

Received: 18 September 2018; Accepted: 7 December 2018; Published: 10 December 2018

Abstract: Injection compression molding (ICM) provides enhanced optical performances of molded polymer optics in terms of birefringence and transmission of light compared to Injection molding (IM). Nevertheless, ICM requires case-dedicated process optimization to ensure that the required high accuracy geometrical replication is achieved, particularly especially in the case of surface micro-features. In this study, two factorial designs of experiments (DOE) were carried out to investigate the replication capability of IM and ICM on a micro structured Fresnel lens. A laser scanning confocal microscope was employed for the quality control of the optical components. Thus, a detailed uncertainty budget was established for the dimensional measurements of the replicated Fresnel lenses, considering specifically peak-to-valley (PV) step height and the pitch of the grooves. Additional monitoring of injection pressure allowed for the definition of a manufacturing signature, namely, the process fingerprint for the evaluation of the replication fidelity under different process conditions. Moreover, considerations on the warpage of parts were related to a manufacturing signature of the molding processes. At last, the global part mass average and standard deviation were measured to correlate local geometrical replication performances with global part quality trends.

Keywords: manufacturing signature; process fingerprint; Fresnel lenses; injection compression molding; injection molding; micro structures replication; confocal microscopy; optical quality control; uncertainty budget; optimization

---

1. Introduction

Fresnel lenses are well-known optical devices with enhanced illumination properties combined with a compact and lightweight design. They are plano-convex optics, where the lens profile curvature is collapsed into a series of discontinuous frusto-conical grooves of reduced thickness. For mobile communication and electronic devices, as well as automotive and medical applications, the dimensions of the grooves lie in the micrometer scale and define the Fresnel lens optical performances [1–4]. Replication technologies represent the state-of-the-art solution to enable mass-manufacturing of polymer optics. The most cost-effective replication processes implemented currently in the industry are molding-based solutions such as injection molding (IM) and injection compression molding (ICM). Even though IM and ICM are established processes, it appears that they are still not clearly understood in the literature with regard to which process conditions provide the optimal results in terms of micro-geometrical replication for complex polymer optical systems.

Despite IM and ICM being interrelated processes, an additional phase in the operations sequence of ICM, i.e., compression, may have a substantial impact on the final product. In fact, while in IM, the polymer melt is injected in a closed mold cavity with almost the same dimensions and geometry of the final part, depending on the material shrinkage (i.e., the ones achieved when the two halves of the mold are forced against each other by the clamping force provided by the IM machine), in ICM, the melt is injected into an "open" cavity with the two mold halves initially being separated from each other. The mold is successively closed during a compression phase at the end of the operations sequence or during the injection phase [5]. The additional gap between the molds is called the compression gap, and it provides the necessary stroke to perform the compressing action. The compression gap is achieved in different ways. One of them consists of the design of molds with a so-called "vertical flash" area (see Figure 1a), in which the entire mold halves are kept separated. Another option employs a "compression frame" into the mold. The frame is built with spring systems, and an additional compression plate is mounted into the movable mold side (see Figure 1b). A more accurate but also more cost-intensive solution consists of the adoption of an independent "compression core" (see Figure 1c). In this configuration, the insert in the movable cavity is directly actuated for cavity closure, and it performs the compression. The selection of the most suitable solution depends on cost, and partly on geometry and target accuracy. The compression action is seen as an additional holding phase that is applied to the material inside the cavity. In IM, holding starts at the so-called switch/over point, i.e., the moment when part filling is considered complete, and the machine control switch from a filling control criterion (injection velocity, screw position, injection pressure, etc.) to a holding control criterion, generally the holding pressure. Holding ensures that the final part volume is equal to the one of the cavity, since the polymer material shrinks during cooling. In IM, holding is consequently a crucial quality step; however, for parts showing long flow length and/or small wall thickness, the required holding pressure to compensate for the pressure drop can be significantly high, as are the resulting residual stresses inside the part. The major advantage of ICM consists of the opportunity to reduce stresses in the part, as compression action provides an in-thickness holding effect on the cavity, ensuring a uniform distribution of stresses inside the cavity, while the part solidifies [6–8]. For this reason, ICM has been extensively favored to IM in the production of optical components. It is proven that it provides enhanced optical functionalities by reducing the birefringence and increasing the transmission efficiency, which are both correlated to the part internal stresses distribution. Birefringence effect is an optical property, which causes preferential light multiple refractions within the optics [9–11]. At the same time, the light transmission measures the effective spectral and power transmittance of the optics, i.e., a combined result of the material absorption of specific wavelengths, reflection, and surface loss. All of these aspects are heavily dependent on the geometrical and dimensional optical design, and on local surface defects such as roundings or drafts that are natural outcomes of the molding process [12,13]. Even though ICM leads to functional benefits, the compression phase increases IM complexity, because additional parameters must be taken into consideration for optimizing the process. Suzuki et al. [14] presented the importance of increasing the compression stroke in order to improve surface replication. On the contrary, Rohde et al. [15] discouraged the increment of the compression gap, as it reduces the transcription ratio of micro-structures. These controversial results in relation to the compression gap require further consideration, and they have received attention in recent research. Masato et al. [16] observed a significant interaction between the compression gap and the injection velocity, confirming that the optimization of the compression gap must take into account the overall polymer flow conditions. The study highlighted the negative impact of a large compression gap with respect to the replication homogeneity. A similar result was achieved by Chen et al. [17], who proved that a smaller compression gap induces a more uniform part shrinkage, and in another study, Chen et al. [18] validated that a larger compression gap increases part birefringence. The different theories regarding the compression gap selection can be justified by the results obtained by Shen et al. [19]. In their work, replication and birefringence improvements were initially noticed by increasing the compression gap. Such a gain was due to a larger compression energy provided to the polymer melt that increased the shear rate and reduced the viscosity. However, the advantage arose from a delay in the compression phase, due to either a large compression gap, or a slow compression speed (or both), that increased the material viscosity as the polymer cooled down before being compressed, generating heat dissipation with the mold and shear rate reduction. In these conditions, the polymer melt front formed a thick solid layer, so-called "skin layer" on the mold wall, limiting the melt flow more than what the compression could favor. This theory was supported by Ho et al. [20] observing an injection pressure and shear rate reduction with increased compression strokes. An additional result on compression parameters was given by Ito et al. [21], who found that the compression starting point and the compression gap are relevant factors affecting both optical performance and internal stresses. The optimal compression start was identified when the cavity injection was completed. The work of Kuo et al. [22] is one of the few studies mentioning the importance of the compression force. In the study, it was found that both IM and ICM parameters affect the replication of micro-features less when the compression force exceeds a certain threshold. Sortino et al. [23] verified the influence on the transcription ratio of IM factors such as holding pressure, injection velocity, and mold temperature in ICM. Their study demonstrated that the statistical effect of the IM factors was reduced in ICM. A confirmation was given by Han et al. [24], who discovered that the holding pressure could be reduced in ICM up to 50% with respect to IM, thanks to a more uniform cavity pressure distribution that is achieved with the compression phase. In the case of micro- or nano-structures replicated by ICM, a significant effect on the replication is also given by the mold temperature, as proved by Rytka et al., Nagato et al., and Chuan-Zhen et al. [25–27]. The effective local mold temperature also supports different replication quality conditions in dependence on the cavity design and the polymer melt flow [28–30]. Understanding the effective replication behavior of surface grooves is of paramount importance, to ensure the designed functionality of polymer optics, such as Fresnel lenses. Moreover, the complexity of the lenses' features demands for dedicated quality control criteria. For example, the so-called "interference by adjacent step" is a Fresnel lens efficiency, loss due to its stepped discontinuous profile [31]. In some cases, it is possible to reduce the efficiency loss by designing total internal reflection (TIR) lenses [32–34]. However, molding-based processes are not always capable of reproducing the ideal design, e.g., because of minimum draft angles required for de-molding [1]. In addition, the sharp edges of the micro-stepped grooves cannot be fully replicated, producing rounded features that reduce the overall optical performance [35–37]. In general, functional optical tests based on photogrammetry ensure the correct optical functionality. Such tests are robust and investigate whether optical aberrations occur while operating the lens. From those results, it is possible to reconstruct the lens geometry when an optical model is available. Nevertheless, such tests do not distinguish whether aberration occurs due to material dependent degradation or geometrical/dimensional imperfections occurring in the manufacturing process. The identification of manufacturing signatures, i.e., the link between a measurable feature of the final part geometry and the individual manufacturing process conditions, allows for a comprehensive understanding of the production steps, and ensures effective and efficient optimization solutions [38,39].

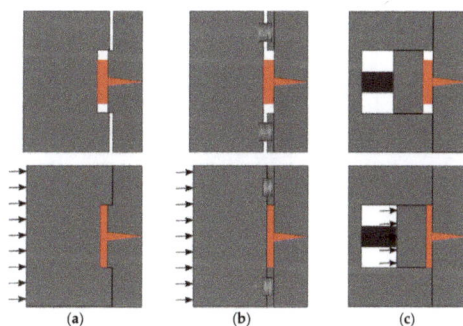

Figure 1. Different injection compression molding (ICM) mold closures before (upper picture) and during compression (lower picture), and their schematic architectures: (**a**) generating a "vertical flash" area; (**b**) using a spring-connected "compression-frame"; (**c**) adopting an actuated compression die.

Dedicated geometrical metrology is needed for the assessment of a manufacturing signature. Tactile measuring equipment is still extensively exploited, even though they can generate scratches on the lenses surface and generally require long set-up time and suffer accuracy loss in PV measurements [40,41]. Alternatively, non-contact optical solutions such as 3D optical microscopes can also be adopted for the scope [42]. Nevertheless, the high transparency of the material prevents the possibility of using focus variation systems or contrast-based microscopes. Similar limitations of these techniques are observed when optical or near-optical surface roughness (i.e., down to single digits to tens of nanometers, respectively) is measured [43,44]. In addition, setting up a scatterometry-based inspection is challenging, as the tips and roots of the Fresnel surface severely manipulate the scattering properties of the specimen [45]. In this study, the low aspect ratio surface micro-grooves of a Fresnel lens were investigated using a confocal microscope. The microscope principle is well known for its flexibility and the possibility to have both lateral and vertical resolutions in the sub-micrometer level. In this work, the identification of different manufacturing signatures in the production of Fresnel lenses is tackled. To do so, an initial metrological procedure with a detailed uncertainty budget is proposed, to evaluate the lens surface micro-feature replication. The methodology is proposed for two different materials, providing robust applicability for the procedure. The objective of this work is to provide a comprehensive methodology for the quantitative evaluation of IM and ICM performances, based on manufacturing signatures that address micro-replication quality. The four different manufacturing signatures (micro-replication accuracy, warpage, injection pressures, and part mass) were applied, providing a methodology for the optimization of IM and ICM. These four manufacturing signatures are employed, and their respective results are compared simultaneously as drivers of the optimization process for micro-structured optical parts manufacturing. The methodology, based on a metrological approach, provides a robust guideline for the effective molding of high precision polymer optics.

2. Materials and Methods

2.1. Device under Investigation

The component under investigation is the aspheric-corrected square Fresnel lens shown in Figure 2. Tolerance specifications are allocated to surface groove peak-to-valley (PV), as presented in a previous study [46]. The materials employed for the experimentation were a cyclo-olefin polymer (COP) commercially available under the trade name Zeonex® E48R, produced by Zeon© company, Tokyo, Japan, and a polymethyl methacrylate (PMMA), traded under Altuglas® V825T, produced by Arkema©, Colombes, France. Their viscosities (a) and pressure-specific volume-Temperature (pvT) curves (b) are reported in Figure 3. The data were collected from the Moldflow® software database, version 2018, by Autodesk® (San Rafael, CA, United States).

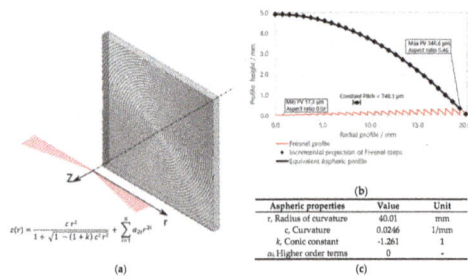

Figure 2. Representation of the studied Fresnel lens: (**a**) 3D view of the flat, squared, 40 mm × 40 mm Fresnel surface aperture, with a section view of the axis-symmetric radial profile, which follows a stepped aspheric profile as described by the conic equation; (**b**) a highlight of the section profile in the z, r plane with indication of the peak-to-valley (PV) nominal specifications in comparison with the equivalent continuous aspheric curvature; (**c**) summary table of the aspheric properties of the considered lens.

(a) Viscosity   (b) pvT

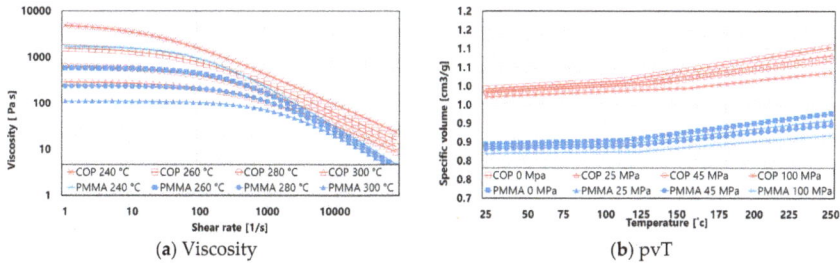

Figure 3. Cyclo-olefin polymer (COP) and polymethyl methacrylate (PMMA) material properties: (**a**) shear rate-dependent viscosity at different temperature values; (**b**) temperature-dependent specific volumes at different pressure values. Data collected from the Moldflow® software database, version 2018, by Autodesk® (San Rafael, CA, USA).

2.2. Injection Molding and Injection Compression Molding Machine

Experiments were performed on an injection molding machine (Negri Bossi©, Cologno Monzese, Milano, Italy) equipped with a reciprocating injection screw having a diameter of 32 mm, capable of a maximum clamping force of 600 kN. The clamping unit was equipped with the toggle clamp system represented in Figure 4a. The unit had a control on the position of the toggle with a repositioning error of 0.1 mm. The closure between the fixed and movable mold plates was measured at the two different positions of the toggle. In Figure 4c, a strong linear correlation (R^2 = 99.8%) was found between the toggle positions and the mold halves measured the closure with a resulting precision on the gap between the plates corresponding to 0.01 mm. In this way, the compression gap was controlled with the same level of precision, which is of central importance since the closing action of the mold compresses the melt inside the cavity. Furthermore, the compression speed is limited by the machine dynamics, and different average compression velocities and times for different starting compression gaps are reported in Figure 4d.

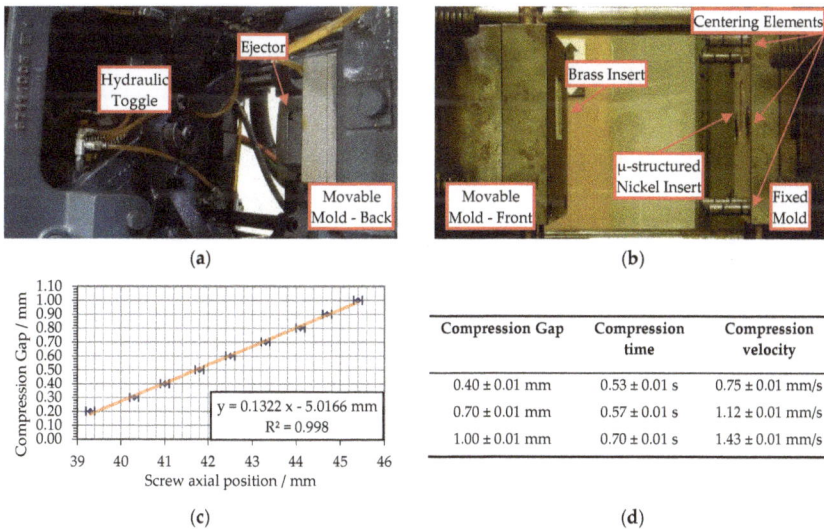

(a)

(b)

(c)

Compression Gap	Compression time	Compression velocity
0.40 ± 0.01 mm	0.53 ± 0.01 s	0.75 ± 0.01 mm/s
0.70 ± 0.01 mm	0.57 ± 0.01 s	1.12 ± 0.01 mm/s
1.00 ± 0.01 mm	0.70 ± 0.01 s	1.43 ± 0.01 mm/s

(d)

Figure 4. Injection compression molding solution: (**a**) toggle clamp unit with its hydraulic circuit; (**b**) movable (left) and fixed (right) mold plates in open configuration; (**c**) linear regression of the toggle unit position against the mold plates' closure to control the compression gap; (**d**) summary table of the compression gap, compression time, and average compression velocity.

2.3. Metrology and Uncertainty

The employed confocal instrument was a commercially available microscope with the trade name Lext OLS4000, manufactured by Olympus, Tokyo, Japan. Measurements were performed in the central location of the lens, as shown in Figure 5, using the microscope 20× standard magnification objective (measurement parameters summarized in Table 1). The lens showed a global squared dimension of 40 mm × 40 mm. The center of the lens was selected, as its near contour allowed for the evaluation of warpage and replication differences, depending on the features geometry, symmetry, and location with respect to the lens center (where the optical axis lies). Stitching of five different images was executed with a 3.712 mm × 0.640 mm sample area of the lens center. In this area, the first groove of the Fresnel lens was assessed. The groove had the lowest aspect ratio in the lens (17.3 μm/748.1 μm); nonetheless, the feature is the shortest in the lens design. The stitching overlapping factor between each image was 20% of the area of a single scan. An image size of 5328 × 913 pixels was achieved, with a resolution below the diffraction limit for the X-axis and above it for the Y one. The overall expanded uncertainty related to measurements was guided by ISO 15530-3:2011 [47] and ISO 14253-2:2007 [48]. The individual uncertainty contributors are reported in Table 2 for step height, and in Table 3 for pitch measurements, and they were calculated as described in a previous study [46]. The traceability was established through calibrated gauge blocks. A step height of (14.45 ± 0.26 μm) was obtained wringing together two gauge blocks, and it was related to the height of the lenses' grooves, while a gauge block with a value of 1500 ± 0.08 μm was related to the pitch measurements. In this last case, the measurements were made by stitching three fields of view in the X direction, both for the measured lenses and the reference standards.

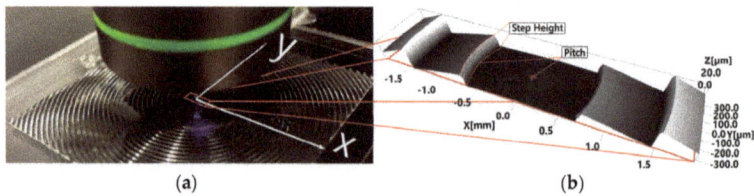

Figure 5. 3D measurements performed with the laser scanning confocal microscope: (a) Measurement set-up using the 20× objective; (b) Resulting 3D view of the sampled image.

Table 1. Measurement set-up of the laser scanning confocal microscope for the studied Fresnel lens.

Objective		Image Properties	
Field of view	640 × 640 μm²	Dimensions	3712 × 640 μm²
Numerical aperture	0.6	Image size	5328 × 913 pixels²
XY diffraction limit	0.412 μm	X pixel size	0.120 μm
Stitching overlap	20%	Y pixel size	0.701 μm
Z spatial resolution	0.030 μm		

Table 2. Step height measurement uncertainty budget for COP and PMMA in injection molding (IM) and ICM, of a nominal grooves' height of 17.3 μm. Measurements are calibrated with a gauge step of 14.45 μm.

Uncertainty Contributor	IM—COP	ICM—COP	IM—PMMA	ICM—PMMA
$u_{cal,z}$ (calibration artefact)	0.26 μm	0.26 μm	0.26 μm	0.26 μm
$u_{p,z}$ (instrument repeatability)	0.03 μm	0.03 μm	0.03 μm	0.03 μm
$u_{b,z}$ (instrument thermal)	0.00 μm	0.00 μm	0.00 μm	0.00 μm
$u_{wp,z}$ (part repeatability)	0.13 μm	0.13 μm	0.11 μm	0.11 μm
$u_{wt,z}$ (part thermal)	0.01 μm	0.01 μm	0.04 μm	0.04 μm
$u_{form,z}$ (form error)	0.17 μm	0.17 μm	0.03 μm	0.05 μm
k (coverage factor)	2	2	2	2
U (exp. Uncertainty)	0.7 μm	0.7 μm	0.6 μm	0.6 μm

Table 3. Pitch measurements' uncertainty budget for COP and PMMA in IM and ICM, of a nominal grooves' width of 748.1 μm. Measurements are calibrated with a gauge block thickness of 1500 μm.

Uncertainty Contributor	IM—COP	ICM—COP	IM—PMMA	ICM—PMMA
$u_{cal,z}$ (calibration artefact)	0.08 μm	0.08 μm	0.08 μm	0.08 μm
$u_{p,z}$ (instrument repeatability)	1.09 μm	1.09 μm	1.09 μm	1.09 μm
$u_{b,z}$ (instrument thermal)	0.01 μm	0.01 μm	0.01 μm	0.01 μm
$u_{wp,z}$ (part repeatability)	1.12 μm	1.12 μm	1.43 μm	1.43 μm
$u_{wt,z}$ (part thermal)	0.45 μm	0.45 μm	0.83 μm	0.84 μm
$u_{form,z}$ (form error)	0.99 μm	1.45 μm	0.73 μm	1.27 μm
k (coverage factor)	2	2	2	2
U (exp. Uncertainty)	3.8 μm	4.4 μm	4.2 μm	4.7 μm

2.4. Design of Experiments

The injected polymer volume and dosage were calibrated with preliminary injection molding short shots experiments, considering a screw injection velocity of 40 mm/s. An optimal switchover was defined for an injection screw position of 10 mm from the end stroke position. The switchover point was kept constant for both COP and PMMA in all process conditions. For all the experiments, the compression phase started after the injection, i.e., when the reciprocating screw reached the optimal switchover position of 10 mm. An initial factorial experimental campaign was performed to investigate the interaction between compression and holding phase, with the aim of analyzing which of the two process phases had a larger influence on the manufacturing signature, for the two analyzed materials:

a.   IM without holding pressure;
b.   IM with holding pressure;
c.   ICM without holding pressure;
d.   ICM with holding pressure.

Compression and post-filling holding phases were alternately switched on and off. Table 4a presents the factorial design. In this case, the compression gap was kept constant at 0.7 mm. All of the other process parameters are presented in Table 4b. A further optimization design was performed for the case of COP material based on the initial experimental campaign results. Aiming to validate the possibility to use the methodology as optimization tool of process parameters settings in ICM, only COP was chosen, as the two materials had different viscosity and processing requirements. Compression gaps and holding pressure levels were varied from 0.4 mm to 1.0 mm, and from 250 bar to 450 bar respectively, as shown in Table 5a. The compression starting point, switchover, and injection velocity were not changed, and kept as in the previous experimentation (see Table 5b). Thermal conditions such as melt and mold temperatures were not varied through the two experimental sessions. The melt temperature for the COP and PMMA materials was set respectively at 280 °C and 260 °C, being the viscosity of the material that was comparable at low shear rates for those temperature values.

The mold's temperature controller was set according to previous experimentation and the material manufacturer suggested values. The measured mold temperature was constant on both fixed and movable sides during the whole experimentations, being (105 ± 3) °C for COP and (93 ± 3) °C for PMMA. The temperature of the mold was not changed, and considered an experimental constant parameter, as it strongly affects the process cycle time.

Table 4. Screening experiments investigating the effect of compression and holding phases as independent process stages of IM and ICM for PMMA and COP materials: (**a**) with three factors on two levels; (**b**) constant parameters.

(a)			(b)	
Factors	Low Level	High Level	Parameters	Value
Compression	OFF	ON	Injection Velocity	40 mm/s
Holding	OFF	ON	Switch/over	10 mm
Material	PMMA	COP	Compression gap	0.7 mm
			Holding pressure	450 bar
			T melt COP	280 °C
			T melt PMMA	260 °C
			T mold COP	105 ± 3 °C
			T mold PMMA	93 ± 3 °C

Table 5. Optimization experiment investigating the effect of compression gap and holding pressure in the ICM of the COP material: (**a**) with two factors on two levels; (**b**) constant parameters.

(a)			(b)	
Factors	Low Level	High Level	Parameters	Value
Compression gap	0.4 mm	1.0 mm	Injection velocity	40 mm/s
Holding pressure	250 bar	450 bar	Switchover	10 mm
			T melt COP	280 °C
			T mold COP	105 ± 3 °C

3. Results

Ten different runs were performed, and three samples were extracted, measured, and averaged for each condition. The quality control was performed on the absolute dimension measurements of the pitch and peak-to-valley (PV) step height of the Fresnel lens' central grooves. The warpage was investigated, considering the central profile of the sampled images. Injection pressure results were analyzed separately for the filling and holding phases. Average and mass standard deviations are presented as part of global quality features.

3.1. Absolute Dimensions

Average absolute dimension measurements and uncertainties are reported in the form of interaction plots. When analyzing the step height (Figure 6a), it is possible to observe that for the case of IM without holding, the use of compression provided a higher step height replication for the COP material. On the contrary, the holding phase increased the step height replication in ICM for PMMA, as shown in Figure 6b. The other shown conditions are statistically equivalent, due to the relatively high measurement uncertainty with respect to the process deviations. In the case of the COP step height, the absolute deviations from nominal specifications ranged from a maximum value of (1.5 ± 0.7) μm, in the case of IM without holding, to a minimum of (0.1 ± 0.7) μm in the case of ICM.

For PMMA, absolute deviations from nominal specifications are at a maximum value of (0.8 ± 0.6) μm in the case of ICM without holding, and at a minimum of (0.1 ± 0.6) μm in the case of IM. Considering the measured values, the compression improved the pitch replication for both COP and PMMA material, as shown in Figure 7a,b. The holding effect was significant and increased the pitch average only for COP in ICM. When employing PMMA, an opposite effect of holding was observed, as shown in Figure 7b. It reduced the average pitch in ICM with statistical significance. In this case, the reduction of pitch was attributed to the higher part of the shrinkage that the PMMA samples underwent in comparison to COP for the considered processing conditions. For all these conditions, the COP maximum absolute deviations from the target values were (13.9 ± 3.8) μm in case of IM without holding, which reduced to a minimum value of (3.3 ± 4.4) μm for ICM. Analyzing PMMA, the absolute maximum deviation from the nominal value was (11.1 ± 4.2) μm for IM, and the minimum one (3.7 ± 4.2) μm for IM without holding. The experimental factors' effects are plotted into the main effect plots in Figure 8a for step height, and Figure 8b for pitch. The addition of compression to IM led to improvements of both pitch and step height replications. PMMA showed an average higher replication than COP, which can be justified by its lower viscosity at the same processing temperature and pressure. The holding phase was responsible for a higher step height replication of the features, while the pitch of the grooves was more dependent on the material and compression.

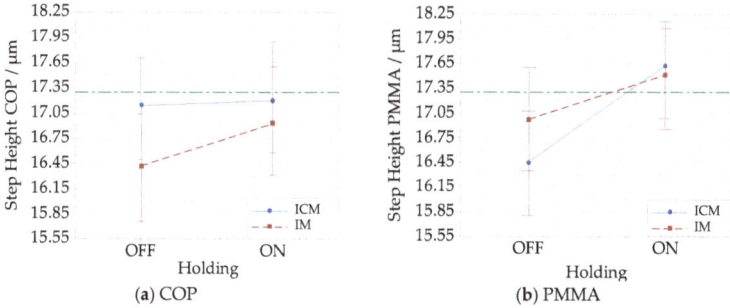

Figure 6. Interaction plots showing the replication of absolute dimensions of the Fresnel lens' grooves with a nominal step height of 17.3 μm. A measurement uncertainty is added to the results obtained for IM without the holding pressure, IM with holding pressure, ICM without holding pressure, and ICM with holding pressure: (**a**) COP material; (**b**) PMMA material.

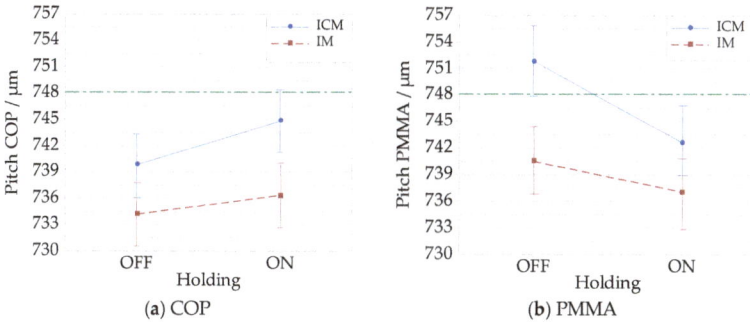

Figure 7. Interaction plots showing the replication of the absolute dimensions of Fresnel lens' grooves with a nominal width of 748.1 μm. The measurement uncertainty is added to the results obtained for IM without holding pressure, IM with holding pressure, ICM without holding pressure, and ICM with holding pressure: (**a**) COP material; (**b**) PMMA material.

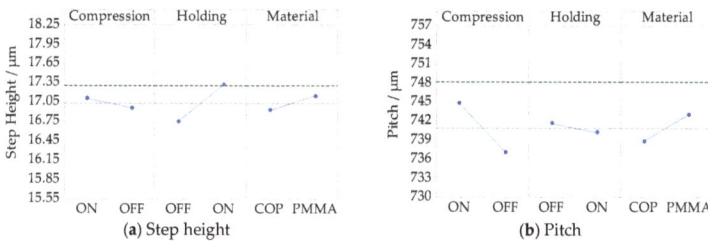

Figure 8. Main effects plot of compression, holding, and material on the average replication of the absolute dimensions of a Fresnel lens' groove with a nominal height of 17.3 μm and width 748.1 μm: (**a**) effects on step height; (**b**) effects on pitch.

A further analysis regarding the interaction between the holding pressure and the compression gap was carried out in the optimization experimental campaign for the COP material. The related results are reported in Figure 9a for step height and Figure 9b for pitch. At a high holding pressure (450 bar), the replication of the step height and pitch was not affected by the compression gap level.

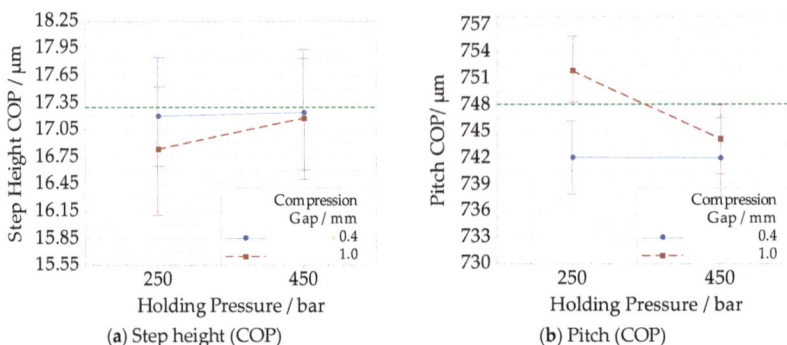

Figure 9. Interaction plots showing the replication of absolute dimensions of the Fresnel lens' groove with a nominal height of 17.30 µm and a width of 748.10 µm. The measurement uncertainty is added for the optimized ICM on compression gaps and holding pressures levels for the COP material: (**a**) to the step height; (**b**) to the pitch.

As shown in Figure 9a, the absolute deviation from the nominal step height is equal to (0.1 ± 0.7) µm for both compression gap levels. For the replicated pitch (see Figure 9b), the absolute deviation was (6.2 ± 4.4) µm with a 0.4 mm compression gap, and (4.1 ± 4.4) µm with a 1.0 mm compression gap. When the holding pressure was lower (250 bar), the effect of the compression gap was higher for both pitch and step height replication. The results also show another trend, that when compression gap was kept at a low level (0.4 mm), the replication difference related to the holding pressure levels was negligible. The reduction of pitch was attributed to the lower shrinkage occurring at high holding pressure level. These combined results indicated that the compression gap and the holding pressure levels should not be considered as independent factors, and that they should be optimized simultaneously to ensure higher replication quality.

3.2. Warpage

Form deviations, named as warpage, are generally associated with uneven cooling and differential shrinkage of the part, as well as deformations induced by the demolding of the part from the mold. In the present case, warpage attributed to unbalanced shrinkage was caused by different process settings, and was investigated along the orthogonal direction with respect to the main symmetry axis of the part, i.e., orthogonally to the melt flow direction (left/right).

Warpage as a manufacturing signature on the produced Fresnel lenses was analyzed on the residuals from the nominal profile geometries (Figures A1–A3 in Appendix A). Residuals were filtered using a median filter, and fitted with second-order polynomials. The selection of the degree of the polynomials was made on the joint inspection of the coefficients of determination (R^2) and the significant coefficients in the regressions of several-order polynomials, eventually choosing the best behavior. Warpage individual results are reported in Table 6.

Table 6. Results summary of part warpage as the maximum absolute deviation in the evaluated area, and as a quadratic coefficient of regression of the evaluated residuals.

Factors	Max Warpage μm	Quadratic Regression Coefficient μm/mm²	Factors	Max Warpage μm	Quadratic Regression Coefficient μm/mm²
IM without holding; COP	2.0	−0.6	ICM; 250 bar, 0.4 mm	1.1	0.3
IM; COP	0.7	0.2	ICM; 250 bar, 1.0 mm	2.0	0.4
ICM without holding; COP	4.3	0.9	ICM; 450 bar, 0.4 mm	1.0	0.2
ICM; COP	2.0	0.4	ICM; 450 bar, 1.0 mm	1.1	0.2
IM without holding; PMMA	1.6	0.4			
IM; PMMA	1.9	0.5			
ICM without holding; PMMA	106.9	39.5			
ICM; PMMA	2.2	0.6			

In the first analysis, for the COP material, a maximum absolute deviation of 4.3 μm was observed for the case of ICM without holding (Figure A1f); for the case of PMMA, the maximum observed deviation was 106.9 μm again for the case of ICM without holding (Figure A2f). In this last case, the form error was not the only defect associated with the part. A further analysis showed that the combined effect of warpage and of an air trap occurred in the central region of the lens. ICM without holding produced the lower form of replication and favored air traps during the part filling. In this case, a higher initial cavity volume ensured the possibility of performing compression; nonetheless, this fact resulted in a higher cavity pressure drop to produce the same filling conditions (as shown in Section 3.3. The shape of the Fresnel lens' grooves promoted the air trap, starting from the lens' center, where the slopes of the grooves changed with respect to the melt front direction, and the air stagnation area was favored (Figure 10).

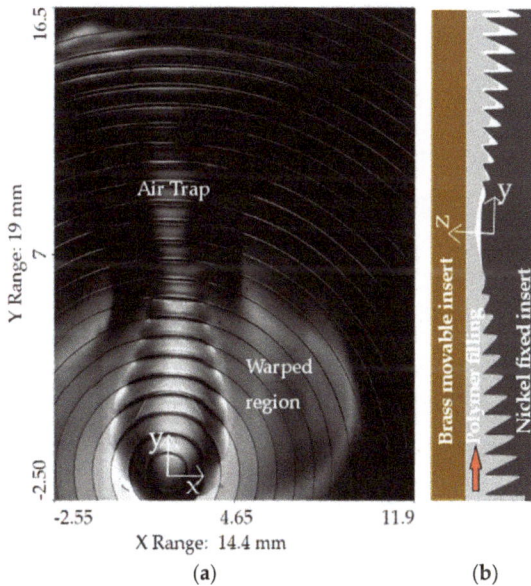

Figure 10. Combination of warpage and air trap in the replication of PMMA in ICM without holding: (**a**) Top view of the defect in the central location of the lens; (**b**) schematic visualization of air trap formation inside the mold cavity during the filling phase.

Warpage was also inspected by observing the quadratic regression coefficients. Since they reduced when considering IM instead of ICM, in the first experiments, compression favored a less stable shrinkage of the parts. This behavior can be recognized in Figure A1d,h for COP, and Figure A2d,h for PMMA, respectively.

When ICM was optimized using a high holding pressure level (450 bar), for both compression gap levels (0.4 mm–1.0 mm), the quadratic regression coefficients were at a minimum value of 0.2 $\mu m/mm^2$ (Figure A3f,h), indicating that warpage could be minimized in ICM with respect to IM. In the optimization case, the effect of compression gap was recognizable only for a low holding pressure (250 bar), as the coefficient increased from 0.3 $\mu m/mm^2$ in the case of a low compression gap (0.4 mm) (Figure A3b), and to 0.4 $\mu m/mm^2$ in the case of a long compression gap (1.0 mm) (Figure A3d).

The interaction of the two factors supports the previous observations, and is in accordance with the step height results described in Section 3.1 by (Figure 9a). Finally, the case of a larger compression gap resulted in a greater pressure drop in the cavity (Section 3.3), not only reducing the absolute replication of the features, but also increasing the warpage of the parts.

3.3. Injection Pressure

Injection pressure profiles over time were investigated as fingerprints of the different process settings [49]. Plastic pressure was measured in the injection chamber before the nozzle with a pressure transducer (Dynisco® Europe GmbH, Heilbronn, Germany, model MDT465C) and a sampling rate of 1 kHz. This pressure was equivalent to the pressure of the hydraulic circuit that moved the ram and was displayed in the machine control user display, and it is shown in Figure 11. The pressure was integrated over time (P_{work}) on a defined time period, according to Equation (1):

$$P_{work} = \sum_{i=1:n-1} \cdot (P_i + P_{i+1}) \cdot \Delta t \cdot 1/2 \tag{1}$$

where n is the total number of sampled pressure points, P_i and P_{i+1} are consecutive injection pressure values, and Δt is the pressure sampling time.

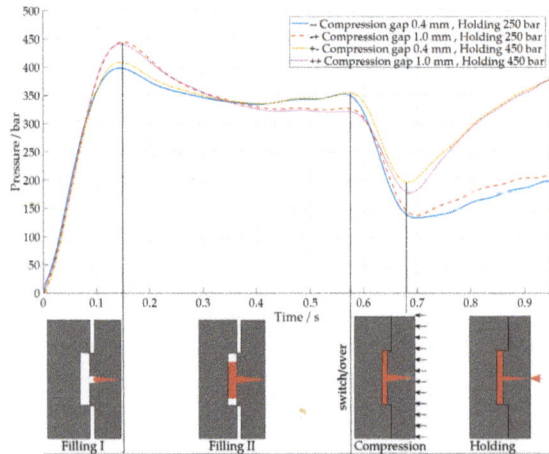

Figure 11. Injection pressure over time for a two-level compression gap (0.4–1.0 mm) and holding pressure (250 bar–450 bar). Compression gap of 1.0 mm and holding pressure of 250 bar; compression gap of 0.4 mm and holding pressure of 450 bar; compression gap of 1.0 mm and holding pressure of 450 bar: before and after switch/over.

P_{work} was calculated in two parts, one before (Figure 12a) and one after (Figure 12b) the switch/over point, which corresponded to an injection time of 0.57 s. Only an initial part of the holding phase was considered, up to an injection time of 0.95 s. The results are presented only for the optimization experiments in ICM using the COP material. Different compression gap levels modified the shapes of the pressure curves before the switchover. During the initial part of the filling, the pressure reached the maximum value depending on the compression gap level. A longer compression gap

(1.0 mm) induced a greater pressure drop, considering the constant injection speed. The maximum injection pressure rose from (398.2 ± 3.7) bar and (408.5 ± 3.8) bar for the low compression gap (0.4 mm), to (444.4 ± 3.4) bar and (440.7 ± 3.5) bar for the high compression gap (1.0 mm), respectively, for low (250 bar) and high holding pressures (450 bar). After this point, the pressure started to decrease as the cavity was further filled. In this case, a higher compression gap also sustained a greater pressure drop. The pressure reached a value at the switchover point of (351.4 ± 3.7) bar and (355.2 ± 3.8) bar for the low compression gap (0.4 mm), which further reduced to (326.7 ± 3.4) bar and (321.0 ± 3.5) bar for a high compression gap in conditions of low and high holding pressures, respectively.

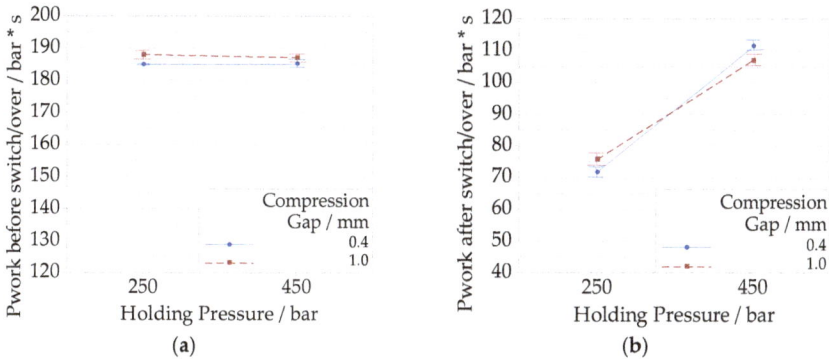

Figure 12. Interaction plots of the integral of pressure over time for the different compression gap and holding pressure levels: (**a**) calculated on a time interval that goes from zero to end of filling (before switch/over); (**b**) calculated on a time interval that goes from the end of injection until end of compression (after switch/over).

The integral of the curves over time (P_{work}) decreased for the low compression gap from (188.0 ± 0.1) bar·s and (187.1 ± 1.3) bar·s to (185.5 ± 1.4) bar·s and (185.2 ± 1.3) bar·s for low and high holding pressure respectively. The overall integration of pressure over time demonstrated that the major contributor of the pressure during filling was given by the compression gap before the switch/over point as cavity size was changed. The interaction plot in Figure 12a shows this result.

After the switch/over, P_{work} decreased from (107.1 ± 2.2) bar·s and (111.4 ± 2.2) bar·s in the case of high holding pressure to (75.7 ± 1.7) bar·s, and (71.7 ± 1.5) bar·s for the low holding pressure level, respectively, for a high and low compression gap. The transient from velocity to pressure control can be seen in Figure 11, as the molding machine increases the pressure to the desired holding control level with a certain lag.

In Figure 12b, the interaction between holding pressure and compression gap levels showed that the compression had a contribution, which depended on the pressure condition at the switchover, and on the energy that is introduced by the compression itself. Nonetheless, this contribution was 10 times lower than the one was given by the holding pressure control, as shown in the last part of Figure 11. As P_{work} could be considered as a direct indicator of the energy stored in the polymer during processing [49]. The maximum energy was achieved when the holding pressure was set at a high level (450 bar) and the compression gap was set to a low level of 0.4 mm.

3.4. Part Mass

The parts' mass was measured on all the 10 specimens per process condition, including the sprue. The interaction plots of the average part mass highlighted that a major effect on the average part mass was given by the holding phase for both COP (Figure 13a) and PMMA (Figure 13b). For COP, in case of IM, the average part mass increased from (13.463 ± 0.024) g to (14.283 ± 0.003) g when holding was performed, and from (13.493 ± 0.014) g to (14.298 ± 0.011) g in the same case for ICM.

Considering PMMA, the increment was from (15.808 ± 0.033) g to (17.168 ± 0.021) g in IM, and from (15.803 ± 0.036) g to (17.187 ± 0.050) g for ICM.

The effect of compression on the average part mass was lower than 30 mg, which was negligible with respect to the effect of the holding pressure. In the case, the standard deviation was considered as a quantitative way to characterize the process precision; for both the materials, precision increased as holding was performed. An interaction between compression and holding was observed. The lowest precision was observed when performing ICM for PMMA (Figure 14b). Meanwhile, the most favorable condition for COP was observed for IM (Figure 14a).

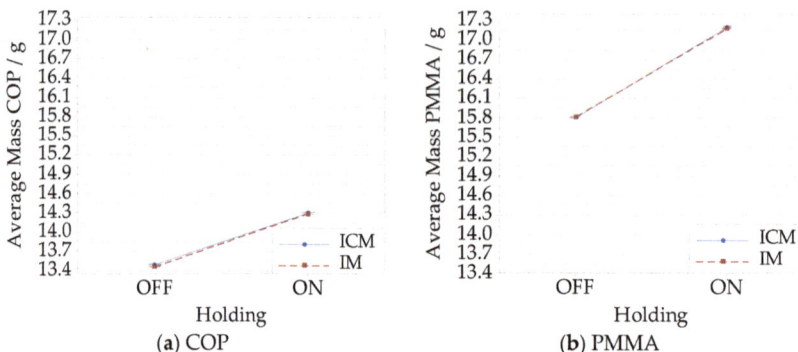

Figure 13. Interaction plots showing the average part mass obtained for IM without the holding pressure, IM and ICM without the holding pressure, and ICM for: (**a**) COP material; (**b**) PMMA material.

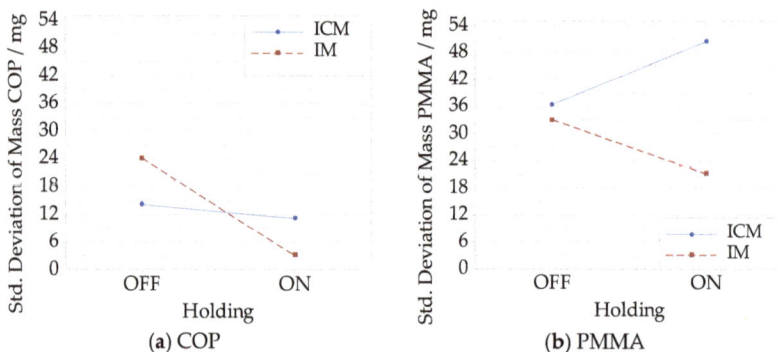

Figure 14. Interactions plots showing the part mass' standard deviation for different compression and holding combinations: IM without the holding pressure, IM with the holding pressure, ICM without the holding pressure, and ICM with the holding pressure: (**a**) COP material; (**b**) PMMA material.

When the process was optimized, the weight variation of ICM parts was reduced to 1 mg when a long compression gap and a low holding pressure were selected. This meant an increment in the precision of three times the previous IM result.

Increasing the holding pressure with a short compression gap resulted in a mass increment from (14.067 ± 0.008) g to (14.282 ± 0.004) g (i.e., +1.5% increase), and from (14.088 ± 0.001) g to (14.298 ± 0.015) g (i.e., +1.5% increase) in the case of long compression gap, see (Figure 15a). The interaction between the compression gap and the holding pressure (Figure 15b) was not negligible when considering the mass standard deviation. An increment of the holding pressure caused a reduction of the mass standard deviation when the compression gap was kept short. The opposite phenomenon was observed with the long compression gap.

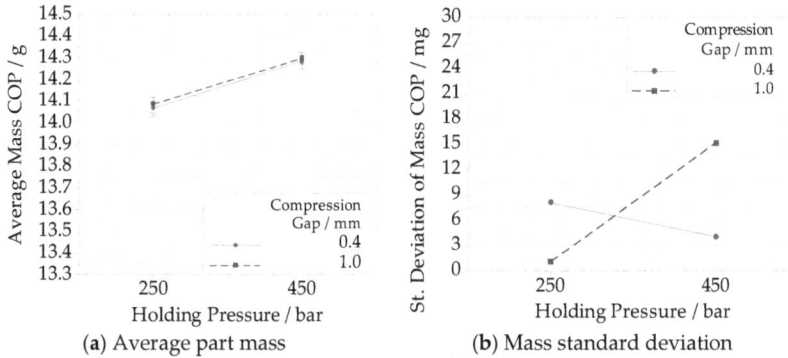

Figure 15. Interaction plots showing for the COP material, and different compression gap and holding pressure levels combinations: (**a**) average part mass; (**b**) mass standard deviation.

During compression, the action on the polymer inside the cavity works as an in-thickness force. On the other hand, during holding, the polymer is pushed from the injection nozzle through the injection point. The two actions on the polymer are of different natures, and the final filling conditions depend on the combined action of the two phases. The significance of the interaction of holding and compression was verified on both at the micro-level and at the global part level with their corresponding quality features.

4. Conclusions

The evaluation of the manufacturing signature of IM and ICM for different quality features in Fresnel lenses manufacturing was performed, investigating the replication of absolute dimensions, part warpage, injection pressure, and part mass in order to identify (both locally, i.e., at the optical micro feature level, and globally, i.e., at the part level). The most suitable quality criteria and manufacturing fingerprints for the optimization of the production have been determined and validated. The methodology is based on a quantitative metrological approach, and it allows for the comparison of the performance of IM and ICM on the experimental case of micro-structured optics. As a result, the adequate selection and setting of process parameters was found, enabling specified quality features to be achieved, in terms of both accuracy and precision. The main conclusions of the research can be summarized as follows:

- The replication of absolute dimensions in terms of groove step height and pitch improved from IM to ICM. A higher replication fidelity was achieved using the PMMA material. The compression phase had a larger influence on pitch values. An optimal condition for ICM was achieved when a higher holding pressure and a lower compression gap were selected.
- The holding phase was of paramount importance in both IM and ICM for the reduction of the warpage. The parts' warpage was described with second-order polynomials and it was related to differential shrinkages of the parts, due to different process conditions. The absence of the holding phase in ICM was detrimental. It increased the warpage and favored the formation of air traps, as shown in the case when processing PMMA. The optimal condition of ICM promoting less warpage occurred when a high holding pressure level was selected.
- The compression phase led to a pressure cavity variation over time, both during and after the filling phase. However, the main driver of pressure variation during the filling phase was the compression gap while its effect during the holding phase was overcome by the holding pressure. P_{work} was used as an indicator of the energy transferred to the polymer part during processing, and its monitoring served as a production manufacturing signature.

- The holding phase was the major contributor to variations in the average part mass. In this case, IM and ICM showed similar process precision. However, process precision, measured as a global part mass standard deviation, can be minimized in the optimized ICM case with a high compression gap and low holding phase levels.

From these conclusions, it was shown that ICM leads to advancements in terms of surface micro-replication, part warpage and process precision with respect to IM. However, particular attention has to be paid in setting the holding phase and the compression gap when carrying out the ICM process, in order to achieve high replication fidelity, as well as the required geometrical accuracy and precision.

Author Contributions: D.L. and G.T. conceptualized the work; performed the investigation, and managed the visualization. D.L., D.Q., M.C., P.P., M.A. and G.T. were all involved in data curation; methodology; writing—review & editing. D.L., D.Q. and G.T. completed the formal analysis. D.L. worked on writing—original draft. D.Q., M.C., P.P., M.A. and G.T. conducted supervision. G.T. was responsible for funding acquisition and project administration.

Funding: This research work was undertaken in the context of the research projects PROSURF and MADE DIGITAL. The PROSURF project ("Surface Specifications and Process Chains for Functional Surfaces", http://www.prosurf-project.eu/) is funded by the HORIZON 2020 program (Project ID: 767589) of the European Commission. MADE DIGITAL, Manufacturing Academy of Denmark (http://en.made.dk/), Work Package WP3 "Digital manufacturing processes", is funded by Innovation Fund Denmark (https://innovationsfonden.dk/en).

Acknowledgments: The collaboration from Eng. Igor Di Vora at Automotive Lighting Italia S.p.A. in connection with the injection and compression molding experiments is greatly acknowledged.

Conflicts of Interest: The authors declare no conflicts of interest.

Appendix A

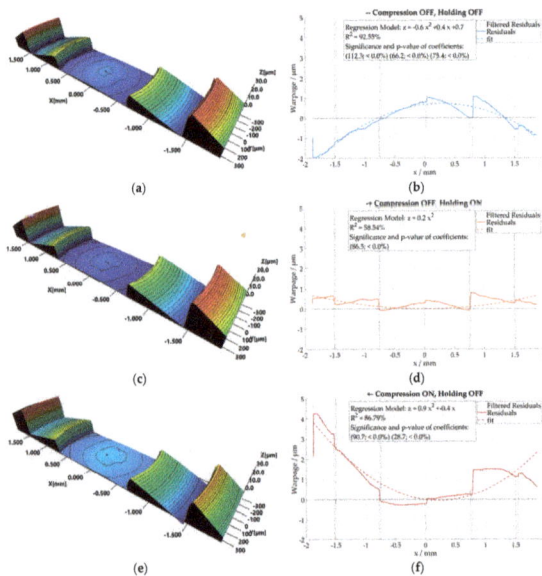

Figure A1. On the left, 3D view of the lens' central location, sampled as shown in Figure 5. On the right, the residuals of the respective x profiles in comparison with nominal geometry, sampled in the image center. The following process conditions are considered for the COP material results: (**a,b**) IM without holding; (**c,d**) IM; (**e,f**) ICM without holding; (**g,h**) ICM.

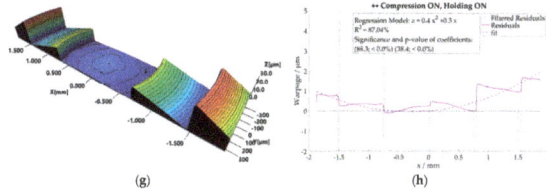

(g)

(h)

Figure A1. *Cont.*

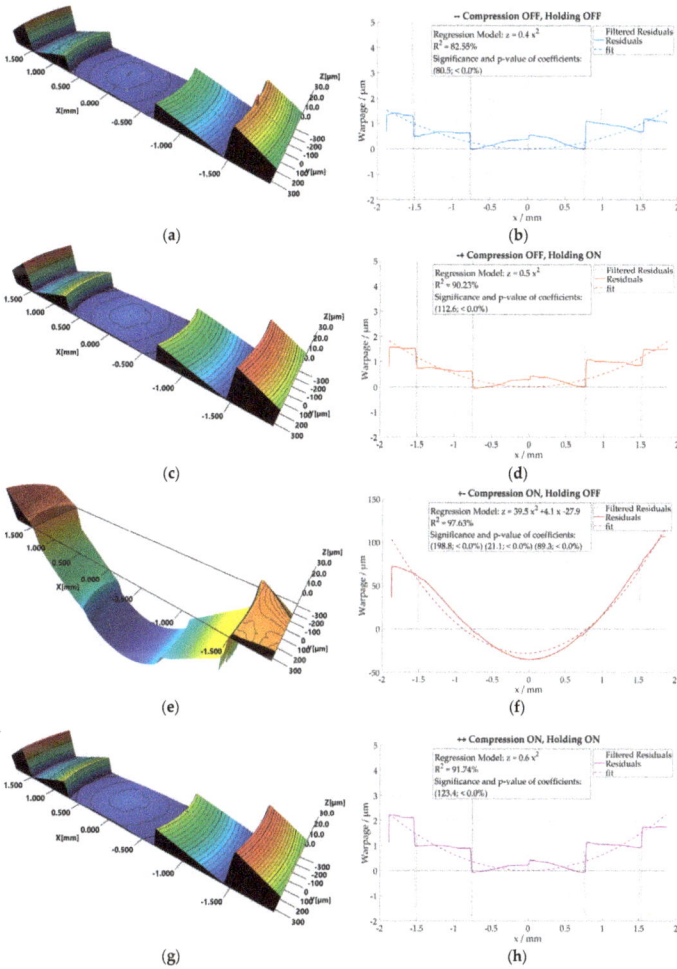

(a)

(b)

(c)

(d)

(e)

(f)

(g)

(h)

Figure A2. On the left, 3D view of the lens' central location, sampled as in Figure 5. On the right, the residuals of the respective x profiles, in comparison with nominal geometry sampled in the image center. The following process conditions are considered for PMMA material results: (**a**,**b**) IM without holding; (**c**,**d**) IM; (**e**,**f**) ICM without holding; (**g**,**h**) ICM.

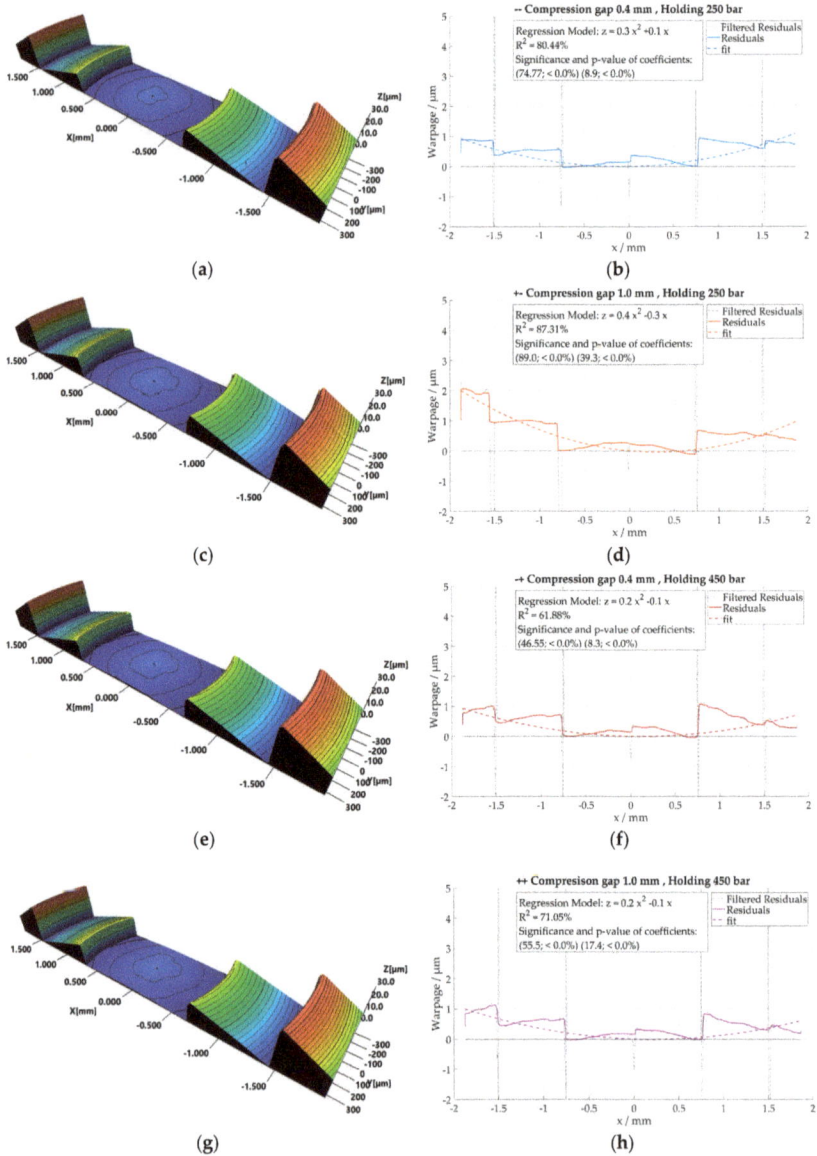

Figure A3. On the left, 3D view of the lens' central location sampled as in Figure 5. On the right, the residuals of the respective x profiles in comparison with nominal geometry, sampled in the image center. The following process conditions are considered for COP material results: (**a,b**) Compression gap 0.4 mm and holding pressure 250 bar; (**c,d**) Compression gap 1.0 mm and holding pressure 250 bar (**e,f**) Compression gap 0.4 mm and holding pressure 450 bar; (**g,h**) Compression gap 1.0 mm and holding pressure 450 bar.

References

1. Davis, A.; Kühnlenz, F. Optical Design using Fresnel Lenses Basic Principles and some Practical Examples. *Opt. Photonik* **2007**, *4*, 52–55. [CrossRef]
2. Aničin, B.A.; Babović, V.M.; Davidović, D.M. Fresnel lenses. *Am. J. Phys.* **1989**, *57*, 312–316. [CrossRef]
3. Egger, J.R. Use of Fresnel lenses in optical systems: some of some advantages and limitations. *Proc. SPIE* **1979**, *193*, 63–69. [CrossRef]
4. Edmund Optics Inc. Modeling of Fresnel lenses in commonly available optical design software. 2001.
5. Brinksmeier, E.; Riemer, O.; Gläbe, R. *Fabrication of Complex Optical Components*; Springer: Berlin/Heidelberg, Germany, 2013; pp. 21–38. ISBN 978-3-642-33000-1.
6. Cao, W.; Min, Z.; Zhang, S.; Wang, T.; Jiang, J.; Li, H.; Wang, Y.; Shen, C. Numerical Simulation for Flow-Induced Stress in Injection/Compression Molding. *Polym. Eng. Sci.* **2015**, *47*, 21–25. [CrossRef]
7. Kulkarni, S. *Robust Process Development and Scientific Molding*, 1st ed.; Carl Hanser Verlag GmbH & Co. KG: Munich, Germany, 2010; ISBN 978-3-446-42275-9.
8. Mayer, R. Precision Injection Molding. In *Optik & Photonik*, 1st ed.; Wiley-VCH Verlag GmbH & Co. KGaA: Weinheim, Germany, 2007; pp. 46–51. ISSN 1863-1460.
9. Lin, C.M.; Hsieh, H.K. Processing optimization of Fresnel lenses manufacturing in the injection molding considering birefringence effect. *Microsyst. Technol.* **2017**, *23*, 5689–5695. [CrossRef]
10. Griffiths, C.A.; Tosello, G.; Dimov, S.S.; Scholz, S.G.; Rees, A.; Whiteside, B. Characterisation of demoulding parameters in micro-injection moulding. *Microsyst. Technol.* **2014**, *21*, 1677–1690. [CrossRef]
11. Ito, H.; Suzuki, H.; Kazama, K.; Kikutani, T. Polymer structure and properties in micro- and nanomolding process. *Curr. Appl. Phys.* **2009**, *9*, 19–24. [CrossRef]
12. Michaeli, W.; Heßner, S.; Klaiber, F.; Forster, J. Geometrical Accuracy and Optical Performance of Injection Moulded and Injection-compression Moulded Plastic Parts. *CIRP Ann. Manuf. Technol.* **2007**, *56*, 545–548. [CrossRef]
13. Davis, A. Fresnel lens solar concentrator derivations and simulations. *Proc. SPIE* **2011**, *8129*, 81290J-1–81290J-15. [CrossRef]
14. Suzuki, H.; Takayama, T.; Ito, H. Replication Behavior for Micro Surface Features With High Aspect Ratio and Structure Development in Injection Compression Molding. *Int. J. Mod. Phys. Conf. Ser.* **2012**, *6*, 563–569. [CrossRef]
15. Rohde, M.; Derdouri, A.; Kamal, M.R. Micro replication by injection-compression molding. *Int. Polym. Process.* **2009**, *24*, 288–297. [CrossRef]
16. Masato, D.; Sorgato, M.; Lucchetta, G. Characterization of the micro injection-compression molding process for the replication of high aspect ratio micro-structured surfaces. *Microsyst. Technol.* **2016**, *23*, 3661–3670. [CrossRef]
17. Chen, S.C.; Chen, Y.C.; Peng, H.S.H.U. Simulation of Injection-Compression-Molding Process. Part 2. *J. Appl. Polym. Sci.* **1999**, *75*, 1640–1654. [CrossRef]
18. Chen, S.C.; Chen, Y.C.; Peng, H.S.; Huang, L.T. Simulation of injection-compression molding process, Part 3: Effect of process conditions on part birefringence. *Adv. Polym. Technol.* **2002**, *21*, 177–187. [CrossRef]
19. Shen, Y.K.; Chang, H.J.; Hung, L.H. Analysis of the Replication Properties of Lightguiding Plate for Micro Injection Compression Molding. *Key Eng. Mater.* **2007**, *329*, 643–648. [CrossRef]
20. Ho, J.-Y.; Park, J.M.; Kang, T.G.; Park, S.J. Three-Dimensional Numerical Analysis of Injection-Compression Molding Process Jae-Yun. *Polym. Eng. Sci.* **2012**, *52*, 901–911. [CrossRef]
21. Ito, H.; Suzuki, H. Micro-Features Formation in Injection Compression Molding. *J. Solid Mech. Mater. Eng.* **2009**, *3*, 320–327. [CrossRef]
22. Kuo, H.C.; Jeng, M.C. The influence of injection molding and injection compression molding on ultra-high molecular weight polyethylene polymer microfabrication. *Int. Polym. Process.* **2011**, *26*, 508–516. [CrossRef]
23. Sortino, M.; Totis, G.; Kuljanic, E. Comparison of injection molding technologies for the production of micro-optical devices. *Procedia Eng.* **2014**, *69*, 1296–1305. [CrossRef]
24. Han, S.; Jin, X. The Three Dimensional Numerical Analysis Of Injection-Compression Molding Process. *Proc. ANTEC Int. Conf.* **2011**, *52*, 1764–1769. [CrossRef]

25. Rytka, C.; Kristiansen, P.M.; Neyer, A. Iso- and variothermal injection compression moulding of polymer micro- and nanostructures for optical and medical applications. *J. Micromech. Microeng.* **2015**, *25*, 065008. [CrossRef]

26. Nagato, K.; Hamaguchi, T.; Nakao, M. Injection compression molding of nanostructures. *J. Vac. Sci. Technol. B* **2011**, *28*, 06FG10-1-4. [CrossRef]

27. Chuan-zhen, Q.; Yong-tao, W.; Jian, Z.; Ya-jun, Z. The Study of Injection Compression Molding of Thin-wall Light-guide Plates with Hemispherical Micro structures. *MEMS* **2012**, 447–450. [CrossRef]

28. Calaon, M.; Tosello, G.; Garnaes, J.; Hansen, H.N. Injection and injection-compression moulding replication capability for the production of polymer lab-on-a-chip with nano structures. *J. Micromech. Microeng.* **2017**, *27*, 105001. [CrossRef]

29. Tosello, G.; Hansen, H.N.; Gasparin, S.; Albajez, J.A.; Esmoris, J.I. Surface wear of TiN coated nickel tool during the injection moulding of polymer micro Fresnel lenses. *CIRP Ann. Manuf. Technol.* **2012**, *61*, 535–538. [CrossRef]

30. Roeder, M.; Schilling, P.; Hera, D.; Guenther, T.; Zimmermann, A. Influences on the Fabrication of Diffractive Optical Elements by Injection Compression Molding. *J. Manuf. Mater. Process.* **2018**, *2*, 5. [CrossRef]

31. Gale, M.T.; Khas, H. Efficiency of Fresnel lenses. *Microelectron. Eng.* **1981**, *9*, 173–183. [CrossRef]

32. Joo, J.Y.; Lee, S.K. Miniaturized TIR fresnel lens for miniature optical LED applications. *Int. J. Precis. Eng. Manuf.* **2009**, *10*, 137–140. [CrossRef]

33. Thanh Tuan, P. A Novel Technique to Design Flat Fresnel Lens with Uniform Irradiance Distribution. *Int. J. Energy Power Eng.* **2016**, *5*, 73–82. [CrossRef]

34. Chen, L.T.; Keiser, G.; Huang, Y.R.; Lee, S.L. A Simple Design Approach of a Fresnel Lens for Creating Uniform Light-Emitting Diode Light Distribution Patterns. *Fiber Integr. Opt.* **2014**, *33*, 360–382. [CrossRef]

35. Holthusen, A.K.; Riemer, O.; Schmütz, J.; Meier, A. Mold machining and injection molding of diffractive microstructures. *J. Manuf. Process.* **2017**, *26*, 290–294. [CrossRef]

36. Tosello, G.; Hansen, H.N.; Calaon, M.; Gasparin, S. Challenges in high accuracy surface replication for micro optics and micro fluidics manufacture. *Int. J. Precis. Technol.* **2014**, *4*, 122–144. [CrossRef]

37. Hansen, H.N.; Hocken, R.J.; Tosello, G. Replication of micro and nano surface geometries. *CIRP Ann. Manuf. Technol.* **2011**, *60*, 695–714. [CrossRef]

38. Mason, R.J.; Rahman, M.M.; Maw, T.M.M. Analysis of the manufacturing signature using data mining. *Precis. Eng.* **2017**, *47*, 292–302. [CrossRef]

39. Zahouani, H.; Mezghani, S.; Vargiolu, R.; Dursapt, M. Identification of manufacturing signature by 2D wavelet decomposition. *Wear* **2008**, *264*, 480–485. [CrossRef]

40. Berger, G.; Wendel, M.; Fair, W.; Cb, B.; Hobson, T. Optical Metrology of Freeforms and Complex Lenses. *Optik Photonik* **2018**, *13*, 40–43. [CrossRef]

41. Bergmans, R.H.; Kok, G.J.P.; Blobel, G.; Nouira, H.; Küng, A.; Baas, M.; Tevoert, M.; Baer, G.; Stuerwald, S. Comparison of asphere measurements by tactile and optical metrological instruments. *Meas. Sci. Technol.* **2015**, *26*. [CrossRef]

42. Nadim, E.H.; Hichem, N.; Nabil, A.; Mohamed, D.; Olivier, G. Comparison of tactile and chromatic confocal measurements of aspherical lenses for form metrology. *Int. J. Precis. Eng. Manuf.* **2014**, *15*, 821–829. [CrossRef]

43. Leach, R. *Optical Measurement of Surface Topography*; Springer: Berlin/Heidelberg, Germany, 2011.

44. Tosello, G.; Haitjema, H.; Leach, R.; Quagliotti, D.; Gasparin, S.; Hansen, H.N. An international comparison of surface texture parameters quantification on polymer artefacts using optical instruments. *CIRP Ann. Manuf. Technol.* **2016**, *65*, 529–532. [CrossRef]

45. Yeh, N. Analysis of spectrum distribution and optical losses under Fresnel lenses. *Renew. Sustain. Energy Rev.* **2010**, *14*, 2926–2935. [CrossRef]

46. Loaldi, D.; Calaon, M.; Quagliotti, D.; Parenti, P.; Annoni, M.; Tosello, G. Tolerance verification of precision injection moulded Fresnel lenses. *Proc. CIRP* **2018**, *75*, 137–142. [CrossRef]

47. ISO 15530-3:2011 Geometrical product specication (GPS) - Geometrical Product Specifications (GPS)—Coordinate Measuring Machine (CMM): techniques for determining the uncertainty of measurement. (ISO 15530-3:2013). (Geneva: International Organization for Standardization).

48. ISO 14253-2:2007 Geometrical product specication (GPS)-Inspection by measurement of workpieces and measuring equipment-Part 2: Guidance for the estimation of uncertainty in GPS measurement, in calibration of measuring equipment and in product verification (ISO 14253-2:2007). (Geneva: International Organization for Standardization).

49. Griffiths, C.A.; Dimov, S.; Scholz, S.G.; Hirshy, H.; Tosello, G. Process Factors Influence on Cavity Pressure Behavior in Microinjection Moulding. *J. Manuf. Sci. Eng.* **2011**, *133*, 031007. [CrossRef]

micromachines

MDPI

Article

Investigation of Product and Process Fingerprints for Fast Quality Assurance in Injection Molding of Micro-Structured Components

Nikolaos Giannekas [1,*], Per Magnus Kristiansen [2,3], Yang Zhang [1] and Guido Tosello [1]

[1]   Department of Mechanical Engineering, Technical University of Denmark, Produktionstorvet,
       Building 427A, DK-2800 Kgs. Lyngby, Denmark; yazh@mek.dtu.dk (Y.Z.); guto@mek.dtu.dk (G.T.)
[2]   Institute of Polymer Nanotechnology (INKA), FHNW University of Applied Sciences and Arts
       Northwestern Switzerland, School of Engineering, Klosterzelgstrasse 2, CH-5210 Windisch, Switzerland;
       magnus.kristiansen@fhnw.ch
[3]   Laboratory for Micro- and Nanotechnology, Paul Scherrer Institute, CH-5232 Villigen-PSI, Switzerland
*   Correspondence: nikgia@mek.dtu.dk; Tel.: +45-4525-4747

Received: 30 October 2018; Accepted: 13 December 2018; Published: 15 December 2018

Abstract: Injection molding is increasingly gaining favor in the manufacturing of polymer components since it can ensure a cost-efficient production with short cycle times. To ensure the quality of the finished parts and the stability of the process, it is essential to perform frequent metrological inspections. In contrast to the short cycle time of injection molding itself, a metrological quality control can require a significant amount of time and the late detection of a problem may then result in increased wastage. This paper presents an alternative approach to process monitoring and the quality control of injection molded parts with the concept of "Product and Process Fingerprints" that use direct and indirect quality indicators extracted from part quality data in-mold and machine processed data. The proposed approach is based on the concept of product and process fingerprints in the form of calculated indices that are correlated to the quality of the molded parts. A statistically designed set of experiments was undertaken to map the experimental space and quantify the replication of micro-features depending on their position and on combinations of processing parameters with their main effects to discover to what extent the effects of process variation were dependent on feature shape, size, and position. The results show that a number of product and process fingerprints correlate well with the quality of the micro features of the manufactured part depending on their geometry and location and can be used as indirect indicators of part quality. The concept can, thus, support the creation of a rapid quality monitoring system that has the potential to decrease the use of off-line, time-consuming, and detailed metrology for part approval and can thus act as an early warning system during manufacturing.

Keywords: precision injection molding; quality control; process monitoring; product fingerprint; process fingerprint

1. Introduction

In recent years, market and consumer needs have led to a shift in the design of complex products by focusing on product design for high volume, cost-effective manufacturing processes in many applications such as automotive components, communication, and medical devices particularly in micro-sized applications [1,2]. Injection molding is one of the processes that can ensure a cost efficient production with short cycle times. Therefore, it has been the favorite process for many manufacturers of cost-effective products and reportedly now accounts for 50% of the produced plastic parts [3].

In many sectors and especially in the medical field, numerous products have micro features with tight tolerances. Satisfying the product specifications and functional requirements for all injection-molded components is a difficult task and requires a highly stable process. Performing frequent metrological inspections is essential to ensure process stability and produced part quality assurance and to approve a production batch. However, metrological inspections require a great deal of resources and effort in comparison to the cost-effectiveness and short cycle time of the injection molding process. Due to the demand for tighter tolerances especially in micro parts and parts with micro features, process monitoring has been the object of many research efforts. The ultimate objective is the monitoring of an optimized process to reveal the occurrence of defects and to ensure that the process produces parts within the specification limits. Out-of-tolerance production can result in significantly increased production costs and a high scrap rate, which reduces the efficiency of the process especially in industrial situations where the components are left to rest for up to 48 h prior to a metrological inspection.

The current paper presents an innovative approach to part quality monitoring and control by proposing indices that serve as part quality indicators (QI) and as "Product and Process Fingerprints" that are based on both process and product data. The proposed approach takes two parallel tracks as follows.

First, the "Product fingerprint" track in which the use of micro-features positioned on the molded part and their replication quality is considered to be directly connected to the overall quality of the component. The correlation of the replication fidelity of these microfeatures on the runner with those of the part itself is explored. Recent research studies have provided examples of part features used for the correlation with part quality. Examples can be found in the use of the weld line positions to assess the quality of the molded part, as described by Tosello et al. [4], and the use of nano features placed on different areas of a component that provides the necessary indicators for rapid part quality assessment as discussed by Calaon et al. [5].

Second, the "Process fingerprint" track explores the use of transient process data that originates from the in-mold sensors and the machine control sensors together with data from the on board quality system of the injection molding machine for process monitoring. Multiple research studies have been conducted with different approaches in the field of sensor technology as a tool for process control and optimization in an attempt to decrease the intensive metrological inspections required for the approval of injection-molded parts. Some of the studies used in/on-mold sensors to regulate and monitor the process with promising results [6–9] with increased tooling costs. Mold separation (MS) monitoring with the use of a linear variable differential transformer (LVDT) is one of those techniques and can provided a reliable indicator for part weight and thickness [9]. Gao et al. [6] used a custom designed multivariate sensor (MVS) to monitor the quality of the injection-molded parts assuming that the part quality indicators (dimensions) can be tightly controlled when the in-mold process parameters are known.

Other studies make use of either data from external sensors placed on the molds or data from in-line measuring equipment to record indirect process parameter data in order to optimize the process with respect to the functional requirements. Johnston et al. have utilized an in-line multivariate optimization system for process control and optimization [10]. Yang et al. make use of digital image processing in-line with the process for defect detection purposes [11]. After detection of a defect, the detection algorithm feeds data to an algorithm for process optimization based on a model-free optimization (MFO) procedure.

Scientific research in this field is not limited to the use of in-line experimental process monitoring of conventional injection molding. In fact, the study by Wang et al. presented a work on warpage optimization with a numerical simulation procedure of dynamic injection molding and sequential optimization based on the Kriging surrogate model [3]. Other studies rely on the application of artificial neural networks (ANN) and genetic algorithms for optimization and monitoring of the process [12].

The previously mentioned approaches focus on the case of a tightly controlled and optimized process. However, the dimensional control of the injection-molded components is not considered directly. The resulting part quality is the main objective of any quality assurance system. Thus, coupling of the dimensional accuracy of the parts with the sensor data is required and this is addressed in the present research.

The approach presented in the current article is based on process and product fingerprints. The paper in divided into three sections. Section 1 is the Introduction. Section 2 includes the Materials and Methods in which the materials and methods used for the experiment, the test geometries, collected data, and analysis methods are presented. A test geometry with two cavities was fitted with a number of both functional and test micro structures to serve as "product fingerprint" candidates. Section 3 includes the Results and Discussion in which several process variables and signals were collected to extract candidates for "process fingerprints" using the analytical methods presented. The results of the analysis are presented and a discussion is commenced with the "product fingerprint" candidates' correlation of the test structure to the functional ones to be assessed. Similarly, the "process fingerprints" candidates were subject to correlation analysis with the features measurements to identify the most suitable ones to act as indicators of the overall quality of the part. The discussion proceeds with a comparison of process and product fingerprints so that the most suitable "fingerprints" for an in-line process monitoring and control system may be selected. Setting up such a system requires the coupling of the selected measurand/product quantity/fingerprint to the proper process fingerprint as shown in the procedure schematically outlined in Figure 1. The last section of the paper summarizes the findings and proposes fingerprints and a way to extract them.

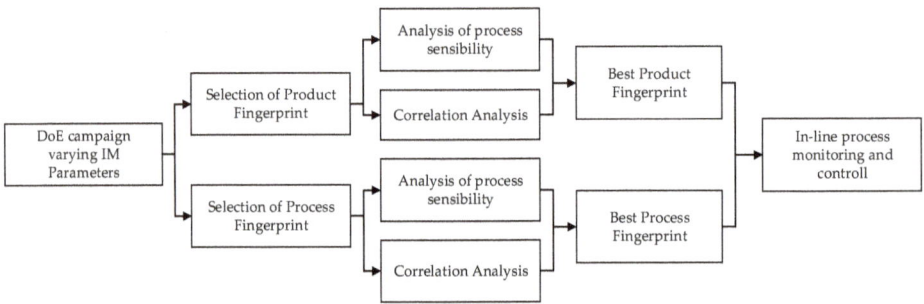

Figure 1. Flow chart for procedure for product and process fingerprint to measurand identification.

2. Materials and Methods

2.1. Case Study and Geometries in Use

The geometry of the molding that was studied was a component disk (see Figure 2) used for testing functional micro-structures and nanostructures. The concept of process and product fingerprint as discussed in the introduction appeared promising but had not yet been tested. In order to examine the viability of the concept, a number of micro-feature geometries were considered as test subjects but only two were selected. Figures 3 and 4 present the two types of micro-features. One consists of different geometries and the measurements were focused mainly on the conical micro pillars positioned in the structure, designated as the "F structure," and the other consists of micro ridges structures designated the "R structures." For the purpose of this paper, the micro features in all positions will be stated as "measurands." It should be noted that the R structures in the particular application are inclined planes created in the mold insert with different angles and mill tool radial engagement, which is shown in Figure 4. The alphanumeric codes for both structures represent the selected measurement positions while Table 1 provides a summary of the characteristics of the micro features and the sensor outputs. More details about the selected structures are provided in Section 2.4 where the measurement strategy for structure characterization is described.

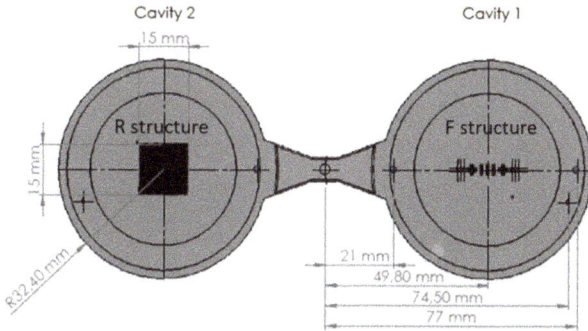

Figure 2. Structures of interest on full molded part. Cavity 1 includes the F structure, which is presented in more detail in Figure 3. Cavity 2 includes the R structure, which is presented in more detail in Figure 4. The F structure is in Cavity1 and the R structure is in Cavity 2.

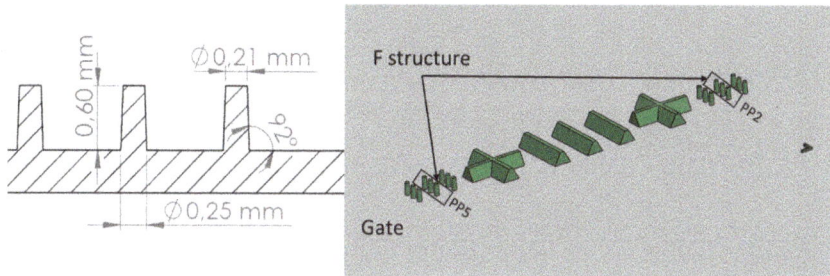

Figure 3. Cavity 1, F structure. Micro pillars used to assess part quality and process fingerprints are marked on the figure. PP5 indicates micro features near the gate while PP2 indicates micro features far from the gate.

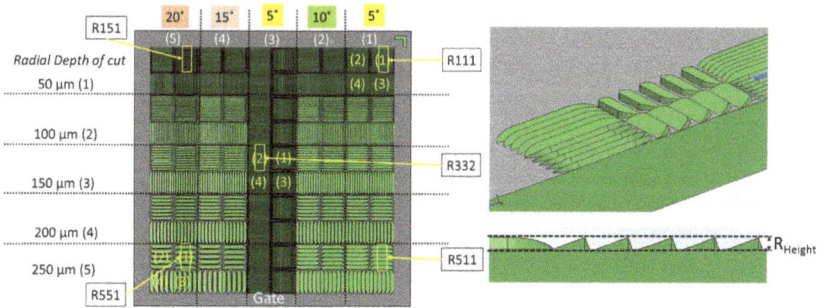

Figure 4. Cavity 2, R structure array created with different radial mill tool engagement (50–250 μm) and inclination angle (5°–20°). Positions of interest used to assess part quality and process fingerprints are marked: R551 and R511 indicate micro features (micro ridges) near the gate on the left and right, respectively. R151 and R111 indicate micro features far from the gate on the left and right, respectively. The position R332 indicates micro features centered in the middle of the part. For this case and for all μ-features of column 3, the inclination and the inclined plane length were kept to 5° and 50 μm, respectively.

Table 1. Product feature characteristics and sensor outputs.

				Product Features		
					Characteristics	
Name	Cavity #	Type	Position from Gate/Lateral	Top Diameter	Bottom Diameter	Pillar Height (P$_{Height}$)
C1PP2	1	Micro pillars	Far	Ø200 μm	Ø250 μm	600 μm
C1PP5	1	Micro pillars	Near	Ø200 μm	Ø250 μm	600 μm
-	-	-	-	Inclination	Incline plane length	R$_{height}$
C2R111	2	Micro ridges	Far/right side	5°	50 μm	2.7 μm
C2R151	2	Micro ridges	Far/left side	20°	50 μm	11.8 μm
C2R332	2	Micro ridges	Middle	5°	50 μm	3.8 μm
C2R511	2	Micro ridges	Near/Right	5°	250 μm	21.6 μm
C2R551	2	Micro ridges	Near/Left	20°	250 μm	81.34 μm
			Sensors			
Name	Cavity #	Type	Position from Gate	Output		
P1a	1	Piezoelectric	Near	Transient pressure		
P1b	1	Piezoelectric	Far	Transient pressure		
P2a	2	Piezoelectric	Near	Transient pressure		
T1C	1	Group N	Far	Transient temperature		
T1vC	1	Group N	In the mold block behind Cavity1	Transient temperature		
T2C	2	Group N	Far	Transient temperature		

2.2. Experimental Setup and Mold Design

After the geometries of the test micro feature geometries had been selected, two mold inserts were machined to include the features on the two cavities of a test mold (Figures 2–6). The inserts were installed on a three-plate mold that was designed to be compatible with both injection and injection compression molding. The plate holding the cavities could slide on secured guideways and the "gap" in the case of ICM was determined by springs at the back side of the plate. For the experiment discussed in this paper, the gap was kept at 0 mm with the springs fully compressed. The ejection of the part was facilitated by the use of four ejector pins at the periphery of the disk part cavity (Ø65 mm), which was larger than the micro-structured area of the mold insert (Ø45 mm).

The mold described above provided in-mold process monitoring capabilities since it contained three N-group thermocouples (4003B) and three piezoelectric pressure sensors (6006BB—Sensitivity of sensors listed in Table 2) from Priamus and was supported by the Priamus Fillcontroll system for in-mold flow front monitoring and process control. Additional thermocouple sensors were positioned in the mold to control the variothermal IM processes. However, this study was conducted under isothermal process conditions. The positions of the sensors in the mold are illustrated in Figure 6.

Table 2. In-mold piezoelectric sensor sensitivity.

Piezoelectric Sensor Sensitivity			
Piezoelectric Sensor	Channel	Sensitivity	Unit
P1a	P2	1.860	pC/bar
P1b	P3	1.880	pC/bar
P2a	P4	1.900	pC/bar

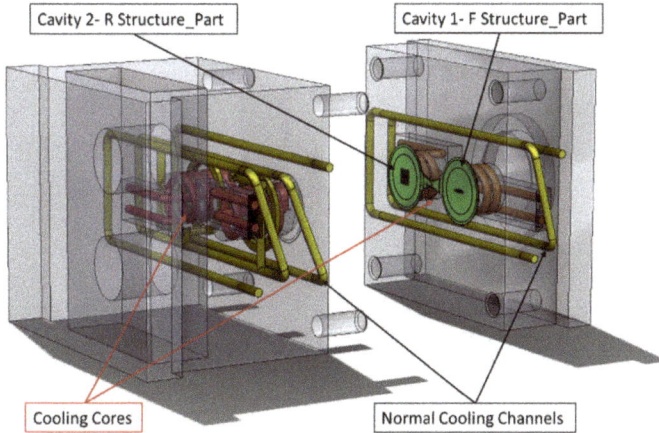

Figure 5. Mold geometry and cooling channels.

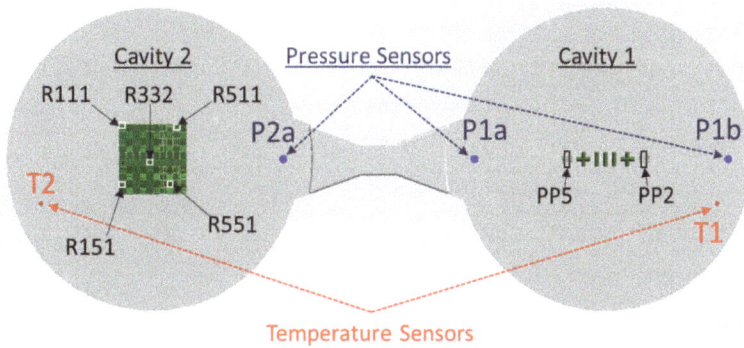

Figure 6. Measurement positions of structures and sensor position on full molding. In this case, the sensor positions are shown in the front side of the molding for easier association of the feature measurement to the sensor recordings. The F structure is in Cavity1 and the R structure is in Cavity 2.

2.3. Experiment Details

A full $2^4 \times 3$ full factorial designed experiment was conducted to examine the validity of the proposed product and process the fingerprint concept. The process parameter of melt temperature "*Tm*" (°C), mold temperature "*Tmld*" (°C), injection speed "*InjSp*" (mm/s), and packing pressure "*PackPr*" (bar) were used in the experiment in which it is proven through well-established research [13–17] that these are the most significant process parameters that affect the part quality in injection molding. The levels of the parameters were set with respect to the process parameter ranges indicated in the material datasheet. For this study, a commercial grade of Acrylonitrile Butadiene Styrene (ABS) (Styrolution Terluran GP-35), which was characterized by a relatively large processing window was used. The limits of the window were used to set the experimental parameter levels, as shown in Table 3, in order to map the effect of processing conditions on the replication of the micro pillars of the F structure and the micro ridges of the R structure. The experiment was performed on an Arburg 320A-Allrounder 600-170 injection-molding machine (Arburg GmbH +Co KG, Lossburg, Germany) with a clamping unit of maximum 600 kN clamping force and a screw diameter of 30 mm. The melt temperature profile was set by using 5 °C intervals in each heating zone of the reciprocating screw to facilitate a gradual heating of the material throughout the barrel. The packing (t_{pack} = 8s) and cooling (t_{cool} = 20s) times were set as values that were high enough to avoid any influence on the experimental results. Figure 7 shows the PvT (7a) and the viscosity (7b) diagrams of the material used in the experiment.

Table 3. Experimental process parameters—Full factorial DOE.

Parameter	Symbol	Unit	Low Level	High Level
Melt temperature	*Tm*	°C	220	260
Mold temperature	*Tmld*	°C	40	60
Injection speed	*InjSp*	mm/s	100	140
Packing pressure	*PackPr*	bar	600	700

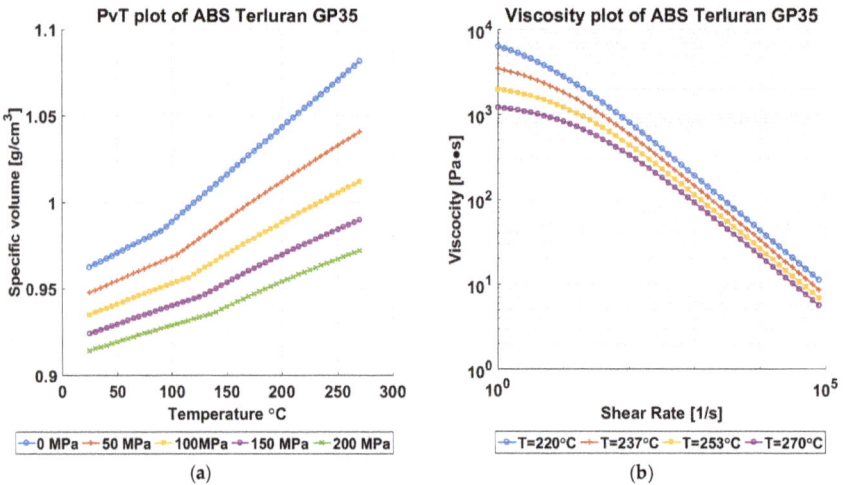

Figure 7. (a) PvT and (b) viscosity plots of material Styrolution Terluran GP-35 (ABS) [18].

Through the use of a DOE approach, two objectives can be met. First, proving that the structures selected were sensitive to the process variation and could be correlated to the part quality, they could be used for quality monitoring as suitable product fingerprints. Second, the data originating from the experiment could be used for the determination of the process fingerprints required to run an in-line process monitoring system as the quality indicators with better correlation with product quality.

2.4. Measurement Strategy and Procedure

From the experiment and for every experimental treatment, the initial 20 molded parts from the start of the process were discarded since the process had not reached stability. The next 10 parts were then collected for assessment with three parts to be measured for the assessment of micro feature replication quality on the parts (μ-pillars (F-structure) in Cavity 1 and μ-ridges (R-Structure) in Cavity 2). It was decided to use the initial, middle, and last part of the collected sample (parts 21, 25, and 29).

As shown in Figures 3, 4 and 6, two positions in Cavity 1 (micro pillars/F structures) and five positions in Cavity 2 (micro ridges/R structures) were selected as possible candidates for product fingerprints. In particular, the five positions were selected with respect to the limits of the inclination and the inclined plane length ranges in order to access the replication fidelity in the two size limits of the R structure. It is important to note that the μ-pillar structures in Cavity 1 were designed as functional microstructures on a microfluidic biochip system, as reported by Marhöfer et al. [19]. However, the structures in Cavity 2 due to their smaller size could be incorporated into the produced component and could easily act as a product fingerprint. In the present case, the two different types of structures were located in different cavities. Since they were symmetrically opposed and balanced, it can be readily assumed that the molding conditions in the two cavities are similar. In all cases for the assessment of the feature quality, the height of the feature was set as the measure and will be denoted by the measurand's name.

Due to the differences in structure type and size, the measurement strategy for the two types of microstructures was not identical but was instead optimized for each case separately, which kept similar setting levels when possible. The feature height dimensional measurements were carried out with the use of a 3D confocal laser scanning microscope (CLSM, Keyence VK-X210 from KEYENCE, Osaka, Japan).

2.4.1. Pillar Dimensional Measurement and Error Evaluation Procedure

Even though the measuring instrument had been designed for the measurement of complex structures, acquiring a full scan of the pillars is still a challenge due to the almost vertical slopes (92°, see Figure 3) of the pillars. Table 4 includes the settings of the microscope that were used for the measurement of the micro-pillars of the F structure.

Table 4. CLSM measurement settings used for μ-pillars (F structure) and μ-ridges (R structure).

Parameter	CLSM Measurement Settings	
	μ-Pillars (F Structure)	μ-Ridges (R Structure)
Objective	50×	50×
Aperture N	0.55	0.95
Lens Working Distance	8.7 mm	0.35 mm
Measurement Range	620 μm	1200 × 200 μm
Brightness	7220	7220
Z pitch	0.36 μm	0.2 μm

In order to assess the stability of the process, the effect of the process parameter changes, and the replication fidelity of the pillar micro features per experimental run, three pillars in each position of interest were scanned to measure the pillar height. The middle μ-pillars in Positions C1PP2 and C1PP5 (Figure 8) were measured five times in order to ensure the repeatability of the measurements (standard deviation in the range of μm) and provided sufficient data for error estimation. The measurement files were consequently processed with the use of SPIP 6.4.1 (SPIP^{TM}, Image Metrology A/S, Hørsholm, Denmark) software to extract four 2D pillar profiles from each scan.

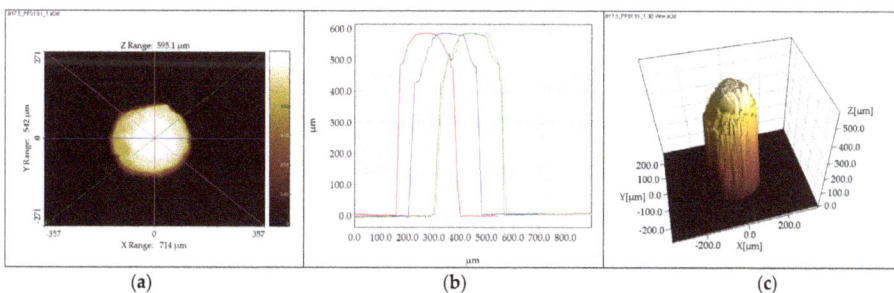

| (a) | (b) | (c) |

Figure 8. Pillar height measurement [20], (**a**) the four planes intersecting the middle of the pillar for the extraction of the cross-section profiles, (**b**) the four cross-section profiles of the pillar, and (**c**) the 3D representation of a pillar.

The average pillar height was calculated by using four profiles (see Figure 8b) that intersected the center of the pillars. The procedure used scans of both the mold and the molded parts to calculate a measure of the replication fidelity of the molding process.

2.4.2. Procedure for Dimensional Measurements of the R Structures

As for the parts produced in Cavity 1, μ-ridges (R structure) on parts produced in Cavity 2 were measured with the use of the same Keyence confocal laser microscope (Figure 9). The CLSM microscope settings are given in Table 4. Similarly to the μ-pillars of Cavity 1, in order to assess the stability of the process, the effect of the process parameter changes and the replication fidelity of the micro-ridges (R structure) as a measuring procedure was devised. Five positions for each part in each experimental run were scanned to access the average area feature height. The measurement positions are illustrated in Figure 4. The measurement files were consequently processed with the use of SPIP 6.4.1 software to extract 2D profiles from the scans of each area.

In Figure 9b, it is important to note the existence of negative "spikes" in the valleys of the surface profile. These surface outliers are not artefacts caused by the laser measurement method. They are physical artefacts that were produced as the negative of the mold insert. The milling strategy on the surfaces of the mold insert that were milled with an angle of 5° produced a large percentage of burr-covered features and consequently the burrs were replicated as deep valleys on the surface of the parts (Figure 9). 3D images of the injection molded micro features provided in Figure 10 illustrate the quality of the features.

Figure 9d shows the measurement procedure used in the Matlab measurement script in order to calculate the average feature height in each area based on the five single profiles extracted from each scan. The profiles are then processed in a Matlab script where the profiles are corrected for tilt and cross-correlated for alignment to the reference profile from the CAD file of the surface. The mean value of the points at the valleys (in specified regions) is calculated (hv1, hv2, hv3, hv4, hv5, ..., hvn) for all five profiles and the point with the maximum value at the region of the peaks is identified (hp1, hp2, hp3, hp4, hp5, ... , hpn) (Figure 9d). To counterbalance the effects of the burrs on the profile and to acquire reliable measurements, the five measurements per peak in each position were subjected to an outlier removal algorithm that used the modified IQR criterion (Equation (1)) [21] to remove the outlier values caused by the burrs. The burr-filtered data were then used to calculate the height of the feature (i.e., Height1 = hp1-|hv1|). Once the height of each peak is calculated, the Chauvenet criterion [21] (Equation (2)) was applied to the five values per peak (five profiles) so that the average height of each peak on the profiles and the average area height could be calculated for all measured parts in each experimental run (Figure 9).

Modified IQR-Range
$$Mod._{IQR} = 1.5\ IQR * \left[1 + 0.1 * \ln\left(\tfrac{n}{10}\right)\right]$$
$where: \ IQR = 3^{rd} Quartile - 1^{st} Quartile,\ \text{of the data}$
$\qquad n: sample\ size$

(1)

Chauvenet Criterion
$$P_t = 2nP_{xL} = \tfrac{1}{2}$$
$where\ P_t: \text{probability band centered on the sample mean}$
$\qquad n: sample\ size$
$\qquad P_{xL} = 1/(4n)\ : \text{probability represented by one tail of the normal distribution}$

(2)

(b)

(a)

(c)

(d)

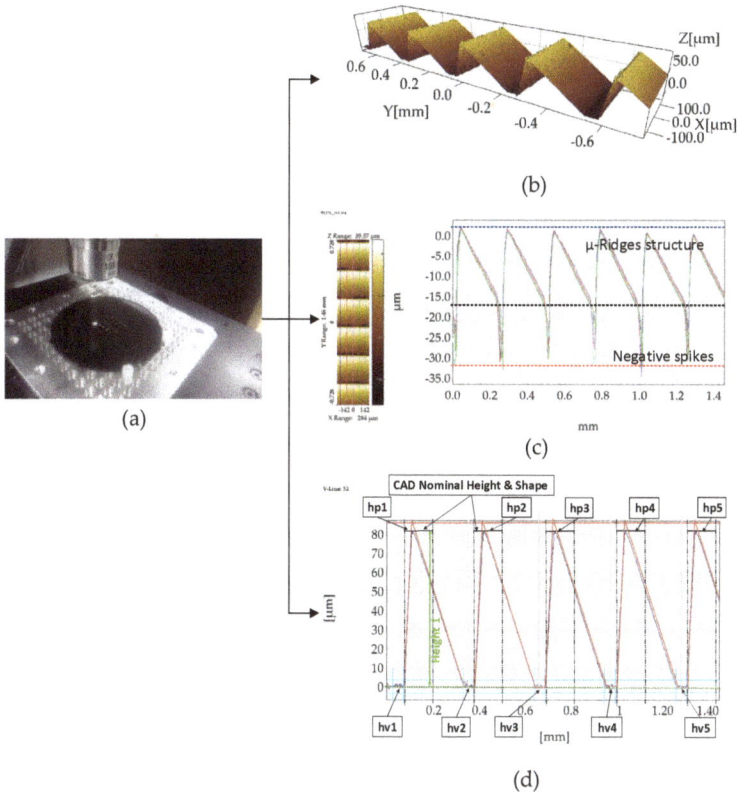

Figure 9. (**a**) Measurement of the R structure on a Keyence CLSM microscope. (**b**) SPIP 3D scan of position C2R551. (**c**) Cross-section profile extraction. In this scenario, the position C2R511 is used to portray the negative "spikes" caused by the replication of the burrs. (**d**) Illustration of a measurement procedure example for peak height (R_{height}) on the profiles of position C2R551.

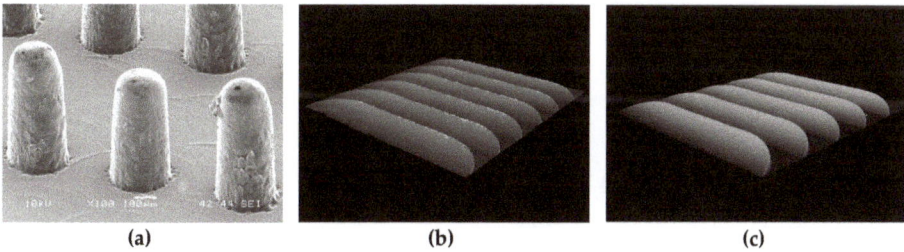

(a)   (b)   (c)

Figure 10. 3D images (**a**) micro pillar (F structure) and micro ridges (R structure) features in positions (**b**) C2R511 with 5° inclination and (**c**) C2R551 with 20° inclination.

2.5. Process Monitoring and Data Collected

In addition to the physical part measurements, a number of process variables were recorded from the injection molding machine and from external sensors. The first type of data (Machine Data) is routinely exported from the injection molding machine's controller and is the easiest to access. The data are used for the quality monitoring subroutine of the machine and can be viewed from the machine monitoring software. The *second type of data* (Machine Signals) is the signals from the sensors used to control the process and the machine. In particular, the injection pressure signal is monitored via a strain gauge transducer positioned at the nozzle of the screw. Injection speed and the screw position are monitored via a linear position sensor. The third type of data (In-Mold Sensors Signals) were recorded with the use of in-mold temperature and piezoelectric pressure sensors with a sampling frequency of 250 Hz (see Section 2.2). A full list of the recorded variables is given in Table 5. The signal data from both the IM machine controller and the Fillcontroll system were recorded separately and were aligned with respect to the time scale since the Fillcontroll system has a maximum delay of 4 ms from the incidence of injection until the start of the recording.

Table 5. List of recorded variables.

Machine Data (Single Value)	Machine Signals	In-Mold Sensor Signals
Injection time (s)	Injection/Pack. Pressure (bar)	T1C (Cavity1) (°C)
Max Pressure (bar)	Injection speed (mm/s)	T1viaC (Cavity1) (°C)
Switch Over Pressure (bar)	Screw position (mm)	T2C (Cavity2) (°C)
Cushion (mm)		P1a (Cavity1-gate) (bar)
Dosage time (s)		P1b (Cavity1-end) (bar)
Cycle time (s)		P2a (Cavity2-gate) (bar)

Before an investigation of the "fingerprint" candidates, an initial analysis of the recorded signals found no unexpected results such as spikes or delays in pressure and temperature signals from the machine and in-mold sensors. Considering the large number of recorded variables and acquired data, many different types of quality indicators were considered as viable process fingerprint candidates and their correlation to the IM parameters was studied. The different types of process fingerprint candidates can be categorized as single value indicators for each molding cycle, which was illustrated in Figure 11.

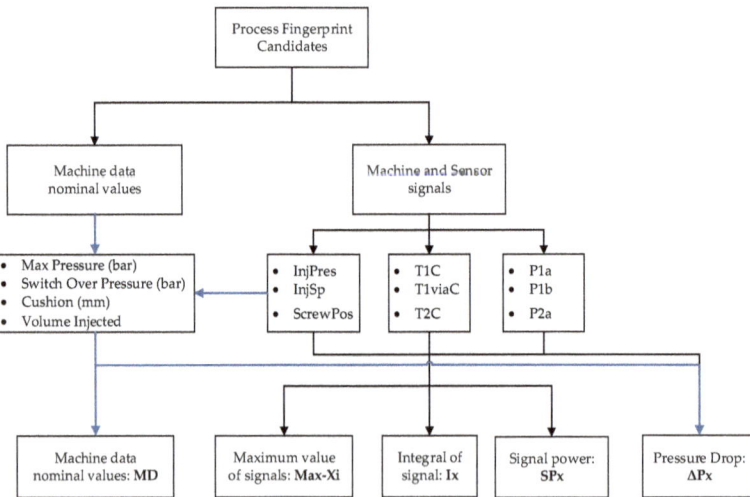

Figure 11. Procedure for the extraction of process fingerprint candidates from the collected data.

In particular, the shape and amplitude of the recorded values and signals were assessed for their viability as process fingerprint candidates. They are explained below.

- **MD**: Machine data nominal values as listed in Figure 11 are single value quantities that originate from the machine controller's quality control subroutine that records quantities during critical points in the process. Such quantities are: the *Max Pressure* (*MaxPr*) in the nozzle occurring for both injection and packing phases, the switch over pressure, and the cushion left in the barrel after the packing phase of the process has been completed and can provide an indication of the filling deviation from cycle to cycle. In addition, cushion and the maximum screw position (*MaxScPos*) are combined to calculate the non-compressed material volume injected (*VolInj*) into the cavity.

- **Max-Xi**: Maximum values from the recorded signals *Xi* originated from the machine controller (*InjPr, InjSp, ScPos*) and in-mold temperature and pressure sensors.
- **Ix**: is the integral of the whole signal *y(t)* recorded from the start of the injection phase ($t_0 = 0$ s) until the end of the packing phase ($t_n = 8$ s), as seen in Equation 3. The integral directly relates to the energy stored in the polymer and can differ from the measured quantity. In particular, the integral calculated from the pressure signals represents the energy stored in the polymer from the melting, compression, and injection of the polymer to the mold cavity.

$$I_x = \int_0^T y(t)dt$$
$$where: \ T = end\ time\ of\ signal\ duration\ (time = 8s)$$
(3)

- **SP-X$_i$**: The power of a signal X_i is the sum of the absolute squares of its time-domain samples divided by the signal length or the square of its RMS level. As for the integral of a signal, the power of the signal relates to the energy of the system for all recorded signal frequencies.

$$SP_x = \lim_{T\to\infty} \frac{1}{T} \int_0^T |y(t)|^2 dt$$
$$where: \ T = end\ time\ of\ signal\ duration\ (time = 8s)$$
(4)

- **ΔP-X$_i$**: refers to the pressure drop from the pressure at the nozzle of the IM machine to the position of the sensor in the mold for both injection phases, which shows the pressure drop at the gate of the cavity *ΔP-P1a* at the switch over point *ΔP-P1b* and during the packing phase *ΔP-PackP1b*.

3. Results and Discussion

3.1. Product Fingerprint Analysis

To assess the suitability of the seven measurands as product fingerprints, the results of the measurement sensitivity analysis were evaluated and are reported in Figures 12 and 13. The figures present the main effect plots and the Pareto plots of the standardized effects. Both types of plots reveal the significance that each of the process parameter has on the response and allows identification of the parameters with the highest significance. The error bars in the main effect plots represent the standard deviations of the respective measurand. Such error bars provide a measure of the process's variation and should be taken into consideration when evaluating the effects of process variations on a measurand. The effects whose variation due to process parameter changes is smaller than the error bar cannot be considered significant and their influence on the measurand is likely to be small or even negligible.

Figure 12a reports the results for the micro-pillar height in position C1PP2 (far from the gate). From the plots, it is evident that the parameter with the greatest influence is the injection speed (*InjSp*). Its increase leads to a 34.9 μm ± 4.2 μm increase in micro-pillar height. In addition, the error bars at the two parameter levels do not overlap. The same can be seen from the Pareto chart. Therefore, the effect of *InjSp* is significant. Similarly, the results for the pillar height in position C1PP5 (near the gate) are illustrated in Figure 12b. From the effect plots, it is evident that the parameter with the greatest influence on the response is the injection speed (*InjSp*) for which an increase of 19.2 μm was observed for the pillar height when increasing *InjSp*. Its effect can be considered significant since the error bars at the two parameter levels do not overlap. Similar conclusions can also be drawn from the Pareto chart as well. In both cases of C1PP2 and C1PP5, only the melt temperature (*Tm*) and the mold temperature (*Tmld*) appears to have an influence among the remaining parameters. However, the error bars at the parameter level do overlap for both parameters, which indicates that the parameters cannot be considered significant.

When the micro features at the two positions are compared, it becomes evident that C1PP2 (far from the gate) is more sensitive to process variations particularly for the increase in *InjSp* and *Tmld*, which results in a larger increase in pillar height at position C1PP2 when compared to position C1PP5 since it is located at the end of the flow path. The reason lies in the rheological behavior of polymers. When higher *Tmld* or *InjSp* is used, the melt viscosity in the cavity is reduced due to either the higher temperature or the shear thinning effect. A lower viscosity level increases the replication fidelity of the micro features, which is beneficial especially for features with a high aspect ratio (2.4-3). In both cases of C1PP2 and C1PP5, the interaction of *Tmld* and *InjSp* appears to have an influence on the responses and is of particular importance for all measurands in Cavity 2. The parameter *Tmld* was the parameter with the largest influence.

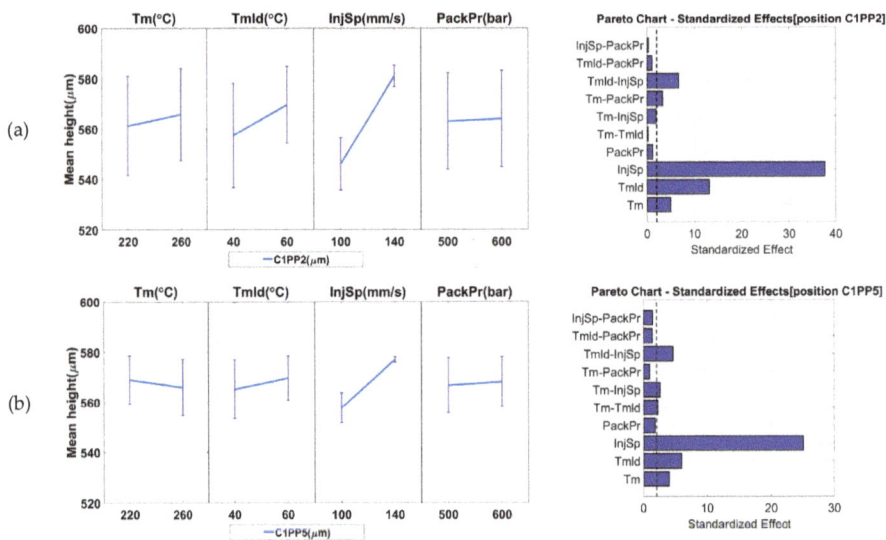

Figure 12. Influence of the IM process parameters on the measurands (pillar height) and possible product fingerprints from cavity: (**a**) C1PP2 (micro pillars far from the gate) and (**b**) C1PP5 (micro pillars near the gate). The figure presents the main effects. The error bars in the main effects plots represent the standard deviations from the respective measurand.

Figure 13a shows the results from the R structure (micro-ridge arrays) height measurements in position C2R111 (far from the gate on the right side). From the effect plots, it can be seen that the parameter with the greatest influence on the response was the mold temperature (*Tmld*). Its increase caused a 0.3 μm increase in height. The same conclusion can be drawn from the Pareto chart. The error bars at the two parameter levels do not overlap and, thus, the effect is considered significant. The rest of the parameters all appear to have had an influence as did the 2-way interaction of *Tmld* and *InjSp*. However, the error bars at the parameter levels of the rest of the parameter effects do overlap, which indicates that the parameters cannot be considered significant.

Figure 13b shows the results for the R structure height measurement in position C2R151 (far from the gate on the left side). The results are similar to what was obtained at position C2R111 where the main effect plots indicate that the parameter with the greatest influence on the response was the mold temperature (*Tmld*). Its increase from 40 °C to 60 °C caused a 2.3 μm increase in feature height. For the rest of the parameters, only the packing pressure (*PackPr*) does not have an influence. However, none can be seen as significant as the error bars in the main effect plot overlap for the two parameter levels.

In comparison to positions C2R111 and C2R151 that are located closer to the end of Cavity 2, the features in positions C2R332 (central position, Figure 13c), C2R511 (near the gate on the right, Figure 13d), and C2R551 (left side, Figure 13d) were less sensitive to process variation than the two previously discussed positions. In the case of C2R511 and C2R551 (Figure 13d), an increase in *Tm*, *Tmld*, *InjSp*, and *PackPr* parameters resulted in a feature height increase of 0.6 μm and 2.1 μm, respectively, which is within the limit of the error bars. This indicated that none of the parameters may be considered significant. Similarly, for the results of the feature height from position C2R332 (Figure 13c), none of the parameters can be considered significant. However, the increase of *Tmld* caused an increase of 0.11 μm in the feature height while an increase in the parameters of *InjSp* from 100 mm/s to 140 mm/s and *PackPr* from 600 bar to 700 bar had the opposite effect. The reason for this behavior lies in a reduction of the melt viscosity. The combination with the geometry of the shallow (3.8 μm) micro-ridges feature and the orientation of the slope perpendicular to the flow path meant that the polymer at higher injection speeds could pass over the features and a frozen surface layer forms before filling the features. When *PackPr* is considered, in such a shallow feature, the already formed frozen layer cannot be deformed by the higher packing pressure to fill the sharp corners at the bottom of the μ-ridges. In all positions of Cavity 2, the parameter with the largest influence was the mold temperature (*Tmld*), which was followed by the melt temperature (*Tm*). From the Pareto charts, it is evident that the two-way interactions existed mainly at the positions further from the gate where the responses were more sensitive to process variations. This behavior originates from the viscosity changes, which were caused by a change in a combination of parameters. As the pressure dropped along the flow path, the effects of viscosity changed and became more prominent further from the gate. The features far from the gate in both cavities (C1PP2, C2R111, and C2R151) are, therefore, considered more suitable options for product fingerprints.

The analysis of the effects of the different IM process parameters on the seven measurands has provided some indication of the most suitable product fingerprints with respect to their sensitivity to process variations. However, a product fingerprint is required to have a high level of correlation with the overall part quality assessed by a measurand. In the current study, the μ-pillars in position C1PP2 were considered representative for the overall quality of the molded part due to their location close to the end of the flow path and are, thus, regarded as a suitable product fingerprint. A correlation analysis was carried out to determine the most suitable product fingerprints from the feature in Cavity 2. For this purpose, the Pearson correlation coefficient ρ was calculated [22].

The calculated |ρ| values for the 49 dataset combinations are shown in Figure 14. It can be seen that the strongest correlations are found for the combinations C2R151/C2R111 (|ρ| = 0.93) in Cavity 2 and C1PP2/C1PP5 (|ρ| = 0.877) in Cavity 1 while the best correlation coefficients between datasets from both cavities were C1PP2/C2R111 (|ρ| = 0.63) and C1PP2/C2R151 (|ρ| = 0.68). The rest of the correlation coefficients varied between |ρ| = 0.43 for measurand combinations within the same cavity and |ρ| = 0.11 for measurand combinations from both cavities.

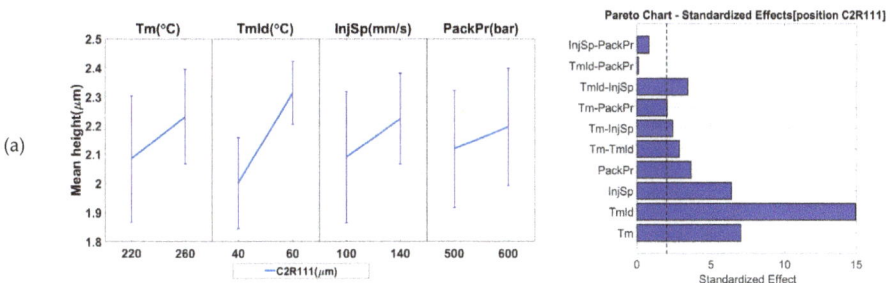

Figure 13. *Cont.*

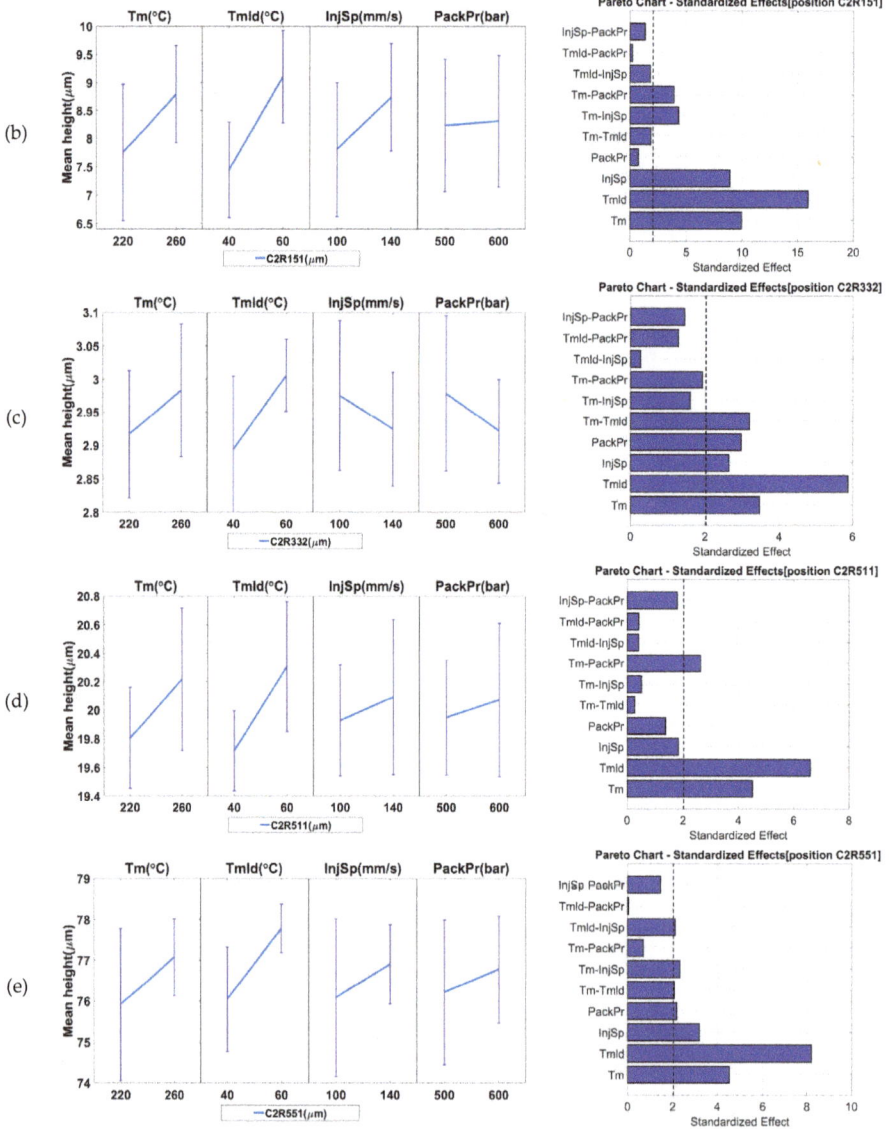

Figure 13. Influence of IM process parameters on the five measurands (micro ridges height—R_{height}) and possible product fingerprints from Cavity 2: (**a**) C2R111, (**b**) C2R151, (**c**) C2R332, (**d**) C2R511, and (**e**) C2R551. The figure presents both the main effects (left column) and the Pareto graphs (right column) of standardized effects. The error bars in the main effect plots represent the standard deviations from the respective measurand. The black dashed line in the Pareto plots represents the significance level at a 95% confidence level.

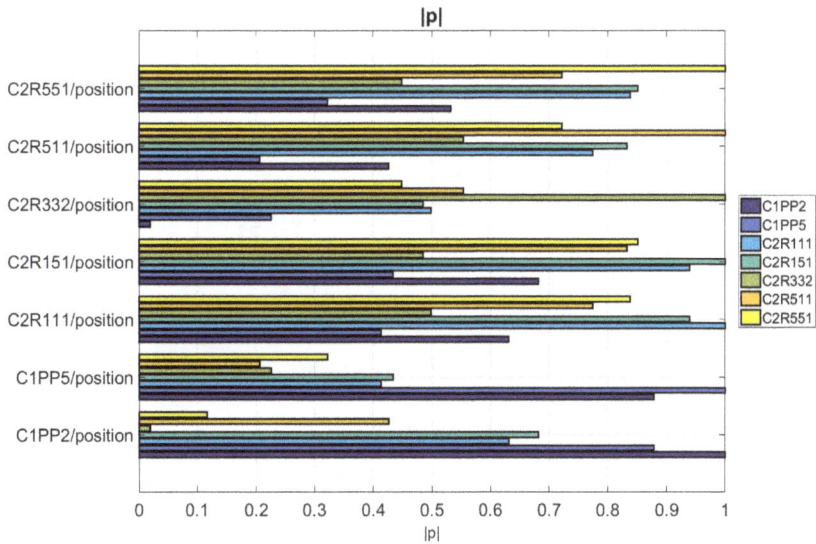

Figure 14. Values of the absolute Pearson correlation coefficient |ρ| calculated from each possible combination of the seven measurands from Cavity 1 and 2. A perfect correlation (|ρ| = 1) exists only for combinations of the same dataset.

3.2. Process Fingerprint Analysis

To compare process quality indicators, a single value was extracted at each cycle either from the molding machine or from the sensors. Then, a similar analysis explained in the previous section was carried out to identify the most suitable process fingerprint, which must have high process sensitivity and be strongly correlated with the physical measurands (i.e., product fingerprints). In this case, the main effect's error bars were calculated from the standard error originating from the datasets associated with a particular level of a single parameter. The process fingerprint candidates (35 candidates) described in Figure 12 were calculated for all seven positions (two positions in Cavity 1: C1PP2 and C1PP5 and five positions in Cavity 2: C2R111, C2R151, C2R332, C2R511, C2R551) yielding a total of 245 combinations of process fingerprint and measurands.

In an initial analysis, the calculated values of the possible process fingerprints were derived by plotting the data and assessing the similarity of the trend in the data, as shown in Figure 15. In most cases, no clearly visible trend exists between the measurement of a measurand position and the calculated process fingerprint candidates. In the case of the integral and signal power of the signals originating from sensor P1b, it is evident that they followed a similar trend to the measurement data from positions C2PP2 and C2PP5 (Figure 15). This preliminary analysis provided an indication of the existence of two combinations of highly correlated datasets (C1PP2 and C1PP5 with *I-P1b* and *SP-P1b*).

The analysis continued with a correlation analysis and screening of the 35 process fingerprint candidates coupled with the height data from the seven measurement positions in the molded components. The process fingerprint candidates were separated into five categories described in Figure 11. Figure 16 illustrates the results of the analysis for the process fingerprint candidates with respect to positions C1PP2 and C1PP5. The candidates that originated from the pressure drop values $\Delta P\text{-}X_i$/position in correlation with any measurand position have a maximum correlation coefficient of |ρ| = 0.32 for the combination $\Delta P\text{-}PackP1b$/C1PP5 ($\Delta P\text{-}PackP1b$ denotes the pressure drop from the nozzle to the location of sensors P1b in the cavity considering both the injection and packing phases). They are, therefore, not considered viable for process fingerprints.

The next category of process fingerprint candidates included the maximum values from the signals that were originally obtained from the IM machine and in-mold sensors. The maximum correlation coefficient for the second category of **Max-X$_i$** was $|\rho| = 0.354$ for the combination *MaxT1C/C2R551*," which denotes the maximum values from the signals from the temperature sensor *T1C* in Cavity 1. The results are similar for signals from sensors *T1vC* and *T2C*. However, due to the low correlation of candidates **Max-X$_i$** with the seven measurands, it indicates that they cannot be considered suitable candidates for process fingerprints.

In addition to the calculated maximum values from the signals, the IM machine's on-board quality control system (i.e., machine data generated, **MD**) provided a number of variables as part of the quality monitoring. Among those variables, the most important ones were: *Max Pressure (MaxPr)*, *Switch Over Pressure (SOPr)*, and *Cushion*. *Cushion* is expressed in millimeters and represents the material volume left in the barrel at the end of the cycle. From *Cushion*, in combination with the maximum screw position calculated from the signal *Screw Position*, the injected material volume *(VolInj)* can be derived. This is only an approximate value since the compressibility of the molten material is not taken into consideration in this calculation. The injected volume *(VolInj)* had a maximum correlation of $|\rho| = 0.57$ to the dataset from position C1PP2. The remaining correlation coefficients for the datasets from all other positions is smaller than 0.44. Similarly, *Cushion* had a maximum correlation coefficient of $|\rho| = 0.61$ with the dataset from position C1PP2 and can, thus, be considered a suitable process fingerprint candidate. Another suitable candidate process fingerprint was the switch over pressure *(SOPr)*, which had a maximum correlation coefficient of $|\rho| = 0.61$ with the dataset from position C1PP5. The height of the features in position C1PP5 was sensitive to changes in *SOPr* since the change of the packing pressure influenced the replication quality of the pillars in this position and can, thus, be considered a suitable process fingerprint. These candidates are particularly interesting since they are directly provided by the IM machine's controller and do not require any additional sensor system, which means they allow process monitoring and control to take place at a minimum additional cost.

The fourth category of potential process fingerprints included the integrals of the recorded signals from both machine and in-mold sensor signals. For pressure signals, the integral directly relates to the energy stored in the polymer from the melting, compression, and injection processes taking place in the mold cavity. The largest correlation coefficients of integral levels to the datasets originated from the signals from the piezoelectric pressure sensor *P1b* and the signals from the temperature sensors (*T1C*, *T1vC*, and *T2C*). However, the values from the temperature sensors were taken into consideration since they are greatly dependent on the mold cavity temperature *Tmld*. Their measurement output is biased, so the focus was set on the pressure *P1b* signals. The maximum correlation coefficient of $|\rho| = 0.758$ was calculated for the combination *I-P1b/C2R151*, which was followed by *I-P1b/C1PP2* ($|\rho| = 0.749$), *I-P1b/C2R111* ($|\rho| = 0.749$), *I-P1b/C2R551* ($|\rho| = 0.658$), and *I-P1b/C2R511* ($|\rho| = 0.655$), which suggests that the *I-P1b* values showed good correlation with all datasets with the exception of dataset C2R332. This dataset contained data from the second nominally smallest micro-ridges of the R structure array and was similar (5° inclination—50 μm inclined plane length) to those in position C2R111. Due to the size of the ridges and their position in the center of the molded part (C2R332), the replication fidelity of the micro features did not deviate significantly between the different experimental treatments in comparison to readings from the pressure sensors *P1b* (at the end of the flow path). Therefore, the correlation of datasets *I-P1b* and C2R322 was very weak.

The fifth category of process fingerprint candidates consisted of the power of the signals calculated from the signal inputs. Similarly to the integral values, the largest correlation coefficients of signal power levels with the datasets occurred for the signals from the piezoelectric pressure sensor *P1b* with a maximum correlation coefficient of $|\rho| = 0.772$ calculated for the combination *SP-P1b/C1PP2*.

From the correlation analysis, a number of promising process fingerprint candidates were selected. The *Cushion, VolInj,* and *SOPr* variables are directly provided by the controller on the IM machine. However, the *I-P1b* and *SP-P1b* process fingerprints were found to have an even higher correlation with the overall part quality, which indicates that the positioning of a pressure sensor at the end of the flow path could actively provide valuable data for fast quality assurance and for a warning system for the detection of production quality issues. However, in order to consider these variables as suitable process fingerprints, the variable must be affected by process variation.

To verify the existence of such effects, a process sensibility analysis was carried out for the most prominent candidates. The results are shown in Figure 17. As in the analysis of the product fingerprints discussed in Section 3.1, Figure 17 shows the main effect plots and the Pareto plots of the standardized effects. Both plot types represent the significance that each of the process parameters have on the responses. They make it possible to identify the parameter with the highest significance. As stated for the product fingerprint, such a measure of process variation should be taken into consideration when evaluating the effects of process variation on a measurand.

Figure 17a shows the results for the response of *VolInj* on changes in process parameters. The main effect plots show that the parameter with the greatest influence on the response is the injection pressure (*InjSp*) whose increase from 100 mm/s to 140 mm/s resulted in an increase of 71 mm^3 in the volume injected (*VolInj*), which was followed by packing pressure (*PackPr*) and the mold temperature (*Tmld*). However, none of these can be considered significant since the error bars at their two levels overlapped. The Pareto chart in the right column shows the significant two-way interactions between *Tm*, *InjSp*, and *PackPr*. Such interactions are expected as the higher level of *Tm* in combination with high levels of either *InjSp* or *PackPr*, reduces the viscosity of the molten polymer, and forces a larger melt volume into the mold cavity. As stated earlier in Sections 2.5 and 3.2, the values of the *VolInj* response were calculated by using the *Cushion* measurements and, as such, there is a direct dependency with *Cushion* levels. A similar but opposite effect can be seen in Figure 17b for the cushion due to the influence of process variation.

Figure 17c shows the results for the response of the switch over pressure *SOPr* as influenced by the process parameter deviation. From the effect plots, it can be seen that changes in melt temperature can have a major influence on the *SOPr*. An increase of *Tm* from 220 °C to 260 °C resulted in a pressure drop of 61.7 bar so the influence of *Tm* must be considered significant.

The other two most prominent process fingerprint candidates were not derived directly from the IM machine but were recorded with the use of an in-mold cavity pressure sensor. Both the integral and power of the signal were dependent on the amplitude and shape of the signal. The integral of the *P1b* signal (Figure 17d) was mostly influenced by an increase in *InjSp* from 100 mm/s to 140 mm/s. This change caused a drop in the amplitude of the pressure signal and delayed the pressure peak. An increase in *PackPr* from 600 bar to 700 bar increased the plateau amplitude for the packing pressure level and, thus, increased the integral of the signal since it was calculated for both injection and packing phases of the process. The signal power and its integral are both measures of the energy stored in the polymer during the process. Similarly to *I-P1b*, the *SP-P1b* process fingerprint was sensitive to process variations. Considering the correlation coefficients, both indicators may be considered suitable for serving as process fingerprints.

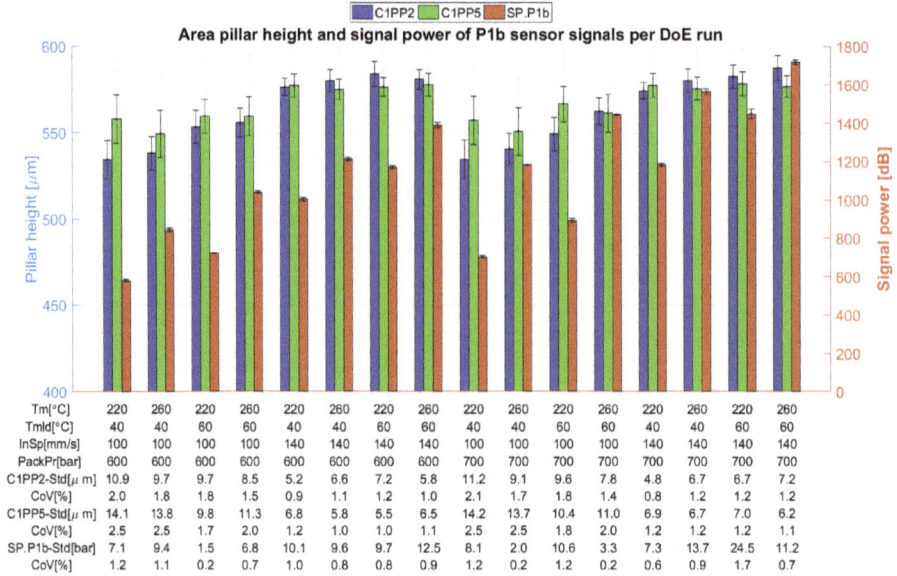

Area pillar height and signal power of P1b sensor signals per DoE run

	C1PP2	C1PP5	SP.P1b

Tm[°C]	220	260	220	260	220	260	220	260	220	260	220	260	220	260	220	260
Tmld[°C]	40	40	60	60	40	40	60	60	40	40	60	60	40	40	60	60
InSp[mm/s]	100	100	100	100	140	140	140	140	100	100	100	100	140	140	140	140
PackPr[bar]	600	600	600	600	600	600	600	600	700	700	700	700	700	700	700	700
C1PP2-Std[μ m]	10.9	9.7	9.7	8.5	5.2	6.6	7.2	5.8	11.2	9.1	9.6	7.8	4.8	6.7	6.7	7.2
CoV[%]	2.0	1.8	1.8	1.5	0.9	1.1	1.2	1.0	2.1	1.7	1.8	1.4	0.8	1.2	1.2	1.2
C1PP5-Std[μ m]	14.1	13.8	9.8	11.3	6.8	5.8	5.5	6.5	14.2	13.7	10.4	11.0	6.9	6.7	7.0	6.2
CoV[%]	2.5	2.5	1.7	2.0	1.2	1.0	1.0	1.1	2.5	2.5	1.8	2.0	1.2	1.2	1.2	1.1
SP.P1b-Std[bar]	7.1	9.4	1.5	6.8	10.1	9.6	9.7	12.5	8.1	2.0	10.6	3.3	7.3	13.7	24.5	11.2
CoV[%]	1.2	1.1	0.2	0.7	1.0	0.8	0.8	0.9	1.2	0.2	1.2	0.2	0.6	0.9	1.7	0.7

Figure 15. Values of the measurands in positions C1PP2 (far from the gate), C1PP5 (near the gate), and the average power of the signal originating from the piezoelectric pressure sensor *P1b* far from the gate in Cavity 1.

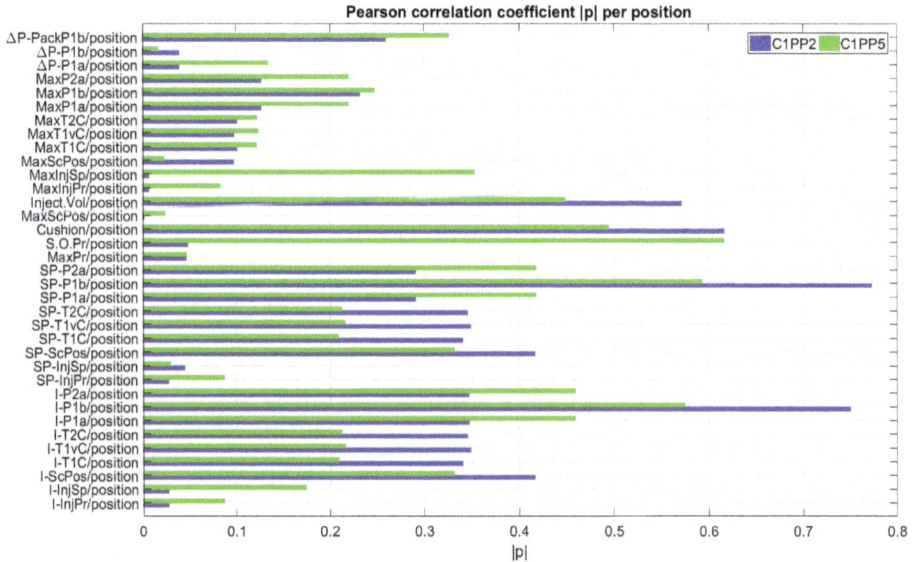

Figure 16. Values of the absolute Pearson correlation coefficient $|\rho|$ calculated from each process fingerprint candidate with respect to the feature height in positions C1PP2 (far from the gate) and C1PP5 (near the gate) of Cavity 1. The process fingerprint candidates are listed in the y axis of the figure.

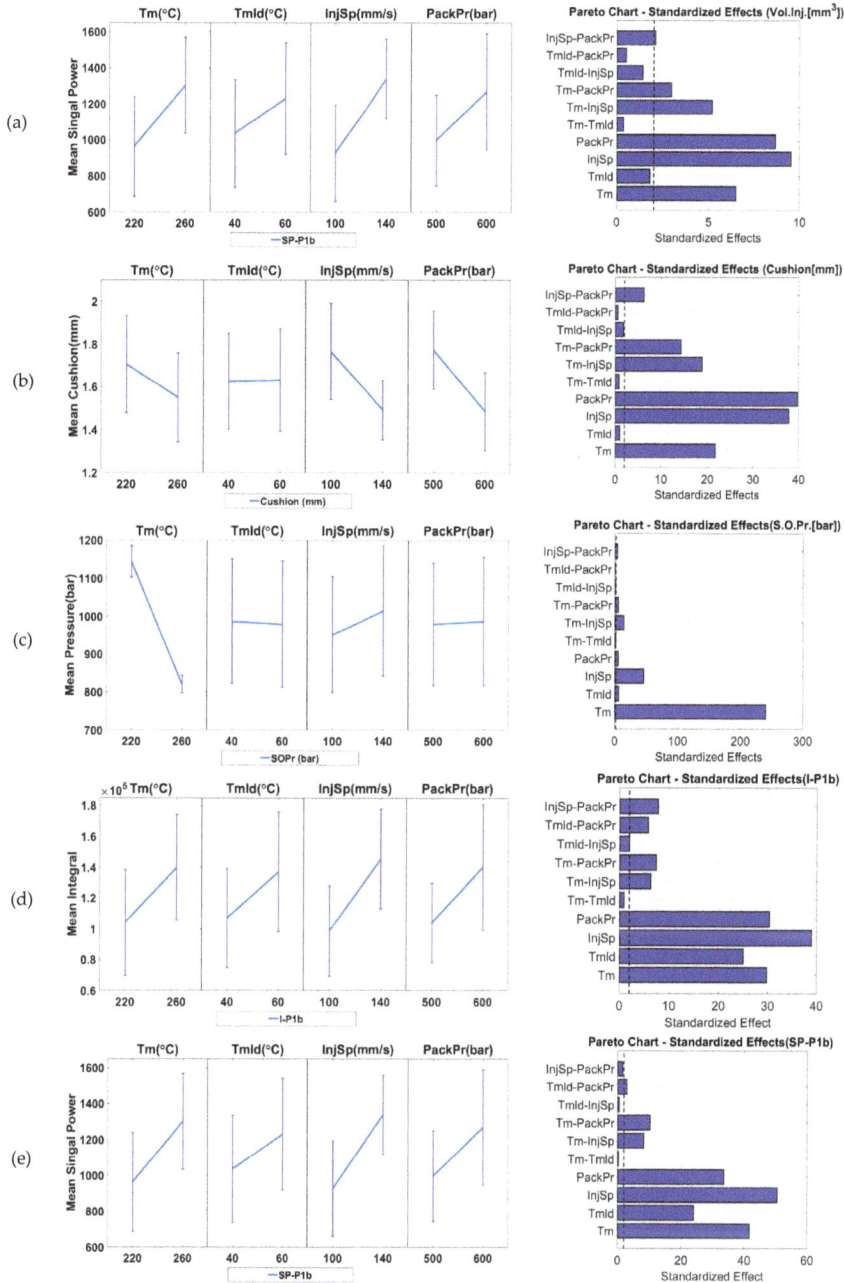

Figure 17. Influence of IM process on the five major process fingerprints candidates: (**a**) *VolInj*, (**b**) *Cushion*, (**c**) *SOPr*, (**d**) *I-P1b*, and (**e**) *SP-P1b*. The figure presents both the main effects (left column) and the Pareto graphs (right column) of standardized effects. The error bars in the main effects plots represent the standard deviations from the dataset of the respective process fingerprint. The black dashed line in the Pareto plots represents the significance level at a 95% confidence level.

4. Conclusions

The present article reports a detailed investigation of the validity of using product and process fingerprints as a new process monitoring concept for fast quality assurance of injection molded components, which adds functional micro features as possible product and process fingerprints. A screening of seven product and 35 process fingerprint candidates was carried out in order to select the most suitable fingerprints based on their sensitivity to process variation and correlation to micro feature replication quality. Topographic analysis of the micro features was conducted off-line by laser scanning confocal microscopy. The signals from machine and external sensors were used for in-line process characterization. Summarizing the previously discussed results, a number of conclusions can be drawn.

- The variation of the IM process parameter settings had an effect on the overall quality and replication of the molded micro-featured surfaces. In particular, the variation affects the quality of the functional micro pillars of the F structure that was positioned in Cavity 1.
- Based on the variation of the injection molding process, the quality of the seven molded structures was assessed and their suitability as a product fingerprint was verified in a correlation analysis of combinations of measurands. For Cavity 1, the dataset C1PP5 (micro pillars near the gate) could be used to predict the quality of the dataset from position C1PP2 (micro pillars far from the gate) since the correlation C1PP2/C1PP5 was $|\rho| = 0.877$. For Cavity 2, the results show that the dataset C2R111 (micro ridges far from the gate) can be used to predict the quality of the dataset from position C2R151 since the correlation C2R151/C2R111 was $|\rho| = 0.93$.
- Additionally, when looking for a product fingerprint in one cavity that could also be representative of the replication in the other cavity, the quality of features in position C1PP2 (micro pillars far from the gate) can be produced by the dataset from position/measurand C2R111 or C2R151 with a correlation C1PP2/C2R111 of $|\rho| = 0.63$ and C1PP2/C2R151 ($|\rho| = 0.68$). This observation suggests that feature C2R151 (micro ridges far from the gate) was the most suitable product fingerprint for both cavities and can be placed on μ-pillar structured components far from the position of the gate in order to be used for the monitoring of the microstructures' quality.
- From the wide variety of process fingerprint candidates assessed for their suitability for monitoring the quality of microstructure replication, a small number of candidates proved to be suitable when considered in combination with specific product measurands. These process fingerprints originated from two data sources including one from the in-mold sensors and the other from the IM machine itself.
- It was concluded that the functional features at position C1PP2 can be successfully monitored with the use of two process fingerprints such as *SP-P1b* (power of signals "**SP**" originated from pressure sensor *P1b*, located far from the gate at the end of the flow path, correlation $|\rho| = 0.772$) and I-P1b (integral "*I*" of signals originated from pressure sensor *P1b*, correlation $|\rho| = 0.749$). These fingerprints are also suitable for the monitoring of a micro feature quality in positions C2R111 ($|\rho| = 0.749$), C2R151 ($|\rho| = 0.758$), C2R511 ($|\rho| = 0.655$), and C2R551 ($|\rho| = 0.658$). Process fingerprints *SP-P1b* and *I-P1b* are more suitable for monitoring the features at positions located closer to the end of the flow path. In comparison, the micro structures in Cavity 2 cannot be monitored reliably by process fingerprints derived from machine data while the larger micro structures in Cavity 1 can.
- The machine process parameter *Cushion* is a suitable process fingerprint and can be used to monitor the quality of micro pillars in position C1PP2 (correlation $|\rho| = 0.61$). *SOPr* is also a suitable process fingerprint and can be used to monitor the quality of micro pillars in position C1PP5 ($|\rho| = 0.61$).
- In summary, *SP-P1b*, *I-P1b*, *Cushion*, and *SOPr* are suitable process fingerprints for the monitoring of the functional micro pillars in positions C1PP2 and C1PP5. The *I-P1b* and *SP-P1b* process fingerprints provide higher correlation to the overall part quality and indicate that the positioning

of a pressure sensor at the end of the flow path can actively provide data for fast quality assurance, which renders high intensity dimensional metrology efforts unnecessary.

Based on the discussed research work, the presented product and process fingerprint concept could be applied to injection molded components with micro-structured geometries similar to the test disk geometries used for this experiment such as plates. Therefore, future work will attempt to validate the concept for different molded components with functional microstructures and to assess the robustness of product and process fingerprint performance in long production runs while implementing mathematical models to predict the quality of the manufactured components.

Author Contributions: N.G., G.T., and Y.Z. conceived and designed the experiments. N.G. performed the experiments, conducted the measurements, and analyzed the data. G.T. and M.K. provided consultation services. N.G. wrote the paper. G.T., Y.Z., and M.K. revised the paper.

Funding: This research was funded by Innovation Fund Denmark grant number 3067-00001B and the APC was funded by the Technical University of Denmark.

Acknowledgments: This paper reports work undertaken within the framework of the project MADE (Manufacturing Academy of Denmark, http://en.made.dk/) Work Package 3 "3D Print and New Production Processes" of the Research Platform MADE SPIR (Strategic Platform for Innovation and Research, http://en.made.dk/spir/). MADE is a collaborative research project supported both by the Danish Manufacturing Industry and by the Innovation Fund Denmark (https://innovationsfonden.dk/en). MADE and Innovation Fund Denmark are thanked for providing financial support to the PhD project "Precision Injection Moulding of Micro Features using Integrated Process/Product Quality Assurance".

Conflicts of Interest: The authors declare no conflict of interest.

References

1. Alting, L.; Kimura, F.; Hansen, H.N.; Bissacco, G. Micro Engineering. *CIRP Ann. Manuf. Technol.* **2003**, *52*, 635–657. [CrossRef]
2. Brousseau, E.B.; Dimov, S.S.; Pham, D.T. Some recent advances in multi-material micro- and nano-manufacturing. *Int. J. Adv. Manuf. Technol.* **2010**, *47*, 161–180. [CrossRef]
3. Wang, X.; Gu, J.; Shen, C.; Wang, X. Warpage optimization with dynamic injection molding technology and sequential optimization method. *Int. J. Adv. Manuf. Technol.* **2015**, *78*, 177–187. [CrossRef]
4. Tosello, G.; Gava, A.; Hansen, H.N.; Lucchetta, G. Study of process parameters effect on the filling phase of micro-injection moulding using weld lines as flow markers. *Int. J. Adv. Manuf. Technol.* **2010**, *47*, 81–97. [CrossRef]
5. Calaon, M. Process Chain Validation in Micro and Nano Replication. Ph.D. Thesis, Department of Mechanical Engineering, Technical University of Denmark, Lyngby, Denmark, 2014.
6. Gao, R.X.; Tang, X.; Gordon, G.; Kazmer, D.O. Online product quality monitoring through in-process measurement. *CIRP Ann. Manuf. Technol.* **2014**, *63*, 493–496. [CrossRef]
7. Kusić, D.; Kek, T.; Slabe, J.M.; Svečko, R.; Grum, J. The impact of process parameters on test specimen deviations and their correlation with AE signals captured during the injection moulding cycle. *Polym. Test* **2013**, *32*, 583–593. [CrossRef]
8. Gao, R.X.; Kazmer, D.O. Multivariate sensing and wireless data communication for process monitoring in RF-shielded environment. *CIRP Ann. Manuf. Technol.* **2012**, *61*, 523–526. [CrossRef]
9. Chen, Z.; Turng, L.-S.; Wang, K.-K. Adaptive Online Quality Control for Injection Molding by Monitoring and Controling Mold Separation. *Polym. Eng. Sci.* **2006**, *46*, 569–580. [CrossRef]
10. Johnston, S.; Mccready, C.; Hazen, D.; Vanderwalker, D.; Kazmer, D. On-Line Multivariate Optimization of Injection Molding. *Polym. Eng. Sci.* **2015**, *55*, 1–8. [CrossRef]
11. Yang, Y.; Yang, B.; Zhu, S.; Chen, X. Online quality optimization of the injection molding process via digital image processing and model-free optimization. *J. Mater. Process. Technol.* **2015**, *226*, 85–98. [CrossRef]
12. Ozcelik, B.; Erzurumlu, T. Comparison of the warpage optimization in the plastic injection molding using ANOVA, neural network model and genetic algorithm. *J. Mater. Process. Technol.* **2006**, *171*, 437–445. [CrossRef]
13. Liu, C.; Manzione, L.T. Process Studies in Precision Injection Molding I Process Parameters and Precision. *Polym. Eng. Sci.* **1996**, *36*, 1–9. [CrossRef]

14. Liu, C.; Manzione, L.T. Process Studies in Precision Injection Molding. II: Morphology and Precision in Liquid Crystal Polymers. *Polym. Eng. Sci.* **1996**, *36*, 10–14. [CrossRef]

15. Park, K.; Ahn, J.-H. Design of experiment considering two-way interactions and its application to injection molding processes with numerical analysis. *J. Mater. Process. Technol.* **2004**, *146*, 221–227. [CrossRef]

16. Attia, U.M.; Alcock, J.R. Evaluating and controlling process variability in micro-injection moulding. *Int. J. Adv. Manuf. Technol.* **2010**, *52*, 183–194. [CrossRef]

17. Attia, U.M.; Alcock, J.R. An evaluation of process-parameter and part-geometry effects on the quality of filling in micro-injection moulding. *Microsyst. Technol.* **2009**, *15*, 1861–1872. [CrossRef]

18. Autodesk. *Moldflow Synergy 2017*; Autodesk: San Rafael, CA, USA, 2017.

19. Marhöfer, D.M.; Tosello, G.; Islam, A.; Hansen Nøgaard, H. Gate design in injection molding of microfluidic compoments using process simulations. In Proceedings of the 4M/ICOMM2015 Conference, Milan, Italy, 31 March–2 April 2015; pp. 546–549. [CrossRef]

20. Giannekas, N.; Tosello, G.; Zhang, Y. A study on replication and quality correlation of on-part and on-runner polymer injection molded micro features. In Proceedings of the World Congress on Micro and Nano Manufacturin, Kaohsiung, Taiwan, 27–30 March 2017; pp. 365–368.

21. Barbato, G.; Garmak, A.; Genta, G. *Measurements for Decision Making: Measurements and Basic Statistics*; Società Editrice Esculapio: Bologna, Italy, 2013.

22. Lee Rodgers, J.; Alan Nice Wander, W. Thirteen ways to look at the correlation coefficient. *Am. Stat.* **1988**, *42*, 59–66. [CrossRef]

micromachines

MDPI

Article

Multi-Response Optimization of Electrical Discharge Machining Using the Desirability Function [†]

Rafał Świercz *, Dorota Oniszczuk-Świercz and Tomasz Chmielewski

Institute of Manufacturing Technology, Warsaw University of Technology, 00-661 Warsaw, Poland;
doo@meil.pw.edu.pl (D.O.-Ś.); t.chmielewski@wip.pw.edu.pl (T.C.);
* Correspondence: rsw@meil.pw.edu.pl; Tel.: +48-22-234-7221
† This paper is an extended version of our paper published in the XIII Internation al Conference Electromachining 2018 that was held in Bydgoszcz, Poland, 9–11 May 2018.

Received: 16 November 2018; Accepted: 14 January 2019; Published: 20 January 2019

Abstract: Electrical discharge machining (EDM) is a modern technology that is widely used in the production of difficult to cut conductive materials. The basic problem of EDM is the stochastic nature of electrical discharges. The optimal selection of machining parameters to achieve micron surface roughness and the recast layer with the maximal possible value of the material removal rate (MRR) is quite challenging. In this paper, we performed an analytical and experimental investigation of the influence of the EDM parameters: Surface integrity and MRR. Response surface methodology (RSM) was used to build empirical models on the influence of the discharge current I, pulse time t_{on}, and the time interval t_{off}, on the surface roughness (Sa), the thickness of the white layer (WL), and the MRR, during the machining of tool steel 55NiCrMoV7. The surface and subsurface integrity were evaluated using an optical microscope and a scanning profilometer. Analysis of variance (ANOVA) was used to establish the statistical significance parameters. The calculated contribution indicated that the discharge current had the most influence (over the 50%) on the Sa, WL, and MRR, followed by the discharge time. The multi-response optimization was carried out using the desirability function for the three cases of EDM: Finishing, semi-finishing, and roughing. The confirmation test showed that maximal errors between the predicted and the obtained values did not exceed 6%.

Keywords: electrical discharge machining; electrical discharge machining (EDM); surface roughness; surface integrity; optimization; desirability function

1. Introduction

Electrical discharge machining is a precision method of manufacturing hard, complex shaped, conductive materials. The removal mechanism of the material in EDM is the result of the electrical discharge, which causes melting and evaporation in the local surface layers of both the workpiece and the working electrode. Owing to the impact of the thermal and chemical processes, the electrical discharge properties of the surface layer of the material are changed [1–3]. Craters form a specific surface texture. Surface roughness parameters directly depend on the discharge current and the time pulse [4]. Heat flux changes the surface integrity. Metallographic images of the machined samples of tool steel show new layers, i.e., the external melting layer (white layer), the heat-affected zones, and the tempered layer. The white layer is non-homogeneous and discontinuous and may vary in thickness on the analyzed sample. Discontinuity of the melted layer is caused by the random occurrence of electrical discharge and the machining conditions. This layer is characterized by high variations in thickness and microstructure defects of the material, such as micro-cracks. Microcracks are an undesirable effect, resulting in reduced fatigue resistance and corrosion resistance. It is important to choose the appropriate parameters and processing conditions to obtain the smallest thickness of the white layer or its complete elimination. The quality of the surface after the EDM process does not always meet expectations [5,6]. Therefore, additional technological operations are used to change the surface integrity. Some of the most applicable operations are electrochemical machining [7,8], laser surface modification [9,10], applying coatings [11,12], or the use of hybrid machining [13,14] or non-conventional finishing [15–18]. However, the use of additional technological operations significantly increases the production costs. Therefore, this work analyzed the impact of the following parameters of the EDM process: discharge current, pulse time, and pulse interval. Furthermore, the optimization of EDM using the desirability function will allow for the selection of the most favorable processing conditions, which reduce the use of additive treatments to the bare minimum.

Electrical discharge machining belongs to a group of non-conventional manufacturing techniques. The material is removed from the workpiece using electrical discharges occurring between the working electrode and the workpiece. Owing to the conducted thermal energy from the electrical discharge, the local temperature increases (in the range of 8000–12,000 °C), leading to the melting and evaporation of a small volume of the surface workpiece and the working electrode. Then, a collapsing plasma channel at the end of the discharge induces high-pressure waves that rinse the molten and evaporated metal [19]. The physics of the material removal phenomenon in EDM is complex. A model of the EDM process proposed by Izquierdo et al. [20], shows that by using the superposition of multiple discharges and calculating the temperature fields inside the workpiece, it is possible to predict the surface roughness parameters and the material removal rate. One of the main problems in modeling is the appropriate determination of the influence of EDM parameters like discharge voltage, discharge current, and pulse time on ionization and growth of the plasma channel. Information about the percentage of discharge energy devoted to the heat flux, the mechanism of plasma channel growth, and the temperature which facilitates material removal, is used to model not only surface roughness, but also to build models of structural changes in the surface layers and their thickness [21]. Ming et al. [22] analyzed the distributions of the energy of the workpiece for different materials like Al 6061, Inconel 718, and SKD11. The authors indicated that the analyses of energy efficiency simultaneously with MRR, could be used in existing thermal–physical models to improve the technical performance of the models. Theoretical models of EDM are significant because the models analyze the processes and their implication on the manufactured material. However, the application of the theoretical model in the manufacturing process is not always possible. Given the complexity in describing the physical phenomena of EDM, such as the random distribution of the electric field and the temperature field, the formation of the plasma channel, and changes in the properties of the dielectric, mathematical models have been built based on the empirical studies. Several process variables promote the application of the optimization method to achieve high productivity.

Optimization of the EDM machining process can be carried out using various methods like response surface methodology [23,24], Taguchi analyses [25], artificial networks [26], grey-based response surface methodology [27,28], and the Deringer desirable [29] or hybrid methods [30,31]. One of the most common manufacturing processes is the response surface methodology (RSM). RSM is extensively used in an analytical and industrial application like turning [32,33], milling [34], welding [35], grinding [36], and erosion machining [37]. Ghodsiyeh et al. [38] used RSM to optimize the wire electrical discharge machining of the titanium alloy Ti-6Al-4 V. Presented results indicated that the pulse time had the highest impact on surface roughness, whilst discharge current had the same role for the wire wear ratio and white layer thickness. Alavi et al. [39] analyzed the influence of the EDM parameters on tool wear, crater size, and microhardness on the titanium alloy Ti-6Al-4V. The presented research indicates that the main effect on the crater size was the discharge voltage, whilst the capacitance was the most important for tool wear and surface micro-hardness. Increasing the capacitance caused a reduction in tool wear and an increase in the micro-hardness of the machined surface. Selvarajan et al. [40] investigated the possibilities of the EDM composite Si_3N_4-TiN. The authors used RSM to determine the optimal parameters for the machining of ceramic composites.

The conducted research showed that the MRR and surface roughness of the manufactured parts depended on the value of the discharge current and the pulse time, and the results were consistent with the results for the processing of tool steel [41]. Kumaran et al. [42] used a grey fuzzy logic approach to optimize the EDM parameters during the machining of carbon fiber reinforced plastic composite. Their research showed that established optimal parameters in ultrasonic-assisted EDM allowed improvements in the deburring rate, with a simultaneous improvement in the tool wear rate (TWR). Gu et al. [43] indicated that the machining of new alloys, which have a high melting point and good thermal conductivity, like titanium–zirconium–molybdenum, require optimizations of the EDM in connection with the analysis. The presented results showed that the crater diameter was much smaller than the plasma-affected zone. To improve the machining performance (*Ra*, MRR), the response surface methodology was used. Dang [44] proposed the Kriging regression model and particle swarm in the optimization of EDM of P20 steel. The authors indicated that the Kriging model could capture the nonlinear characteristics and was better able to obtain the optimum parameters for the MRR, tool wear, and surface roughness. Mohanty et al. [45] pointed out that the choice of the electrode material should be considered in the optimization of the EDM process. The authors found that in the machining of Inconel 718, the material removal rate and tool wear could be improved by using a graphite electrode and to improve surface integrity, the better choice was a brass and copper electrode. Using the utility concept and the quantum particle swarm optimization (QPSO) algorithm, the optimal parametric setting was developed with the objectives to maximize the MRR and minimize tool wear, surface roughness, and radial overcut. Research presented by Maity et al. [46] on the influence of EDM parameters on the thickness of the recast layer, material removal rate, and overcut on the machining of Inconel 718, showed that the optimization parameters using the RSM and Artificial Bee Colony algorithm gave an average prediction accuracy of about 3.5% in relation to confirmation tests. The predictive efficiency of neural networks may be affected by different factors like noise corruption, spatial distribution, and the size of the data used to construct the artificial neural network (ANN) model. Tripathy et al. [47], in order to optimize the powder mixed (SiC) electro-discharge machining of H-11 die steel, which seeks to maximize the MRR and minimize electrode wear and surface roughness, used a different method of optimization. The authors used the grey relational analysis and the technique for the order of preference by similarity, where the Technique for Order of Preference by Similarity to Ideal Solution (TOPSIS) solution achieved a similar effect of an improved performance of the quality characteristics. Nguyen et al. [48] investigated the influence of powder mixed (Ti) electro-discharge machining of SKD61, SKD11, SKT4 steel on the surface roughness (SR), MRR, and microhardness. The presented results showed that the addition of Ti powder in dielectric resulted in reduced SR and increased microhardness. The authors indicated that optimization with the Taguchi–TOPSIS made it difficult to select the optimal parameters. The presented research showed that the measured distance could lead to confusion in selecting the best alternative. A fuzzy analytic hierarchy process (AHP) and fuzzy TOPSIS method were used by Roy et al. [49] to optimize multiple responses of the material removal rate, tool wear, and tool overcut in EDM based on various process parameters. Kandpal et al. [50] investigated the influence of EDM parameters on the MRR, tool wear, and overcut of aluminum matrix composites. The authors showed optimization with the utility concept, which provided the collective optimization of both responses for improving the mean of the process. D'Urso et al. [51] proposed the optimization of EDM micro drilling using a cost index which combined two opposite effects of the material removal rate and tool wear. The minimization of the cost index enabled optimal working conditions. Parsana et al. [52] indicated that in the optimization process of EDM drilling, an important point to consider was the roundness of the holes. Using the RSM and passing vehicle search algorithm, normalized weights proved to be useful in obtaining the Pareto fronts for a combination of different objectives at a time. Research carried out by Hadad et al. [53] showed that the analysis of the optimal parameters of EDM machining, in addition to the electrical parameters, should also include the initial roughness of the working electrodes, which has significant effects on the machining performance during the finishing, semi-rough, and rough EDM processes.

The published literature indicates that few studies have reported on the optimization of the EDM, which considers the surface 3D roughness parameter, white layer thickness, and the MRR in the three stages of machining: roughing, semi-finishing, and finishing, with the desirable function. Therefore, in this paper, a multi-response optimization of the EDM of tool steel 55NiCrMoV7 was conducted. This material has a wide range of industry applications on die matrices, matrix inserts, and hydraulic and mechanical press dies. The surface roughness was investigated using the 3D roughness parameter. The parameter, Sa, gives more information about the surface properties. In electrical discharge machining, the surface roughness is obtained by overlapping the craters of individual discharges. EDM with parameters corresponding to the roughing and semi-finishing operations results in surfaces which have different profiles on the cross-sections of samples. The calculated Ra parameters on the profile cross-sections may provide inaccurate results.

The optimization of parameters was performed using the desirable function. Optimization of electrical discharge machining was divided into three cases: finishing, semi-finishing, and roughing. In each case, different goals were set. For finishing, the goal was to minimize the surface roughness (Sa) simultaneously whilst minimizing the thickness of the white layer (WL), with a possible maximized MRR. In the case of semi-finishing, the EDM goal was to obtain a specific value of the MRR whilst possibly minimizing the roughness (Sa), simultaneously minimizing the thickness of the white layer (WL). In the last case of roughing, the goal was to maximize the material removal rate with the possibility of minimizing the roughness (Sa) and the thickness of the white layer (WL).

The topic of the article focuses on describing the changes occurring in the material as a result of the local thermal processes due to electric discharges. Experimental studies allowed for a better understanding of the relationship between the changes in surface integrity of tool steel 55NiCrMoV7 and how to optimize the process to achieve a micron roughness and recast layer, with the possibility of achieving the maximal value of the material removal rate.

2. Materials and Methods

Industry applications of electrical discharge machining are limited by the obtained specific surface integrity and low material removal rate. The purpose of the research was to develop the multi-response optimization of the EDM process of the tool steel 55NiCrMoV7 for three cases: finishing, semi-finishing, and roughing. Tool steel 55NiCrMoV7 was chosen because of its wide industry applications on die matrices, matrix inserts, and hydraulic and mechanical press dies. This material is characterized by high dimensional stability and crack resistance, with dynamically changing pressures and rapid heating and cooling during operation. Heat-treated samples of the tool steel (55 HRC) had the dimensions of 12 mm × 12 mm × 3 mm. Experimental studies were conducted on the electrical discharge machine, Charmilles Form 2LC ZNC (GF Solutions, Geneva, Switzerland). The electrode used was graphite (EDM-3 POCO), and the EDM fluid 108 MP-SE 60 was used as the dielectric. The present paper was focused on the selection of optimal parameters for EDM, which led to minimum surface roughness, as well as the thickness of the white layer and maximum productivity.

The main object of the study was the optimization of the EDM process using statistical models on the influence of EDM parameters on surface roughness (Sa), the thickness of the white layer, and the material removal rate. To achieve this goal, experimental research was carried out using a completely orthogonal design of the experiment, three-level three parameters full factorial design. Choice of this type of experiment design allowed the reduction in the number of experimental runs required to generate sufficient information for a statistically adequate result. A schematic diagram of the experimental set-up is shown in Figure 1. Investigation of the surface roughness parameters after the EDM was carried out on a Taylor–Hobson FORM TALYSURF Series 2 scan profilometer (Taylor Hobson, Leicester, United Kingdom). The roughness parameter (Sa), the arithmetic mean of the deviations from the mean, was measured on a surface area of 2 mm × 4 mm with a discretization step (10 µm) in the X-axis and Y-axis. The Sa (average value of the absolute heights over the entire surface) parameter responded to the 2D roughness profile parameters Ra. This may be obtained by adding the individual height values, without regard to sign, and dividing the sum by the number of the data matrix, where M is the number of points per profile, N is the number of profiles, and z, x, y are the heights of the profile at a specific point.

$$Sa = \frac{1}{NM} \sum_{x=0}^{N=1} \sum_{y=0}^{M=1} |z_{x,y}| \tag{1}$$

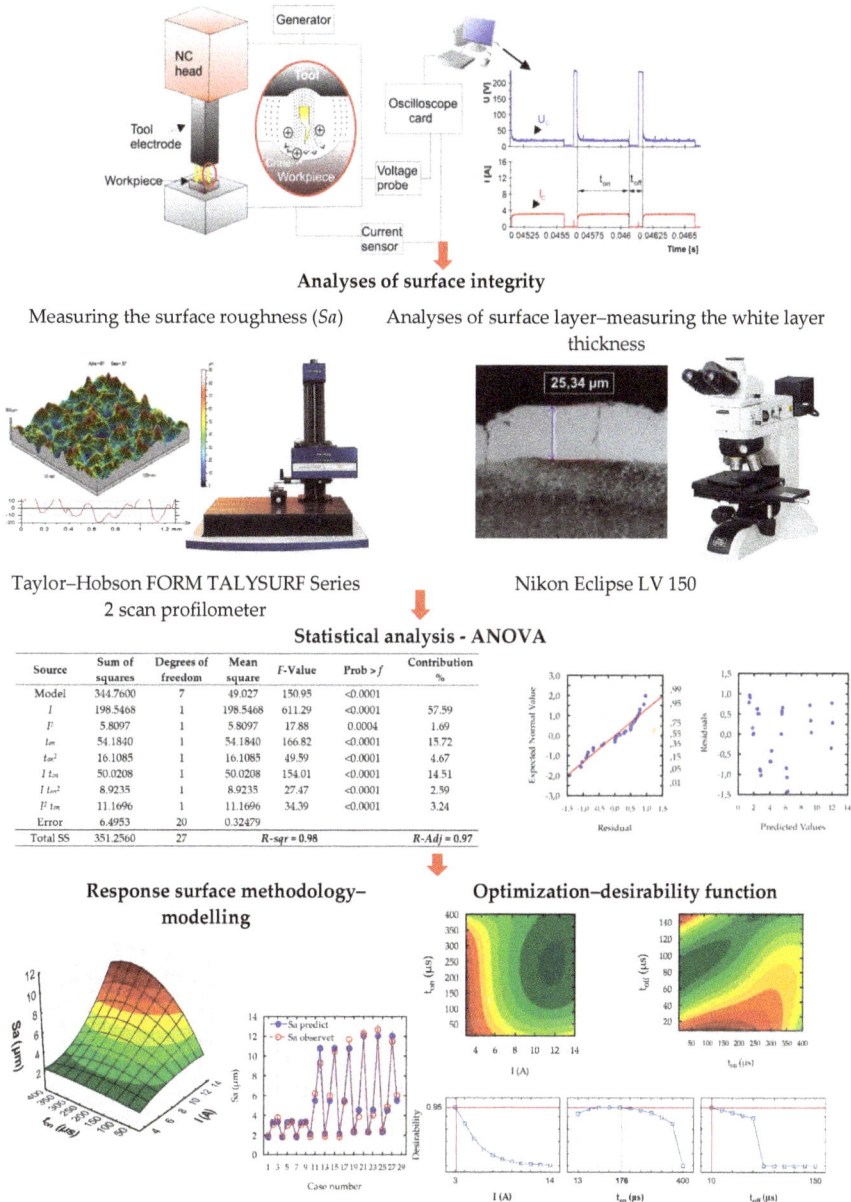

Figure 1. The schematic diagram of the experimental set-up.

Metallographic surface structure studies were performed using a Nikon Eclipse LV 150 optical microscope (Nikon, Tokyo, Japan), coupled to an NIS-Elements BR 3.0 image analyzer (Nikon). Specimens were included in the resin, and were then machined with grinding and polishing. Micro-etching was performed with nital (5%) to reveal the microstructure of the material. The maximum thickness of the white layer in sections was measured for each sample.

2.1. Uncertainty Evaluation Procedure

To verify the quality of the measurements, an uncertainty evaluation was carried out. The measurements were carried out inside a metrological laboratory with a 20 °C ± 0.5 °C controlled temperature. The thermal deformation of the samples was neglected. The surface topography measurements were carried out using a Taylor–Hobson FORM TALYSURF Series 2 scan profilometer (Taylor Hobson, Leicester, United Kingdom). The raw surface acquisitions were post-processed with the dedicated image metrology software TalyMap (Taylor Hobson, Leicester, United Kingdom). The calibration of the scanning profilometer was performed with a calibrated roughness artifact (nominal value: $Ra = 810$ nm). According to Reference [54], it is possible to establish uncertainties for surface roughness Sa measurements following ISO 15530-3 [55].

The uncertainties for surface roughness measurements when using the induction scanning profilometer U_{PROF} was calculated according to the following equation:

$$U_{PROF} = \sqrt{u^2_{cal} + u^2_p + u^2_{res, PROF}}$$ (2)

where u_{cal} is the standard calibration uncertainty of the roughness standard; u_p is the standard uncertainty related to the measurement procedure and is calculated as a standard deviation of ten repeated measurements on the calibrated standard; and $u_{res,PROF}$ is the resolution standard uncertainty related to the declared 3 nm vertical resolution of the scanning profilometer for measuring range 0.2 mm.

The expanded uncertainty of the surface roughness Sa measurement was calculated as follows:

$$U_{95,Sa} = k \sqrt{U^2_{PROF} + U^2_{Sa,EDM}}$$ (3)

where k is the coverage factor, equal to 2 for a 95% confidence interval, $u_{Sa,EDM}$ is calculated using the standard deviation of repeated Sa measurements on the electrical discharge machining sample. Table 1 presents the uncertainty budget.

Table 1. The uncertainty contributions for the Sa roughness measurements on the scanning profilometer.

Uncertainty Contributions (nm)					
u_{cal}	u_p	$u_{res,PROF}$	U_{PROF}	$U_{Sa, EDM}$	$U_{95,Sa}$
20	2	3	20.5	5	42

The measurements of the thickness of the white layer were carried out using a Nikon Eclipse LV 150 optical microscope, coupled to an NIS-Elements BR 3.0 image analyzer (Nikon). The uncertainty was calculated using the above method. The calibrated slide with 10 μm division was selected as the calibrated artifact.

The uncertainties for the thickness of the white layer when using optical microscope U_{OM} was calculated according to the following equation:

$$U_{OM} = \sqrt{u^2_{cal} + u^2_p + u^2_{res, OM}}$$ (4)

where u_{cal} is the standard calibration uncertainty of the calibration slide, u_p is the standard uncertainty related to the measurement procedure and is calculated as the standard deviation of ten repeated measurements on the calibrated slide; and $u_{res,OM}$ is the resolution standard uncertainty related to the objective magnification (50x). The expanded uncertainty of the thickness of the white layer measurement was calculated as follows:

$$U_{95,WL} = k \, U_{OM} \qquad (5)$$

where k is the coverage factor, equal to 2 for a 95% confidence interval. Table 2 presents the uncertainty budget.

Table 2. The uncertainty contributions for the thickness of the white layer measurements on microscope.

Uncertainty Contributions (µm)				
u_{cal}	u_p	$u_{res,OM}$	U_{OM}	$U_{95,WL}$
0.060	0.048	0.312	0.321	0.6

The material removal rate (*MRR*) was calculated based on the volume of material removed from the workpiece divided by the machining time:

$$MRR = \frac{m_1 - m_2}{\rho \, \Delta t} \left[\frac{mm^3}{min} \right] \qquad (6)$$

where m_1 is the sample weight before processing, m_2 is the sample weight after processing, ρ is specific material density, Δt is a time of manufacturing.

Each sample was weighed before manufacturing on a precision electronic balance (Radwag, Radom, Poland). The samples after the EDM process were cleansed with the compressed air and then weighed again. The measurement uncertainties for the weight measurements were calculated according to the following equation:

$$U_B = \sqrt{u^2_{m1} + u^2_{res} + u^2_{i} + u^2_{ie}} \qquad (7)$$

where u_{m1} is the standard uncertainty related to the measurement procedure and is calculated as the standard deviation of the ten repeated measurements of the sample, u_{res}, is the resolution uncertainty related to the declared resolution of the balance with a readability 0.01 mg for the measuring samples of a maximum capacity 50 g, u_i is the uncertainty related to a balance indication error, and u_{ie} is the uncertainty on a determining indication error.

The expanded uncertainty of the weight measurement was calculated as follows:

$$U_{95,W} = k \, U_B \qquad (8)$$

where k is the coverage factor, equal to 2 for a 95% confidence interval. Table 3 reports the uncertainty budget.

Table 3. The uncertainty contributions for the weight measurements on the precision balance.

Uncertainty Contributions (mg)					
u_{m1}	u_{res}	u_i	u_{ie}	U_B	$U_{95,W}$
0.02	0.0029	0.0058	0.01	0.023	0.046

2.2. Analyses of Current and Voltage Waveforms.

The complexity of the physical phenomena of the EDM process and its conditions caused considerable difficulties in describing and identifying the impact of individual machining parameters on the surface roughness, integrity, and the MRR. In the first stage of the conducted research, the measurement circuit was developed to determine the current–voltage characteristics of the generator machines. A primary test was conducted to investigate a range of stability discharges for different values of the discharge current, pulse time, and time interval. The measurement of the current and voltage waveforms in the EDM process conditions was done using a National Instruments NI5133 oscilloscope card (National Instruments, Austin, TX, USA). An application was developed in the LabView environment, which enabled the control of the work of the oscilloscope card. The current measurement was done using the indirect method as the voltage drop on the non-inductive current sensor. The maximum value of the voltage drop for the set current values did not exceed 3V, so the signal was fed directly to the oscilloscope card. The measurement of the voltage during the electric discharge was done with the Tektronix probe (Tektronix UK Ltd., Berkshire, UK). The sampling rate was 100 MS/s, 2-Channel registration. Analyses of the obtained data were performed in DIAdem (National Instruments).

Exemplary current and voltage waveforms registered for the investigated EDM are shown in Figure 2. The workpiece was machined at the moment when the supply voltage Uz dropped to the discharge voltage Uc, with the increase of the discharge current I, during the pulse t_{on}. Then, during the t_{off} interval, the conditions in the gap stabilized, and the process was cyclically repeated.

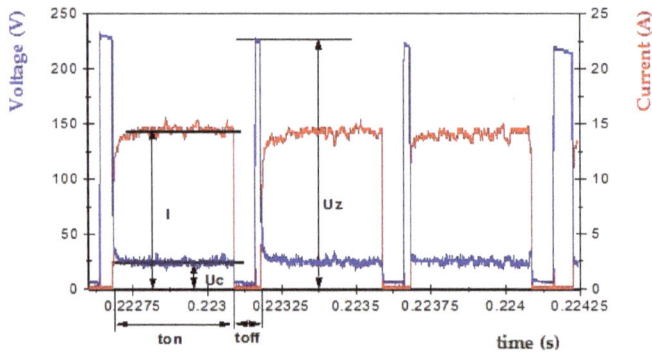

Figure 2. The recorded current and voltage waveforms.

The following parameters can characterize the voltage–current waveforms (Figure 2):

- I = the height of the peak current during discharging,
- Uz = open circuit voltage, this is the system voltage when the EDM circuit is in the open state, and the energy has been built up for discharge,
- Uc = discharge voltage,
- t_{on} = pulse time, the time required for the current to rise and fall during discharging,
- t_{off} = time interval, this is the time from the end of one pulse to the beginning of the next pulse with the current.

Analysis of the obtained voltage and current waveforms showed that at the highest adjustable currents, pulse duration, and minimum values of the break times, in most cases, arc discharges or short circuits occurred. For short time intervals, t_{off}, the plasma channel may not be completely deionized, which increases the probability of another discharge being in the same place. Furthermore, the ineffective removal of the products of erosion from the gap causes a reduction in the dielectric resistance and destabilization of the conditions. There is a high probability of a short circuit. The machine's control system resists the phenomena described above by increasing the gap (temporarily raising the electrode), whilst at the same time extending the break time (Figure 3).

The proper operation of the generator control system ensures the energy repeatability of discharges. Therefore, the presented disturbances will not have a significant impact on the quality of the treated surfaces. Nevertheless, unfavorably selected ranges of the set parameters will significantly affect the efficiency of the process [56]. The following are examples of stable $U(t)$, $I(t)$ waveforms, which enabled the selection of the range of variability of parameters used in the experimental research (Figure 4).

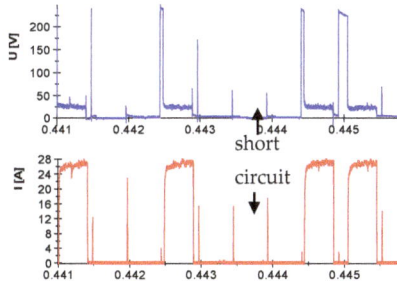

Figure 3. The recorded current and voltage waveforms.

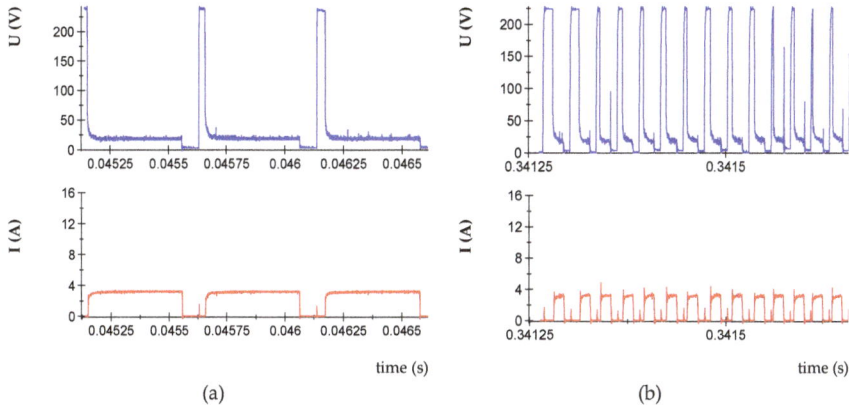

Figure 4. The recorded voltage and current waveforms for the following parameters: (**a**) U_c = 25 V, U_z = 230 V, I = 3 A, t_{on} = 400 μs, t_{off} = 100 μs; (**b**) U_c = 25 V, U_z = 230 V, I = 3 A, t_{on} = 13 μs, t_{off} = 13 μs.

Analysis of the recorded voltage and current waveforms enabled the selection of stable parameters in the EDM process for roughing, semi-finishing, and finishing machining. Experimental studies were conducted using the orthogonal full factorial design DOE (design of experiments) methodology: Orthogonal full factorial design. Preliminary experiments were conducted to obtain the stable discharges in all ranges of the design matrix. After analysis of the results, the following machining conditions were selected: discharge current in the range I = 3–14 A, pulse time in the range t_{on} = 10–400 μs, and time interval t_{off} = 10–150 μs, with the following constants :open voltage U_0 = 225 V, discharge voltage Uc = 25 V. Table 4 shows the levels of machining parameters carried out in the experimental design.

Table 4. The levels of machining parameters carried out in the experimental design.

EDM Parameters	Level 1	Level 2	Level 3
discharge current I (A)	3	8.5	14
pulse time t_{on} (μs)	13	206	400
time interval t_{off} (μs)	9	80	150

3. Results and Discussion

3.1. Analysis of Surface Integrity

The primary factor that affected the properties of the machined parts was the surface texture. Experimental investigations showed that parameters such as surface roughness (*Sa*), directly depended on the applied machining parameters. Surface topography after EDM was the result of the overlapping of craters from single discharges and it had an isotropic structure (Figure 5).

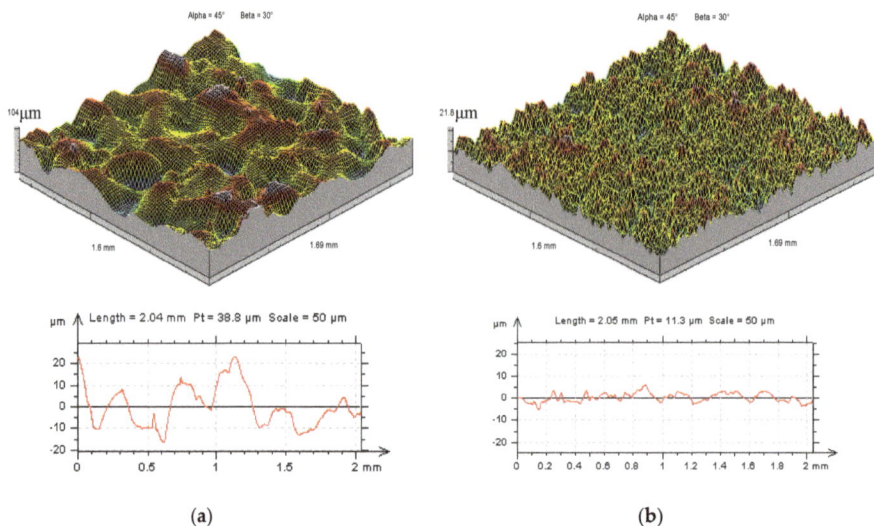

Figure 5. The surface texture of the tool steel 55NiCrMoV7 after electrical discharge machining (EDM): (a) $U_c = 25$ V, $I = 8.5$ A, $t_{on} = 400$ μs, $t_{off} = 150$ μs; (b) $U_c = 25$ V, $I = 3$ A, $t_{on} = 400$ μs, $t_{off} = 80$ μs.

Owing to rapid local thermal processes during electrical discharge machining, phase changes occurred on the surface layer of the workpiece. The analysis of images of the metallographic structure of tool steel 55NiCrMoV7 showed the occurrence of three characteristic sublayers for the whole range of the investigated machining parameters (Figure 6).

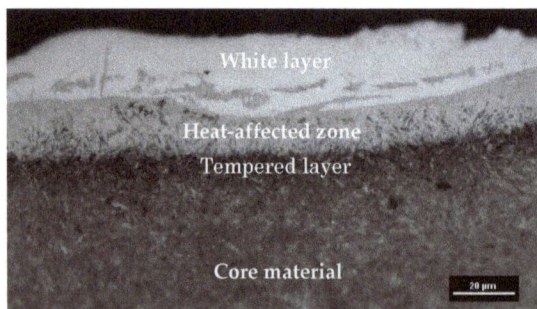

Figure 6. The metallographic structure of tool steel 55NiCrMoV7 after (EDM).

An external molten layer (commonly referred to as a white layer), was formed by melting and rapidly solidifying a thin layer of metal not removed from the surface of the crater during an electric discharge. The white layer in its structure may have chemical decompositions from both the core material and the working electrode. The heat-affected zone (HAZ), which is located directly under the melted layer (visible as a light structure), was characterized by an increased hardness around the core material. The last observed layer was the tempered layer, which was visible as a dark streak immediately below the heat-affected zone layer. The thickness of each observed layer depended on the investigated EDM parameters. The white layer, unlike the heat-affected zone and the tempered layer, was characterized by local discontinuities and thickness changes (Figure 7). The industrial application of the results of the conducted experiments requires information on the largest thickness of the white layer. This information allows for the correct selection of machining allowances for semi-finishing and finishing manufacturing.

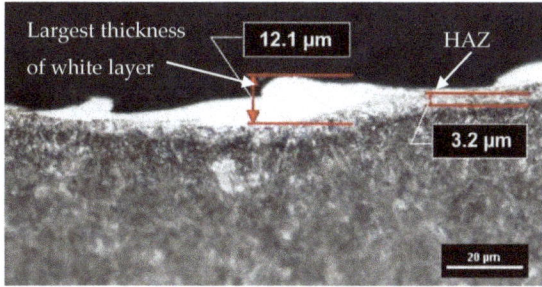

Figure 7. The metallographic structure of tool steel 55NiCrMoV7 after EMD: $Uc = 25$ V, $I = 14$ A, $t_{on} = 13$ μs, $t_{off} = 10$ μs.

Electrical discharges caused the local melting and evaporation of material. The thermal processes of removing material and the rapid re-solidification of the molten metal which was not removed from the discharge crater generated thermal stress. Exceeding the maximum tensile strength of the material caused the generation of micro-cracks. Micro-cracks are an undesirable effect, resulting in reduced fatigue resistance and corrosion resistance. In most cases, micro-cracks propagate to the end of the white layer (Figure 8). In rare cases, the propagation of a crack has been observed to penetrate the core of the material. Micro-cracks can be observed directly on the machining surface (Figure 9).

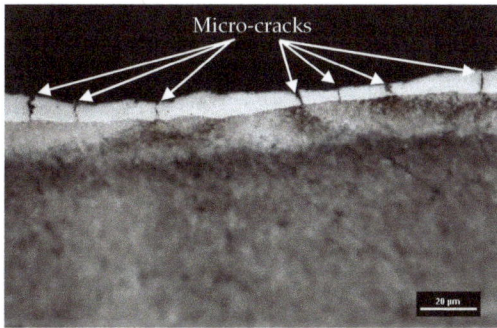

(a)

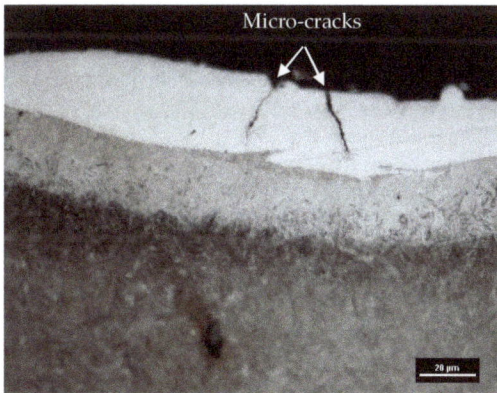

(b)

Figure 8. The metallographic structure of tool steel 55NiCrMoV7 after EMD: (**a**) $Uc = 25$ V, $I = 3$ A, $t_{on} = 206$ μs, $t_{off} = 80$ μs; (**b**) $Uc = 25$ V, $I = 14$ A, $t_{on} = 400$ μs, $t_{off} = 80$ μs.

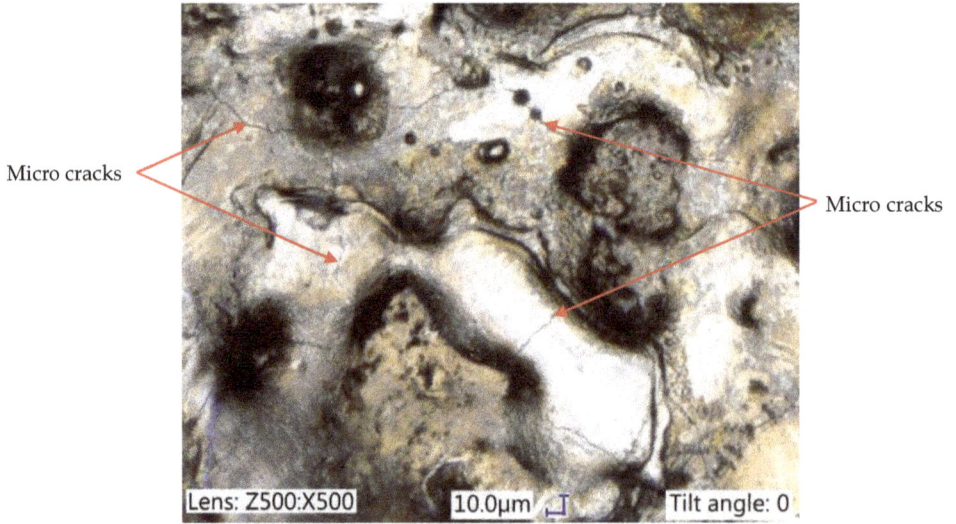

Figure 9. The structure of the surface after EMD: $Uc = 25$ V, $I = 14$ A, $t_{on} = 400$ μs, $t_{off} = 150$ μs.

3.2. Response Surface Methodology

Experimental investigation of the influence of the EDM parameters on the surface roughness (Sa), the thickness of the white layer, and the MRR was carried out using response surface methodology. In RSM, the dependence between the desired response and the independent variables can be represented by the following:

$$Y = f(I, t_{on}, t_{off}) \pm \varepsilon \tag{9}$$

where Y is the response; f is the response function; ε is the experimental error. I the discharge current, t_{on} (μs) the pulse time, and t_{off} (μs) the time interval are independent parameters. In the study, the polynomial regression model was chosen to fit the response function to the experimental results.

The experimental investigation was carried out based on the full factor orthogonal experiment design: three-level three-parameter. The study of the influence of the input factors on three equidistant levels of variation allows for the determination of regression equations with a high degree of correlation and a small spread of values. According to the full factor orthogonal design plan, twenty-eight samples, with one additional replication in the center point, were manufactured and measured. Based on the experimental data, an empirical model of the influence of the discharge current I, pulse time t_{on}, and time interval t_{off} was built. The results of the experimental studies are presented in Table 5. The surface roughness (Sa) was in the range of 1.88 μm to 12.7 μm. The maximal thickness of the white layer was in the range of 5.5 μm to 33.5 μm. The material removal rate was in the range 0.1 mm^3/min to 29.19 mm^3/min. The obtained value of roughness (Sa), the maximal thickness of the white layer, and the *MRR* corresponded to the finishing and roughing machining.

Analysis of variance (ANOVA) was used to check the significance of each independent variable in the response function. The ANOVA test was conducted at a 5% significant level. The *F*-value corresponded to a continuous probability distribution. If this probability (Prob $> f$) value for each factor was less than 0.05, this indicated that the model factor was significant (i.e., at a 95% confidence level). Values of Prob $> f$ higher than 0.05 indicated that a model factor was non-significant.

Table 5. The design of the experimental matrix.

Exp. no.	EDM Parameters			Observed Values		
	Discharge Current I (A)	Pulse Time t_{on} (μs)	Time Interval t_{off} (μs)	Surface Roughness Sa (μm)	Maximal Thickness of the White Layer (μm)	MRR (mm^3/min)
1	3	13	10	2.0	5.5	0.54
2	8.5	13	10	3.1	11.5	3.47
3	14	13	10	3.8	12	11.06
4	3	13	80	1.9	6	0.17
5	8.5	13	80	3.0	12	1.18
6	14	13	80	3.4	11.5	3.21
7	3	13	150	1.9	6	0.10
8	8.5	13	150	3.0	11.5	0.55
9	14	13	150	3.3	12	1.31
10	3	206	10	1.9	7	0.51
11	8.5	206	10	6.2	22	8.09
12	14	206	10	9.3	25.4	28.46
13	3	206	80	1.9	10	0.36
14	8.5	206	80	6.0	24	5.77
15	14	206	80	10.5	28	19.23
16	3	206	150	1.8	10	0.29
17	8.5	206	150	5.4	25	4.68
18	14	206	150	11.7	32	15.48
19	3	400	10	2.4	12	0.37
20	8.5	400	10	3.9	17	6.58
21	14	400	10	12.3	28	29.19
22	3	400	80	2.4	13.5	0.34
23	8.5	400	80	4.0	20	5.61
24	14	400	80	12.7	29	24.84
25	3	400	150	2.5	14	0.28
26	8.5	400	150	4.9	18.4	2.56
27	14	400	150	11.5	33.5	20.31
28	8.5	206	80	6.1	24.5	5.88

The ANOVA results for the *Sa*, white layer thickness, and the MRR are shown in Tables 6–8, respectively. Table 6 shows the ANOVA results for surface roughness (*Sa*). The calculated contribution indicated that the discharge current had the most influence on the surface roughness (*Sa*) (57.6%). Second, the affecting variable was pulse time (15.7%) and the interaction of the discharge current with pulse time (14.5%). Other variables and their interactions had a significant influence on the surface roughness (*Sa*), but each of the contributions did not exceed 5%. The ANOVA results presented in Table 7 indicated that the most significant influence on the maximal thickness of the white layer was the discharge current (47.5%), followed by the pulse time (27.8%) and the squared pulse time (9.2%). The contribution of other variables on the *Wl* was significant but less important. Table 8 presents the ANOVA results for the MRR. The calculated contributions indicated that the discharge current (55.9%) had the most influence on the MRR, followed by the interaction of the discharge current with the pulse time (12.7%) and also the pulse time (11.6%). Other variables and their interactions were significant, but their contributions were smaller and contained in the range (1.6%–5.6%). From the presented ANOVA Tables 6–8, calculated Fisher coefficients for the models *Sa*, WL, and MRR were 150.95, 141.26, and 208.65, respectively. The results implied that all the developed models were significant at a 95% confidence level.

Table 6. The analysis of variance (ANOVA) table for the *Sa* (after elimination).

Source	Sum of Squares	Degrees of Freedom	Mean Square	F-Value	Prob > f	Contribution %
Model	344.7600	7	49.027	150.95	<0.0001	-
I	198.5468	1	198.5468	611.29	<0.0001	57.6
I^2	5.8097	1	5.8097	17.88	0.0004	1.7
t_{on}	54.1840	1	54.1840	166.82	<0.0001	15.7
t_{on}^2	16.1085	1	16.1085	49.59	<0.0001	4.7
$I\,t_{on}$	50.0208	1	50.0208	154.01	<0.0001	14.5
$I\,t_{on}^2$	8.9235	1	8.9235	27.47	<0.0001	2.6
$I^2\,t_{on}$	11.1696	1	11.1696	34.39	<0.0001	3.2
Error	6.4953	20	0.32479	-	-	-
Total SS	351.2560	27		R-sqr = 0.98		R-Adj = 0.97

Table 7. The ANOVA table for the maximal thickness of the white layer (after elimination).

Source	Sum of Squares	Degrees of Freedom	Mean Square	F-Value	Prob > f	Contribution %
Model	1886.366	11	171.48	141.26	<0.0001	-
I	896.656	1	896.656	738.59	0.0022	47.5
I^2	16.041	1	16.041	13.21	<0.0001	0.8
t_{on}	524.880	1	524.880	432.35	<0.0001	27.8
t_{on}^2	174.366	1	174.366	143.62	0.0002	9.2
t_{off}	27.406	1	27.406	22.57	<0.0001	1.4
$I\,t_{on}$	90.750	1	90.750	74.75	0.0004	4.8
$I\,t_{on}^2$	61.584	1	61.583	50.72	0.0031	3.3
$I^2\,t_{on}$	37.210	1	37.210	30.65	0.0071	2.0
$I^2\,t_{on}^2$	44.018	1	44.017	36.25	<0.0001	2.3
$t_{on}\,t_{off}$	5.603	1	5.603	4.61	<0.0001	0.3
$t_{on}^2\,t_{off}$	7.860	1	7.859	6.47	0.0473	0.4
Error	19.424	16	1.2140	738.59	0.0216	-
Total SS	1905.799	27		R-sqr = 0.99		R-Adj = 0.98

Table 8. The ANOVA table for the material removal rate (MRR) (after elimination).

Source	Sum of Squares	Degrees of Freedom	Mean Square	F-Value	Prob > f	Contribution %
Model	2243.49	9	247.881	208.65	<0.0001	-
I	1253.287	1	1253.287	1055.08	<0.0001	55.9
I^2	126.243	1	126.243	106.27	<0.0001	5.6
t_{on}	260.655	1	260.655	219.43	<0.0001	11.6
t_{on}^2	56.489	1	56.489	47.55	<0.0001	2.5
t_{off}	101.381	1	101.381	85.34	<0.0001	4.5
$I\,t_{on}$	285.948	1	285.948	240.72	<0.0001	12.7
$I\,t_{on}^2$	36.030	1	36.030	30.33	<0.0001	1.6
$I^2\,t_{on}$	44.106	1	44.106	37.13	<0.0001	2.0
$I\,t_{off}$	79.350	1	79.350	66.80	<0.0001	3.5
Error	21.381	18	1.188	-	-	-
Total SS	2264.87	27		R-sqr = 0.99		R-Adj = 0.99

Regression analysis with a backward elimination process was performed. For each equation, we calculated the coefficient of determination, *R-squared*, and the adjusted coefficient of determination, *R-Adj*. The coefficients represented the percentage of variance explained by the model. When the value of the *R-sqr* and *R-Adj* approaches unity, a more accurate fit of the regression equation to the research results would be obtained.

After eliminating the non-significant factors in the response equations for the surface roughness (*Sa*), maximal white layer thickness (WL), and MRR, this was described by the following polynomial function:

$$Sa = 0.38 + 0.54\,I - 0.027\,I^2 - 0.0004\,t_{on} + 0.00002\,t_{on}^2 + 0.00006\,I\,t_{on} \\ - 0.000007\,I\,t_{on}^2 + 0.0003\,I^2\,t_{on} \tag{10}$$

$$WL = 1.422 + 1.697\,I - 0.075\,I^2 - 0.101\,t_{on} + 0.0003\,t_{on}^2 - 0.002\,t_{off} + 0.035\,I\,t_{on} - 0.0001\,I\,t_{on}^2 \\ - 0.0014\,I^2\,t_{on} + 0.000005\,I^2\,t_{on}^2 + 0.00027\,t_{on}\,t_{off} - 0.000001\,t_{on}^2\,t_{off} \tag{11}$$

$$MRR = -1.2087 + 0.342\,I + 0.02967\,I^2 - 0.00817\,t_{on} + 0.00004\,t_{on}^2 + 0.02287\,t_{off} \\ + 0.00096\,I\,t_{on} - 0.00001\,I\,t_{on}^2 + 0.00057\,I^2\,t_{on} - 0.00668\,I\,t_{off} \tag{12}$$

Analyses of the results of the MRR showed that the values of the *R*-squared for surface roughness (*Sa*), the maximal thickness of the white layer, and the MRR were over 98%, 99%, and 99%, respectively. This result indicated that the regression models provided an excellent explanation of the relationship between the independent variables and the response *Sa*, WL, and MRR. Differences between the *R*-squared and the *R*-adjustable were smaller than 0.2, which indicated that the established model was adequate in representing the process. The developed models can be used to predict the values of surface roughness (*Sa*), the maximal value of white layer thickness (WL), and the material removal rate (MRR). Comparisons between the results of experimental studies and the values calculated based on the developed models for *Sa*, WL, and MRR are shown in Figure 10. The results indicated that the predicted values were very close to the experimental data.

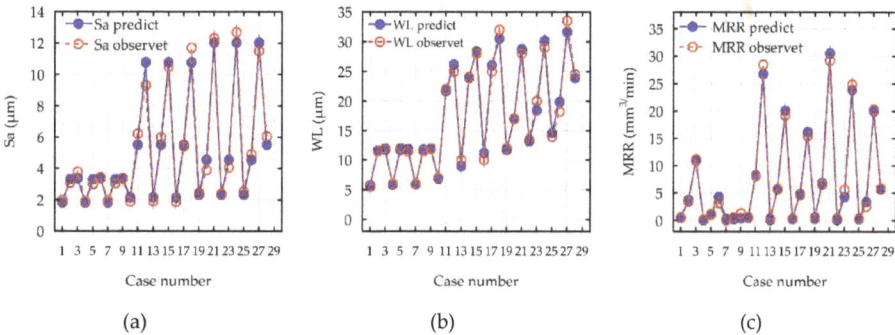

Figure 10. The comparison between the results of experimental studies to the values calculated based on the developed models for (**a**) surface roughness (*Sa*); (**b**) maximal thickness of white layer (WL); (**c**) material removal rate (MRR).

Developed models for the *Sa*, WL, and MRR were checked using additional statistical tests, which confirmed the basic assumptions using the ANOVA. Residuals had a normal distribution, constant variance, and were independent of an order of data. The assumption of constant variance was checked by plotting the residuals versus the predicted values. The normality assumption and independence of residuals were checked by plotting the expected normal value versus the residuals and the residuals versus the order of data, respectively. Analysis of the residual normal probability plots (Figures 11a, 12a and 13a) showed that the residuals had normal distributions. Plots of the residuals versus the predicted values (Figures 11b, 12b and 13b); and the residuals versus the case number values (Figures 11c, 12c and 13c) showed that the residuals had a stochastic nature. The analysis of the plotted residuals versus the case values indicated that the error terms were independent of one another. The analyses of the residuals confirmed that the developed models were adequate.

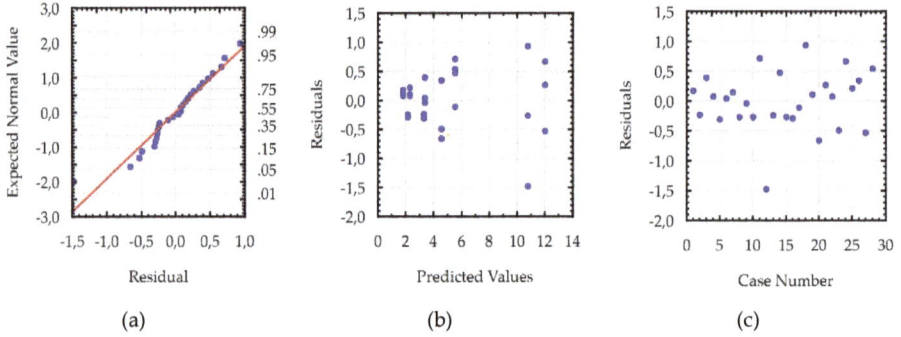

Figure 11. The plots to check the model for surface roughness (*Sa*): (**a**) the normal plot of residuals; (**b**) the residuals versus the predicted values; and (**c**) the residuals versus the case values.

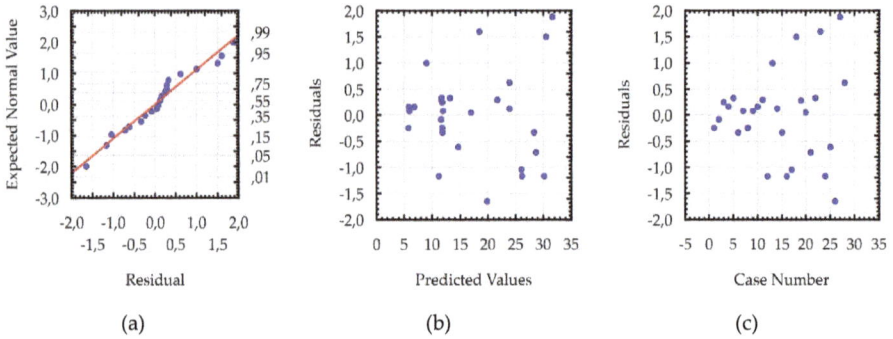

Figure 12. The plots to check the model for maximal white layer thickness: (**a**) the normal plot of residuals; (**b**) the residuals versus the predicted values; and (**c**) the residuals versus the case values.

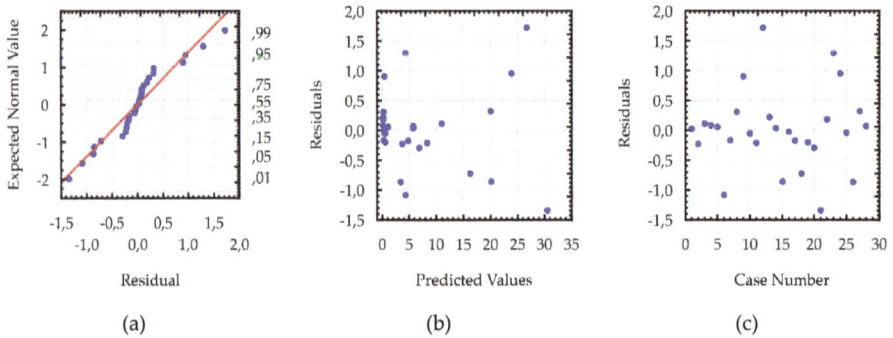

Figure 13. The plots to check the model MRR: (**a**) the normal plot of residuals; (**b**) the residuals versus the predicted values; and (**c**) the residuals versus the case values.

To better understand the influence of the EDM parameters on surface roughness (*Sa*), the maximal thickness of white layer, and the material removal rate, the response surface plots were estimated. Based on the regression models (Equations (10)–(12)), the influence of the discharge current *I*, pulse time t_{on}, and time interval t_{off} on the *Sa*, WL, and MRR is shown in Figures 14–16, respectively.

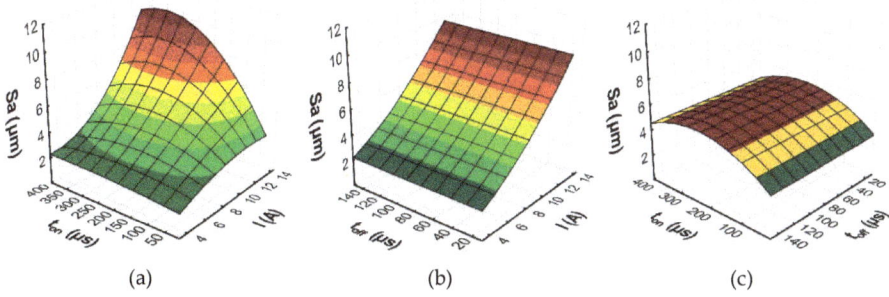

Figure 14. The estimated response surface plot for roughness (*Sa*): (**a**) constant t_{off} = 80 µs; (**b**) constant t_{on} = 206 µs; and (**c**) constant *I* = 8.5 A.

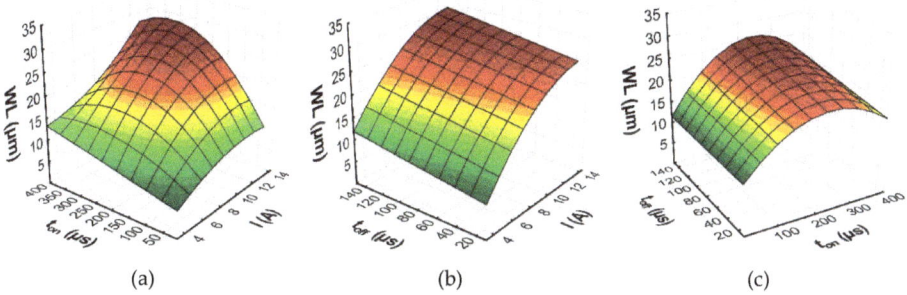

Figure 15. The estimated response surface plot for the WL: (**a**) constant t_{off} = 80 µs; (**b**) constant t_{on} = 206 µs; and (**c**) constant *I* = 8.5 A.

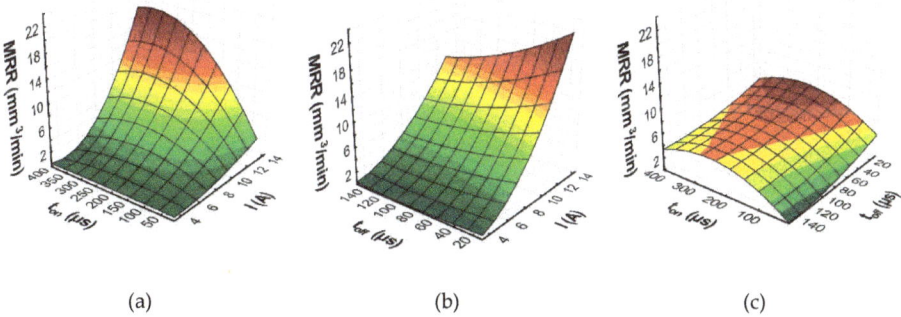

Figure 16. The estimated response surface plot for the MRR: (**a**) constant t_{off} = 80 µs; (**b**) constant t_{on} = 206 µs; and (**c**) constant *I* = 8.5 A.

The results of the experimental studies indicated that the main parameters that influenced surface roughness (*Sa*), were the discharge current and the pulse time (Figure 14). Time interval, in the case of stability discharges, does not have a significant impact on the surface texture properties. The surface roughness (*Sa*) increases with the growth of the discharge current and the pulse time. These two parameters, with constant voltage, determine the amount of energy of the electrical discharge. At the lowest value of the discharge current, the changing of the pulse time does not generate a crater with greater depth. The surface roughness (*Sa*) does not change significantly. With the increase of the discharge current, the amount of energy delivered to the workpiece causes the melting and evaporation of a higher volume of material, which generates a crater with a larger depth. The presented dependence also has effects on the material removal rate. The MRR, similar to surface roughness, is influenced by the volume of material which is removed in single discharges, and it mainly depends on the discharge current (Figure 16). The time interval is responsible for the stabilization of the conditions in the gap after discharges. In a stable EDM process, increases of the time interval result in the decrease of the material removal rate. Discharge energy devoted to the heat flux, the mechanism of growing the plasma channel, and the removal process of the material implies the thickness of the white layer. Figure 15 shows the estimated response surface plot for the white layer thickness in relation to the EDM parameters. The increase in the pulse time and current resulted in an increase in the amount of melted and evaporated material from the discharge zone. However, more material which was melted in the single crater was not removed from the surface of the workpiece and it re-solidified on the core.

In the industrial application of the developed models, splitting the electrical discharge machining into several steps should be considered. In the first stage, the material will be removed from the workpiece using the highest discharge energy (roughing technology). In the roughing step, the applied parameters should ensure the maximum removal rate. In the next step, the parameters of the process should be changed to achieve proper surface layer properties and a low roughness for the machined surface of the manufactured parts. The EDM process will be conducted with the semi-finishing and finishing step, with respectively lower discharge energies. In the last step—the finishing treatment—it is vital to obtain the appropriate surface roughness and thickness of the white layer. However, the finishing can take more time than the roughing. The result is that a significant increase in the cost of production may be observed. In the finishing machining, a combination of minimum surface roughness with the minimum value of white layer thickness, and with a possibly maximum *MRR* is desirable. In the case of EDM, the simultaneous achievement of these three goals is conflicting. For that reason, in considering the properties of the EDM optimization, it should be based on the desirability technique. This method uses the Derringer's [57] desirability function, which in the case of the same importance of each response can be described by the equation:

$$D = (d_1 \times d_2 \times \ldots \times d_n)^{1/n} = \left(\prod_{i=1}^{n} d_i \right)^{\frac{1}{n}} \tag{13}$$

where n is the number of responses in the measure.

The desired function is established for each investigated response, $d_i(\hat{y}_i)$, and it has a range from zero to one (one being the most desirable). Different desirable functions can be built, depending on the adopted optimization criteria which determine the desirable value, maximal (upper-U_i) or minimal (lower-L_i). If the response for the investigated parameter should be minimized, then $d_i(\hat{y}_i)$ can be calculated according to the following equation:

$$d_i(\hat{y}_i) = \begin{cases} 1 & \hat{y}_i < L_i \\ \left(\frac{U_i - \hat{y}_i}{U_i - L_i} \right)^t, & L_i \leq \hat{y}_i \leq U_i \\ 0 & \hat{y}_i > U_i \end{cases} \tag{14}$$

If the desirable function should be maximized, then it can be expressed by the following equation:

$$d_i(\hat{y}_i) = \begin{cases} 0 & \hat{y}_i < L_i \\ \left(\frac{\hat{y}_i - L_i}{U_i - L_i} \right)^s, & L_i \leq \hat{y}_i \leq U_i \\ 1 & \hat{y}_i > U_i \end{cases} \tag{15}$$

Calculations of the desirable function consider the extent to which the estimated values ($\hat{y}_i$) are close to the minimum or maximum. Figure 17 presents a graphical interpretation of the desirability functions with the "importance" levels s and t. When considering the case when the "importance" t

(for minimum) and s (for maximum) is large, the desirability is low unless the response moves close to the target. For low-value parameters, t and s desirability has a high value for a wide range of responses. It means that it is possible to achieve satisfactory desirability not only in the target value (minimum L_i or maximum U_i).

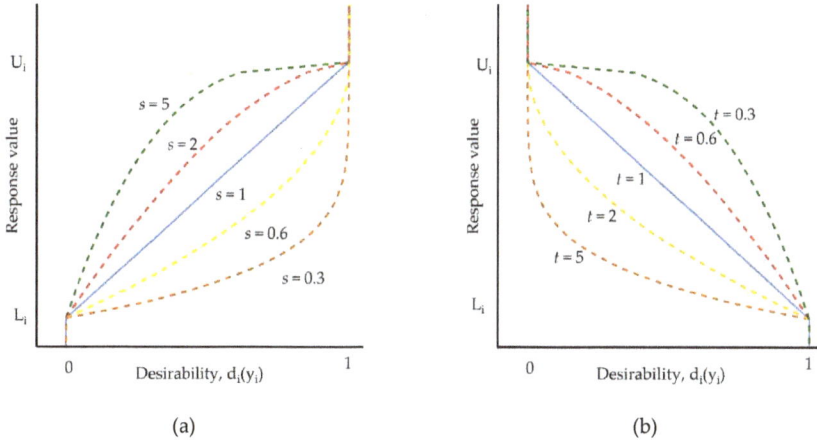

(a)   (b)

Figure 17. The desirability functions for the target value: (**a**) maximized; (**b**) minimized criteria.

The multi-response optimization was divided into three cases. In the first case, we examined the optimal parameters of the finishing EDM. For this task, the goal was defined as minimizing the surface roughness and white layer thickness, whilst maximizing the maximum material removal rate. The second case was to find the optimal parameters for semi-finishing EDM. In this task, we aimed to achieve an average value of the MRR (about 14.5 mm^3/min), with the possibility of minimizing the surface roughness and white layer thickness. In the last case, the optimal parameters for roughing EDM were considered. In this task, the goal of optimization was to achieve the maximum MRR possible, with the possibility of minimizing the surface roughness and white layer thickness. These three cases of the optimization of electrical discharge machining were carried out using the desirability function, based on the regression Equations (10)–(12). It should be maintained that the success of the optimization with the desirability function mainly depends on the quality of the regressions models. In this study, each established model had a coefficient of determination, *R-squared*, that was over 98%, and the differences between the R-squared and the R-adjustable were smaller than 0.2, which indicated that the models were adequate in representing the process. For each EDM parameter (discharge current, pulse time, time interval) there was a simultaneous analysis of every combination of the factors for each of the nine responses (Figures 14–16). A multi-response optimization procedure was performed for the global desirability function. The ranges for the constraints and factors for optimization are shown in Table 9.

Table 9. The goals and factor range for optimization.

Factors	Goal	Lower Limit	Upper Limit	Weight	Importance		
					Finishing EDM	Semi-Finishing	Roughing
I (A)	In range	3	14	1	-	-	-
t_{on} (µs)	In range	13	400	1	-	-	-
t_{off} (µs)	In range	10	150	1	-	-	-
Sa (µm)	Minimize	1.85	12.7	1	$t = 5$	$t = 3$	$t = 0.3$
WL (µm)	Minimize	5.5	33.5	1	$t = 5$	$t = 3$	$t = 0.3$
MRR (mm^3/min)	Maximize	0.01	29.19	1	$s = 0.3$	$s = 3$	$s = 5$

The results of the multi-response optimization procedure of the global desirability function for the finishing, semi-finishing, and roughing operations are shown in the contour plots in Figures 18–21, and in Table 10. The desirable function in the first case (finishing EDM) reached 0.95 (Figure 19). The semi-finishing and roughing operations reached 0.98 and 0.99, respectively. The optimal EDM parameters for finishing electrical discharge machining were a discharge current I = 3 A, pulse time t_{on} = 176 μs, and pulse interval t_{off} = 10 μs. The predicted surface roughness (1.7 μm) and the white layer thickness (6 μm) after optimization were close to the results obtained in the experimental studies for sample number four. Nevertheless, the material removal rate grew almost seven times and reached an MRR = 1.1 mm³/min. The increase of the MRR was achieved, along with the minimization of the surface roughness and the white layer thickness, which has a significant effect on productivity.

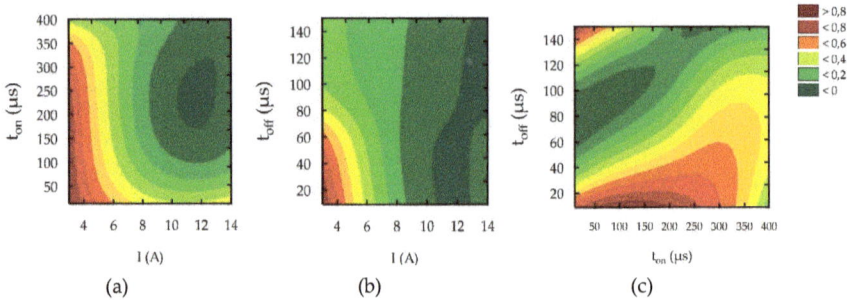

Figure 18. The contour plots of the desirability function for the finishing EDM optimization.

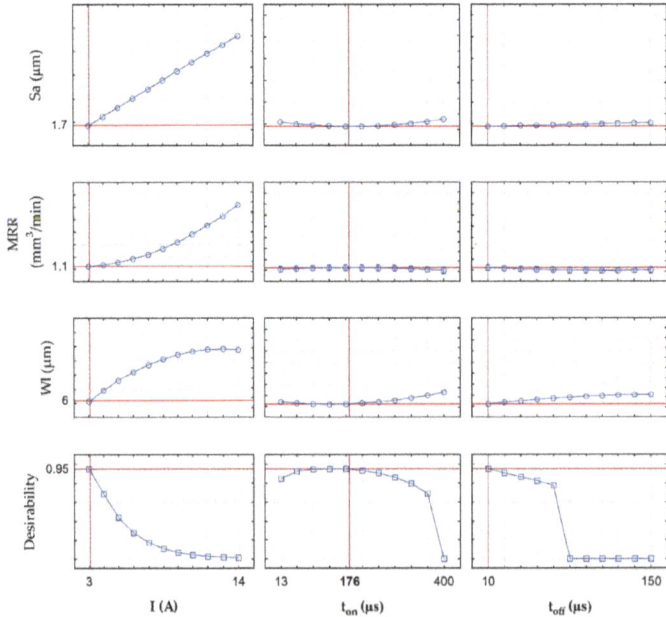

Figure 19. The example of profiles for the predicted values and desirability for the finishing EDM.

Figure 20 and Table 10 present the results of the optimization for semi-finishing EDM. Values of the optimal EDM parameters were as follows: discharge current I = 14 A, pulse time t_{on} = 52 μs, and pulse interval t_{off} = 24 μs. In this case, if the material removal rate reached 14.5 mm³/min (i.e., the average value of the MRR from experimental studies), the optimized surface roughness and the white layer thickness were as follows: Sa = 5.2 μm and WL = 15 μm. Figure 21 and Table 10 present the results of the optimization for roughing EDM. Values of the optimal EDM parameters as follows: I = 14 A, t_{on} = 361 μs, and t_{off} = 24 μs. The optimized material removal rate, surface roughness, and white layer thickness were as follows: MRR = 29.2 mm^3/min, Sa = 12.1 μm, WL = 28.8 μm.

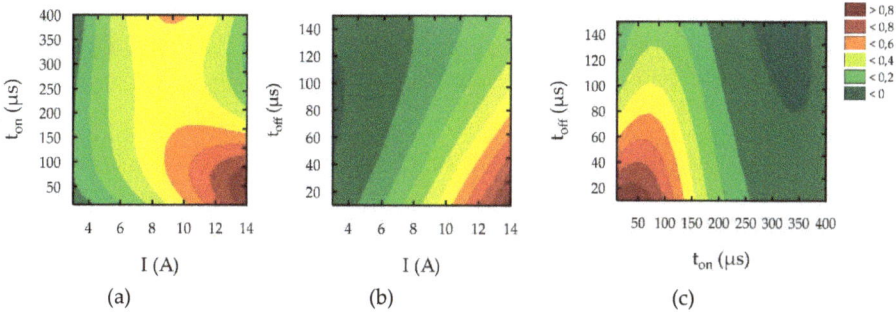

Figure 20. The contour plots of the desirability function for semi-finishing EDM optimization.

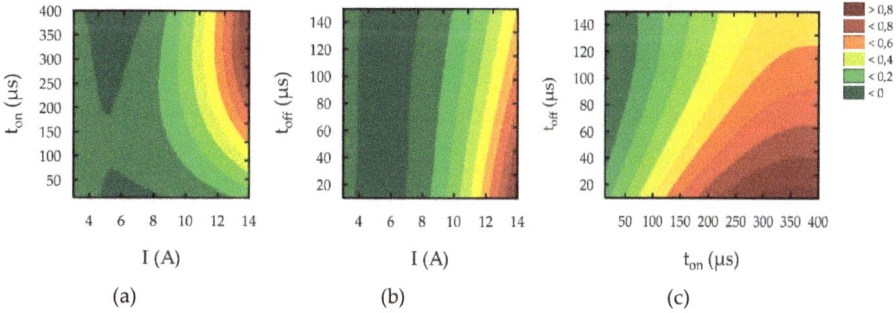

Figure 21. The contour plots of the desirability function for roughing EDM optimization.

In the last stage of the experimental investigations for the optimal EDM parameters, a confirmation test was conducted on the EDM machine Charmilles Form 2LC ZNC (Bern, Switzerland). Table 10 presents the results of the validations of multi-response optimizations. The maximal errors between the predicted and the obtained values were 6%, which could be considered a very good result. The calculated measurement uncertainty (at 95% confidence) of the Sa and MRR is much lower than the measurement uncertainty of the thickness of the white layer. Moreover, prediction errors are in the range of maximal measurement uncertainty.

Table 10. The experimental validations of the multi-response optimizations.

Optimal EDM Parameters		Summary of Values Obtained in Optimization			
		Response	Predicted	Experimental Verification	Error%
Finishing	I = 3 A	Sa (μm)	1.7	1.8	6
	t_{on} = 176 μs	WL (μm)	6	6.3	5
	t_{off} = 10 μs	MRR (mm^3/min)	1.13	1.06	6
Semi- finishing	I = 14 A	Sa (μm)	5.2	5.4	4
	t_{on} = 52 μs	WL (μm)	15	15.8	5
	t_{off} = 24 μs	MRR (mm^3/min)	14.5	15	3
Roughing	I = 14 A	Sa (μm)	12.1	12.7	5
	t_{on} = 361 μs	WL (μm)	28.8	30.5	6
	t_{off} = 24 μs	MRR (mm^3/min)	29.2	28.1	4

4. Conclusions

In this study, the EDM of tool steel 55NiCrMoV7 was analyzed and optimized for three cases: roughing, semi-finishing, and finishing machining. Based on the theoretical analyses and experimental research, the following conclusions were obtained:

1. Experimental research on the influence of discharge current, pulse time, and pulse interval on the surface roughness (*Sa*), white layer thickness, and the MRR showed that the discharge current had the main effect on *Sa*, WL, and the MRR. With an increase in the discharge current and pulse time, the amount of energy delivered to the workpiece caused the melting and evaporation of a higher volume of material, which generated craters with a larger depth and diameter. However, more material which melted in the single crater was not removed from the surface of the workpiece and it re-solidified on the core. The time interval between pulses did not significantly affect the change in surface integrity and the MRR, but it played an important role in the stability of the process.

2. The desirability function was used in the multi-response optimization of three functions: *Sa*, WL, and MRR. For the three cases of EDM—finishing, semi-finishing, and roughing operations—the optimal parameters were established. The confirmation tests for the established optimal parameters showed that the maximal errors between the predicted and the obtained values did not exceed 6%, which could be considered as a very good result.

3. The developed regression equations could be used in electrical discharge machining as a guideline for the selection of EDM parameters.

Author Contributions: Conceptualization, R.Ś.; Methodology, R.Ś., D.O.-Ś. and T.C.; Software, D.O.-Ś.; Validation, R.Ś., D.O.-Ś. and T.C.; Formal Analysis, R.Ś. and T.C.; Investigation, R.Ś. and D.O.-Ś.; Resources, R.Ś.; Data Curation, R.Ś. and D.O.-Ś.; Writing-Original Draft Preparation, R.Ś.; Writing-Review & Editing, RŚ, D.O.-Ś., and T.C.; Visualization, D.O.-Ś.; Supervision, R.Ś.; Project Administration, R.Ś.; Funding Acquisition, R.Ś.

Funding: This research was funded by using the statutory subsidy of the Faculty of Production Engineering of the Warsaw University of Technology in 2018.

Acknowledgments: Thanks are given to the assistance from the technical specialist Eng. Adrian Kopytowski and Eng. Rafał Nowicki.

Conflicts of Interest: The authors declare no conflict of interest.

References

1. Hsu, W.-H.; Chien, W.-T. Effect of electrical discharge machining on stress concentration in titanium alloy holes. *Materials* **2016**, *9*, 957. [CrossRef] [PubMed]
2. Ramulu, M.; Spaulding, M. Drilling of hybrid titanium composite laminate (HTCL) with electrical discharge machining. *Materials* **2016**, *9*, 746. [CrossRef] [PubMed]
3. Górka, J. Assessment of steel subjected to the thermomechanical control process with respect to weldability. *Metals* **2018**, *8*, 169. [CrossRef]
4. Świercz, R.; Oniszczuk-Świercz, D.; Dąbrowski, L.; Zawora, J. Optimization of machining parameters of electrical discharge machining tool steel 1.2713. *AIP Conf. Proc.* **2018**, *2017*, 020032. [CrossRef]
5. Liang, J.F.; Liao, Y.S.; Kao, J.Y.; Huang, C.H.; Hsu, C.Y. Study of the EDM performance to produce a stable process and surface modification. *Int. J. Adv. Manuf. Technol.* **2018**, *95*, 1743–1750. [CrossRef]
6. Valíček, J.; Držík, M.; Hryniewicz, T.; Harničárová, M.; Rokosz, K.; Kušnerová, M.; Barčová, K.; Bražina, D. Non-contact method for surface roughness measurement after machining. *Meas. Sci. Rev.* **2012**, *12*, 184–188. [CrossRef]
7. Klocke, F.; Zeis, M.; Klink, A. Interdisciplinary modelling of the electrochemical machining process for engine blades. *CIRP Ann.* **2015**, *64*, 217–220. [CrossRef]
8. Kozak, J.; Zybura-Skrabalak, M. Some problems of surface roughness in electrochemical machining (ECM). *Procedia CIRP* **2016**, *42*, 101–106. [CrossRef]
9. Gusarov, A.V.; Grigoriev, S.N.; Volosova, M.A.; Melnik, Y.A.; Laskin, A.; Kotoban, D.V.; Okunkova, A.A. On productivity of laser additive manufacturing. *J. Mater. Process. Technol.* **2018**, *261*, 213–232. [CrossRef]

10. Wyszyński, D.; Ostrowski, R.; Zwolak, M.; Bryk, W. Laser beam machining of polycrystalline diamond for cutting tool manufacturing. *AIP Conf. Proc.* **2017**, *1896*, 180007. [CrossRef]

11. Chmielewski, T.; Golański, D.; Włosiński, W.; Zimmerman, J. Utilizing the energy of kinetic friction for the metallization of ceramics. *Bull. Pol. Acad. Sci. Tech. Sci.* **2015**, *63*, 201–207. [CrossRef]

12. Rokosz, K.; Hryniewicz, T.; Matýsek, D.; Raaen, S.; Valíček, J.; Dudek, Ł.; Harničárová, M. SEM, EDS and XPS analysis of the coatings obtained on titanium after plasma electrolytic oxidation in electrolytes containing copper nitrate. *Materials* **2016**, *9*, 318. [CrossRef] [PubMed]

13. Skoczypiec, S.; Ruszaj, A. A sequential electrochemical–electrodischarge process for micropart manufacturing. *Precis. Eng.* **2014**, *38*, 680–690. [CrossRef]

14. Gołąbczak, M.; Święcik, R.; Gołąbczak, A.; Nouveau, C.; Jacquet, P.; Blanc, C. Investigations of surface layer temperature and morphology of hard machinable materials used in aircraft industry during abrasive electrodischarge grinding process. *Materialwissenschaft und Werkstofftechnik* **2018**, *49*, 568–576. [CrossRef]

15. Kelemesh, A.; Gorbenko, O.; Dudnikov, A.; Dudnikov, I. Research of wear resistance of bronze bushings during plastic vibration deformation. *East.-Eur. J. Enterp. Technol.* **2017**, *2*, 16–21. [CrossRef]

16. Golański, D.; Dymny, G.; Kujawińska, M.; Chmielewski, T. Experimental investigation of displacement/strain fields in metal coatings deposited on ceramic substrates by thermal spraying. *Solid State Phenom.* **2016**, *240*, 174–182. [CrossRef]

17. Salacinski, T.; Winiarski, M.; Chmielewski, T.; Świercz, R. Surface finishing using ceramic fibre brush tools. In Proceedings of the 26th International Conference on Metallurgy and Materials, Brno, Czech Republi, 24–26 May 2017; pp. 1220–1226.

18. Guo, J.; Wang, H.; Goh, M.H.; Liu, K. Investigation on surface integrity of rapidly solidified aluminum RSA 905 by magnetic field-assisted finishing. *Micromachines* **2018**, *9*, 146. [CrossRef]

19. Kunieda, M.; Lauwers, B.; Rajurkar, K.P.; Schumacher, B.M. Advancing EDM through fundamental insight into the process. *CIRP Ann. Manuf. Technol.* **2005**, *54*, 64–87. [CrossRef]

20. Izquierdo, B.; Sánchez, J.A.; Plaza, S.; Pombo, I.; Ortega, N. A numerical model of the EDM process considering the effect of multiple discharges. *Int. J. Mach. Tools Manuf.* **2009**, *49*, 220–229. [CrossRef]

21. Izquierdo, B.; Sánchez, J.A.; Ortega, N.; Plaza, S.; Pombo, I. Insight into fundamental aspects of the EDM process using multidischarge numerical simulation. *Int. J. Adv. Manuf. Technol.* **2011**, *52*, 195–206. [CrossRef]

22. Ming, W.; Zhang, Z.; Wang, S.; Huang, H.; Zhang, Y.; Zhang, Y.; Shen, D. Investigating the energy distribution of workpiece and optimizing process parameters during the EDM of Al6061, Inconel718, and SKD11. *Int. J. Adv. Manuf. Technol.* **2017**, *92*, 4039–4056. [CrossRef]

23. Gulbinowicz, Z.; Świercz, R.; Oniszczuk-Świercz, D. Influence of electrical parameters in electro discharge machining of tungsten heavy alloys on surface texture properties. *AIP Conf. Proc.* **2018**, *2017*, 020007. [CrossRef]

24. Salcedo, A.T.; Arbizu, I.P.; Pérez, C.J.L. Analytical modelling of energy density and optimization of the EDM machining parameters of Inconel 600. *Metals* **2017**, *7*, 166. [CrossRef]

25. Rahang, M.; Patowari, P.K. Parametric optimization for selective surface modification in EDM using taguchi analysis. *Mater. Manuf. Process.* **2016**, *31*, 422–431. [CrossRef]

26. Vagaská, A.; Gombár, M. Comparison of usage of different neural structures to predict AAO layer thickness. *Tehnički Vjesnik* **2017**, *24*, 333–339. [CrossRef]

27. Wojciechowski, S.; Maruda, R.W.; Królczyk, G.M. The application of response surface method to optimization of precision ball end milling. *MATEC Web Conf.* **2017**, *112*, 01004. [CrossRef]

28. Adalarasan, R.; Santhanakumar, M.; Rajmohan, M. Optimization of laser cutting parameters for Al6061/SiCp/Al2O3 composite using grey based response surface methodology (GRSM). *Measurement* **2015**, *73*, 596–606. [CrossRef]

29. Vera Candioti, L.; De Zan, M.M.; Cámara, M.S.; Goicoechea, H.C. Experimental design and multiple response optimization. Using the desirability function in analytical methods development. *Talanta* **2014**, *124*, 123–138. [CrossRef]

30. Ayesta, I.; Izquierdo, B.; Sanchez, J.A.; Ramos, J.M.; Plaza, S.; Pombo, I.; Ortega, N. Optimum electrode path generation for EDM manufacturing of aerospace components. *Robot. Comput.-Integr. Manuf.* **2016**, *37*, 273–281. [CrossRef]

31. Abidi, M.H.; Al-Ahmari, A.M.; Siddiquee, A.N.; Mian, S.H.; Mohammed, M.K.; Rasheed, M.S. An investigation of the micro-electrical discharge machining of nickel-titanium shape memory alloy using grey relations coupled with principal component analysis. *Metals* **2017**, *7*, 486. [CrossRef]

32. Rubio, E.M.; Villeta, M.; Valencia, J.L.; Sáenz de Pipaón, J.M. Experimental study for improving the repair of magnesium–aluminium hybrid parts by turning processes. *Metals* **2018**, *8*, 59. [CrossRef]

33. Chabbi, A.; Yallese, M.A.; Meddour, I.; Nouioua, M.; Mabrouki, T.; Girardin, F. Predictive modeling and multi-response optimization of technological parameters in turning of Polyoxymethylene polymer (POM C) using RSM and desirability function. *Measurement* **2017**, *95*, 99–115. [CrossRef]

34. Kilickap, E.; Yardimeden, A.; Çelik, Y.H. Mathematical modelling and optimization of cutting force, tool wear and surface roughness by using artificial neural network and response surface methodology in milling of Ti-6242S. *Appl. Sci.* **2017**, *7*, 1064. [CrossRef]

35. Rogalski, G.; Fydrych, D.; Łabanowski, J. Underwater wet repair welding of API 5L X65M pipeline steel. *Pol. Marit. Res.* **2017**, *24*, 188–194. [CrossRef]

36. Unune, D.R.; Mali, H.S. Parametric modeling and optimization for abrasive mixed surface electro discharge diamond grinding of Inconel 718 using response surface methodology. *Int. J. Adv. Manuf. Technol.* **2017**, *93*, 3859–3872. [CrossRef]

37. Hlaváč, L.M.; Krajcarz, D.; Hlaváčová, I.M.; Spadło, S. Precision comparison of analytical and statistical-regression models for AWJ cutting. *Precis. Eng.* **2017**, *50*, 148–159. [CrossRef]

38. Ghodsiyeh, D.; Golshan, A.; Izman, S. Multi-objective process optimization of wire electrical discharge machining based on response surface methodology. *J. Braz. Soc. Mech. Sci. Eng.* **2014**, *36*, 301–313. [CrossRef]

39. Alavi, F.; Jahan, M.P. Optimization of process parameters in micro-EDM of Ti-6Al-4V based on full factorial design. *Int. J. Adv. Manuf. Technol.* **2017**, *92*, 167–187. [CrossRef]

40. Selvarajan, L.; Manohar, M.; Kumar, A.U.; Dhinakaran, P. Modelling and experimental investigation of process parameters in EDM of Si3N4-TiN composites using GRA-RSM. *J. Mech. Sci. Technol.* **2017**, *31*, 111–122. [CrossRef]

41. Świercz, R.; Oniszczuk-Świercz, D. Influence of electrical discharge pulse energy on the surface integrity of tool steel 1.2713. Proceedings of 26th International Conference on Metallurgy and Materials, Brno, Czech Republic, 24–26 May 2017; pp. 1452–1457.

42. Kumaran, S.T.; Ko, T.J.; Kurniawan, R. Grey fuzzy optimization of ultrasonic-assisted EDM process parameters for deburring CFRP composites. *Measurement* **2018**, *123*, 203–212. [CrossRef]

43. Gu, L.; Zhu, Y.; Zhang, F.; Farhadi, A.; Zhao, W. Mechanism analysation and parameter optimisation of electro discharge machining of titanium-zirconium-molybdenum alloy. *J. Manuf. Process.* **2018**, *32*, 773–781. [CrossRef]

44. Dang, X.-P. Constrained multi-objective optimization of EDM process parameters using kriging model and particle swarm algorithm. *Mater. Manuf. Process.* **2018**, *33*, 397–404. [CrossRef]

45. Mohanty, C.P.; Mahapatra, S.S.; Singh, M.R. An intelligent approach to optimize the EDM process parameters using utility concept and QPSO algorithm. *Eng. Sci. Technol. Int. J.* **2017**, *20*, 552–562. [CrossRef]

46. Maity, K.; Mishra, H. ANN modelling and Elitist teaching learning approach for multi-objective optimization of μ-EDM. *J. Intell. Manuf.* **2018**, *29*, 1599–1616. [CrossRef]

47. Tripathy, S.; Tripathy, D.K. Multi-response optimization of machining process parameters for powder mixed electro-discharge machining of H-11 die steel using grey relational analysis and topsis. *Mach. Sci. Technol.* **2017**, *21*, 362–384. [CrossRef]

48. Nguyen, H.-P.; Pham, V.-D.; Ngo, N.-V. Application of TOPSIS to Taguchi method for multi-characteristic optimization of electrical discharge machining with titanium powder mixed into dielectric fluid. *Int. J. Adv. Manuf. Technol.* **2018**, *98*, 1179–1198. [CrossRef]

49. Roy, T.; Dutta, R.K. Integrated fuzzy AHP and fuzzy TOPSIS methods for multi-objective optimization of electro discharge machining process. *Soft Comput.* **2018**. [CrossRef]

50. Kandpal, B.C.; Kumar, J.; Singh, H. Optimization and characterization of EDM of AA 6061/10%Al$_2$O$_3$ AMMC using Taguchi's approach and utility concept. *Prod. Manuf. Res.* **2017**, *5*, 351–370. [CrossRef]

51. D'Urso, G.; Giardini, C.; Quarto, M.; Maccarini, G. Cost index model for the process performance optimization of micro-EDM drilling on tungsten carbide. *Micromachines* **2017**, *8*, 251. [CrossRef]

52. Parsana, S.; Radadia, N.; Sheth, M.; Sheth, N.; Savsani, V.; Prasad, N.E.; Ramprabhu, T. Machining parameter optimization for EDM machining of Mg–RE–Zn–Zr alloy using multi-objective Passing Vehicle Search algorithm. *Arch. Civil Mech. Eng.* **2018**, *18*, 799–817. [CrossRef]

53. Hadad, M.; Bui, L.Q.; Nguyen, C.T. Experimental investigation of the effects of tool initial surface roughness on the electrical discharge machining (EDM) performance. *Int. J. Adv. Manuf. Technol.* **2018**, *95*, 2093–2104. [CrossRef]

54. Baruffi, F.; Parenti, P.; Cacciatore, F.; Annoni, M.; Tosello, G. On the application of replica molding technology for the indirect measurement of surface and geometry of micromilled components. *Micromachines* **2017**, *8*, 195. [CrossRef]

55. ISO. *IOS 15530-3: Geometrical Product Specifications (GPS)—Coordinate Measuring Machines (CMM): Technique for Determining the Uncertainty of Measurement*; IOS: Geneva, Switzerland, 2011.

56. Xin, B.; Li, S.; Yin, X.; Lu, X. Dynamic observer modeling and minimum-variance self-tuning control of EDM interelectrode gap. *Appl. Sci.* **2018**, *8*, 1443. [CrossRef]

57. Gopalakannan, S.; Senthilvelan, T. Optimization of machining parameters for EDM operations based on central composite design and desirability approach. *J. Mech. Sci. Technol.* **2014**, *28*, 1045–1053. [CrossRef]

![micromachines logo] *micromachines*

MDPI

Article

Charged Satellite Drop Avoidance in Electrohydrodynamic Dripping

Lei Guo [1], Yongqing Duan [1], Weiwei Deng [2], Yin Guan [3], YongAn Huang [1,*] and Zhouping Yin [1,*]

[1] State Key Laboratory of Digital Manufacturing Equipment and Technology, Huazhong University of Science and Technology, Wuhan 430074, China; hustgl@hust.edu.cn (L.G.); duanyongqing@hust.edu.cn (Y.D.)
[2] Department of Mechanics and Aerospace Engineering, Southern University of Science and Technology, Shenzhen 518055, China; dengww@sustc.edu.cn
[3] School of Energy and Power Engineering, Huazhong University of Science and Technology, Wuhan 430074, China; yinguan@hust.edu.cn
* Correspondence: yahuang@hust.edu.cn (Y.H.); yinzhp@mail.hust.edu.cn (Z.Y.); Tel.: +86-27-87558207 (Y.H.)

Received: 16 January 2019; Accepted: 25 February 2019; Published: 1 March 2019

Abstract: The quality of electrohydrodynamic jet (e-jet) printing is crucially influenced by the satellite drop formed when the primary drop detaches from the meniscus. If the satellite drop falls onto the substrate, the patterns on the substrate will be contaminated. The electric charge carried by the satellite drop leads to more complex satellite/meniscus interaction than that in traditional inkjet printing. Here, we numerically study the formation and flight behavior of the charged satellite drop. This paper discovered that the charge relaxation time (CRT) of the liquid determines the electric repulsion force between the satellite drop and meniscus. The satellite drop will merge with the meniscus at long CRT, and fail to merge and deteriorate the printing quality at short CRT. The simulations are adopted to discover the mechanism of generation and flight behavior of charged satellite drops. The results show that the critical CRT decreases with the dielectric constant of the liquid and the supplied flow rate. Namely, for small dielectric constant and fixed CRT, the satellite drop is less likely to merge with the meniscus, and for high flow rate, the satellite drop is prone to merge with the meniscus due to the delay of necking thread breakup. These results will help to choose appropriate parameters to avoid the satellite drop from falling onto the substrate.

Keywords: satellite drop; electrohydrodynamic jet printing; charge relaxation time

---

1. Introduction

Electrohydrodynamic jet (e-jet) printing has received much attention recently due to its high printing resolution, ink compatibility, and process flexibility [1–3]. It can adopt nozzles much smaller than those used in traditional inkjet printing [4,5], and can work in versatile manners such as the cone-jet mode, dripping mode, or microdripping mode [6–11]. E-jet printing is suitable for a variety of inks and has found widespread applications in printed electronics [12–15], DNA microarrays [16], protein microarrays [17], photonic devices [18,19], 3D structures [20–22], solar cells [23,24], and others [25].

In cone-jet mode, the charge level of the printed drop may exceed the Rayleigh limit and modest evaporation of the drop will lead to Coulombic fission and blurry printed patterns when printing on dielectric substrates. In contrast, the dripping mode has lower charge level and Coulombic fission can be prevented. However, satellite drops may appear in the dripping mode. It is known that satellite drops are undesirable for inkjet printing as they may deteriorate the resolutions of printed patterns. Existing experimental and numerical studies on e-jet printing in dripping mode [26–30] mainly focused on the breakup of the primary drop instead of the subsequent satellite drop formation and trajectories. The satellite drop formation in e-jet printing is similar to that in traditional inkjet printing [31–35], except that the electric forces play an important role in the satellite drop formation and interaction between the satellite drop and meniscus. Huo et al. [36] visualized the satellite flight behaviors in electrohydrodynamic dripping mode through high-speed imaging and found that the satellite drop merged with the meniscus when the applied voltage was sufficiently low. The liquid used in the experiment was ethanol. However, the mechanism of this phenomenon was not reported.

The electric field and electric stress play important roles in the satellite formation and flight behavior. The charge relaxation time (CRT or $\tilde{t}_e = \varepsilon_L/K$), which is the ratio between the dielectric constant (ε_L) and the conductivity (K) of the liquid, has a big influence on the electric field distribution and surface charge along the drop surface during electrohydrodynamic dripping. CRT describes the time scale required for the surface charge responses to the change of electric field. Whether the satellite drop will merge or separate from the meniscus can be predicted by a critical CRT which also depends on the dielectric constant of the liquid and the supplied flow rate. Here, we investigated the effect of CRT on the satellite drop merging and separation. The paper is organized as follows. First, the numerical method and governing equations are introduced. Second, the influence of the CRT on the generation and flight of the satellite drop is described and discussed. Finally, the influence of dielectric constant and flow rate on the critical CRT as well as the underlying mechanisms are discussed.

2. Numerical Methods

Figure 1a shows the schematic of the experimental setup. The liquid was supplied by pushing the syringe by a syringe pump. The high voltage was applied between the nozzle and the substrate. The high voltage was generated by a signal generator and a high-voltage amplifier. A Taylor cone formed at the front of the nozzle under the applied voltage. Figure 1b shows two representative flight behaviors of satellite drops. In both cases, a long and thin liquid thread formed when the primary drop detached from the meniscus that suspended on the nozzle, and the breakup of this liquid thread led to the generation of the satellite drop. For the upper images, the satellite drop moved toward the meniscus and eventually merged with the meniscus. Thus, the satellite drop did not influence the printed patterns. For the lower images, the satellite drop moved downward and separated from the meniscus. The satellite drop may oscillate between the meniscus and the main drop due to the electric repulsion between the satellite drop and the meniscus/main drop. Finally, the satellite drop may deposit on the substrate and form unwanted marks. Therefore, to achieve clean and high-resolution e-jet printing in dripping mode, it is necessary to understand the mechanism of satellite drop merging and separation. As shown in Figure 1c, the surface of the drop suffers normal electric stress, tangential electric stress, and surface tension stress during the breakup process. The stresses influence the breakup process of the drop and the subsequent satellite flight behavior. Once the satellite drop forms, it suffers electric repulsion force from the meniscus and the main drop. Finally, the satellite drop may merge with the meniscus or separate from the meniscus.

Figure 1d shows the axisymmetric numerical domain and the boundary conditions of the model, which consists of a nozzle of radius $\tilde{R}$ kept at potential $\tilde{\phi}$ in a cylindrical domain of air. The no-slip boundary condition is applied at the inner wall of the nozzle. In all the expressions, $(\cdot)$ denotes the dimensionless form of a quantity and $(\tilde{\ })$ denotes the dimensional form of a quantity. The electrical potential is applied at the nozzle and the top boundary. The zero potential and the outflow boundary condition are applied at the bottom. The symmetric boundary conditions are set for the velocity and potential at the left boundary and the symmetric boundary condition is applied at the right boundary for the electric potential. The governing equations are non-dimensionalized for the simulation (see Methods). At the nozzle inlet, a parabolic velocity profile is applied,

$$v = v_0\left(1 - r^2\right) \tag{1}$$

Here, v is the axial component of the velocity, r is the radial coordinate. The magnitude of the velocity can be tuned by changing v_0. Both the positive and negative charge densities ($c^{\pm}$) are set as constant at the nozzle inlet. The dimensionless inner and outer radii of the nozzle are 1 and 1.25, respectively. The dimensionless nozzle-to-substrate distance is 20. We used *Gerris*, an open-source code based on volume of fluid method to simulate the electrohydrodynamic dripping process [30,37]. The physics are based on the leaky dielectric model [38].

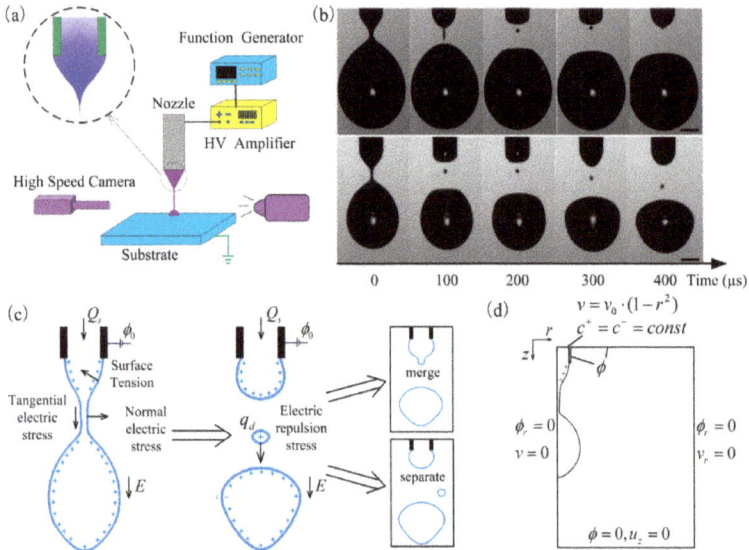

Figure 1. (**a**) The schematic of the experimental setup. (**b**) The two different satellite drop flight behaviors. For the upper images, the liquid was ethanol and the applied voltage was 1350 V. For the lower images, the liquid was a mixture of glycerin and water with the volume ratio of 1:2 and the applied voltage was 2300 V. The nozzle-to-substrate distance was 2.6 mm, which was 20 times the nozzle radius for both experiments. Scale bars: 200 µm. (**c**) The stress on the surface of the drop during breakup and the stress on the satellite drop during satellite drop motion. (**d**) The simulation domain and the boundary conditions for the simulation.

The governing equations of the leaky dielectric model are established for electrohydrodynamic dripping as follows. The liquid is assumed to be incompressible and Newtonian, and the mass conservation equation can be expressed as

$$\nabla \cdot u = 0 \tag{2}$$

where u is the velocity of the flow. The liquid flow of e-jet printing is governed by the modified Navier-Stokes equation [30]:

$$\frac{\partial u}{\partial t} + u \cdot \nabla u = -\nabla p + Oh \cdot \nabla^2 u + F_e - 2H\delta_s n \tag{3}$$

where p is the pressure of the liquid, F_e is the electric stress on the liquid, $-2H$ represents the surface tension stress, δ_s is the Dirac function which indicates that the surface tension stress only exists at the surface of the liquid. The surface tension stress is transformed into a volume stress in the liquid by the continuum surface force (CSF) model. The electric stress can be expressed as

$$F_e = \rho_e E - \frac{1}{2}E^2 \nabla \varepsilon_r \tag{4}$$

where ρ_e is the volume charge density, E is the electric field. The first term on the right-hand side of the equation represents the electrostatic stress due to the free charge under electric field. The second term represents the electric polar stress caused by the polarity of the liquid under electric field.

The motion of the charge in the liquid is governed by

$$c_t^\pm + \nabla \cdot (c^\pm u) = \nabla \cdot (D^\pm \nabla c^\pm) \mp \nabla \cdot (\Lambda^\pm c^\pm E) \tag{5}$$

The relation of the electric field, the electric potential, and the free charge is

$$\nabla \cdot (\varepsilon_r E) = -\nabla \cdot (\varepsilon_r \nabla \phi) = \rho_e \tag{6}$$

$$\rho_e = \frac{ez^+ c^+ + ez^- c^-}{\varepsilon_0 \tilde{\phi} / \tilde{R}^2} \tag{7}$$

where $z^\pm$ are the valences of the positive and negative charge species, respectively. The conductivity of the liquid can be expressed as

$$K = \omega^+ z^+ c^+ e + \omega^- z^- c^- e \tag{8}$$

where $\omega^\pm$ are the mobilities of the positive and negative charge species, respectively, and e represents the elementary charge.

Figure 2 shows the comparison of the experiment and the simulation. The satellite drop position in the simulation is the same as in the corresponding picture of the experiment. It can be seen that the satellite drop in the simulation falls down as that in the experiment, but the falling speed of the satellite drop in the simulation is smaller than that in the experiment. Maybe this is caused by the inaccuracy in calculating the electric stress between the meniscus and the satellite drop in the simulation.

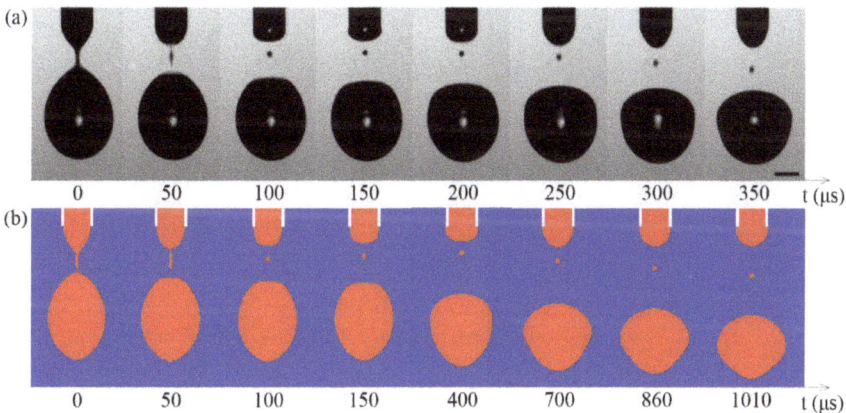

Figure 2. The comparison of (**a**) the experiment and (**b**) the simulation. The conductivity of the liquid is 2.3 µS/cm. The viscosity of the liquid is 3.5 mPa·s. The supplied flow rate is 223.5 µL/min. Scale bar = 100 µm.

3. Results and Discussion

There are 10 parameters influencing the electrohydrodynamic dripping process [39]:

$$\rho,\ \gamma,\ \varepsilon_L,\ K, \mu, \varepsilon_G, \tilde{R},\ \tilde{L},\ \tilde{\phi},\ \tilde{Q}_s$$

where ρ is the density of the liquid, γ is the surface tension of the liquid, ε_L is the dielectric constant of the liquid, K is the conductivity of the liquid, μ is the viscosity of the liquid, ε_G is the dielectric constant of air, $\tilde{R}$ is the inner radius of the nozzle, $\tilde{L}$ is the nozzle-to-substrate distance, $\tilde{\phi}$ is the applied voltage,

$\tilde{Q}_s$ is the supplied flow rate. After non-dimensionalization, there are six dimensionless numbers describing the electrohydrodynamic dripping process (see Appendix A):

$$\varepsilon_r, \ t_e, \ Oh, \ L, \ \phi, \ Q_s.$$

This study only considers the flow with low viscosity ($Oh = 0.05$). The dimensionless voltage ϕ is 6.

3.1. The Influence of CRT on the Flight of Satellite Drop

Figure 3 shows the motion of satellite drops for liquids with different CRTs. By altering the conductivity of the liquid, the CRT can be varied while keeping other dimensionless parameters constant. When the CRT is long enough, the satellite drop moves upward and merges with the meniscus (Figure 3a). When the CRT is short, the satellite first moves up and then goes down due to electric repulsion force (Figure 3b,c). The satellite drop goes farther away from the meniscus for liquid with smaller CRT within the same duration after breaking up from the bottom of the neck. Figure 3d shows the satellite drop position with time. The satellite drop goes upward and eventually merges with the meniscus for the liquid with CRT of $t_e = 2.5$. The satellite drop first moves up and then goes down for the liquid with CRT of $t_e = 0.667$ and $t_e = 0.333$. In order to avoid unwanted deposition of satellites, liquids with longer CRT should be adopted. Therefore, in order to prevent the satellite drop from falling onto the substrate, the CRT of the liquid should be sufficiently long.

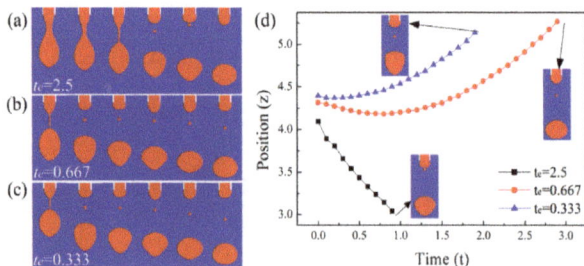

Figure 3. The satellite flight behaviors for liquids with CRTs of (**a**) $t_e = 2.5$, (**b**) $t_e = 0.667$, (**c**) $t_e = 0.333$. The dimensionless time between adjacent images is 1. (**d**) The position of the satellite drops with time for liquids with different CRTs. The other parameters are $Oh = 0.05$, $\phi = 6$, $Q_s = 0.314$, $\varepsilon_r = 50$.

The electric repulsion force between the meniscus and the satellite drop is important for the satellite flight behavior. Since the satellite drop and the meniscus have the same type of charge, the meniscus will electrostatically repel the satellite drop and alter the satellite motion, and the satellite drop acceleration is:

$$a = \frac{E_d}{\rho} \cdot \frac{q_d}{V} \tag{9}$$

where V is the volume of the satellite drop, a is the deceleration rate of the satellite drop, E_d is the average electric field on the satellite drop, q_d is the charge on the satellite drop. Equation (9) suggests the deceleration rate mainly depends on the charge-to-volume ratio (q_d/V). Figure 4a shows the amount of charges carried by the satellite drops and the satellite drop volumes for liquids with different CRTs. It can be seen that the charges carried by the satellite drops decrease with CRT and increases a little when the charge relaxation time is long. The satellite drop volume increases with CRT. The charge-to-volume ratio decreases with CRT as shown in Figure 4b. Since the charge-to-volume ratio is small for the liquid with long CRT, the deceleration rate of the satellite drop is small. Thus, the satellite drop can merge with the meniscus for the liquid with long CRT.

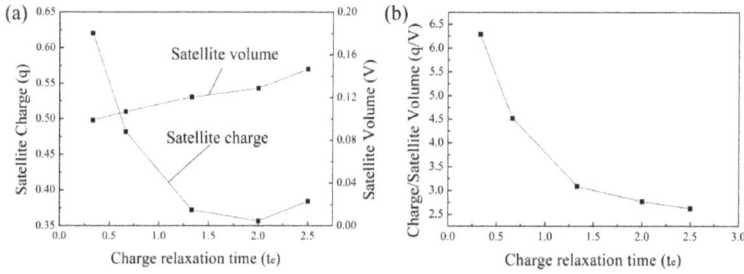

Figure 4. The satellite drops and their flight behavior with different CRTs. (**a**) The correlation between the satellite drop charge and CRT and the correlation between satellite drop volume and CRT. (**b**) The correlation between the charge-to-volume ratio of the satellite drop and CRT. The other parameters are $Oh = 0.05$, $\phi = 6$, $Q_s = 0.314$, $\varepsilon_r = 50$.

The initial speed of the satellite drop upon its formation also has a significant influence on the satellite drop motion. If the initial upward speed of the satellite drop is high, it is prone to merge with the meniscus. The initial speed is 0.833, 0.308, and 0.132 for fluids with charge relaxation time of 2.5, 0.667, and 0.333, respectively. The initial speed is higher for the satellites with longer CRT. The initial position is closer to the meniscus for the liquid with longer CRT. The initial position and initial speed of the satellite drop are determined by the breakup process of the drop. The breakup process is influenced by the electric field and electric stress distribution along the drop surface. Figure 5 shows the evolution of the droplet shape for liquids with different CRTs during the electrohydrodynamic dripping.

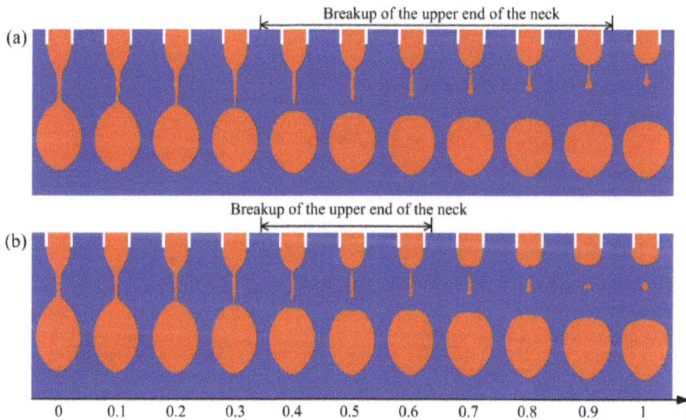

Figure 5. The evolution in time of the drop shape for liquids with CRTs of (**a**) $t_e = 2.5$ and (**b**) $t_e = 0.333$. The other parameters of the liquids are $\varepsilon_r = 50$, $Oh = 0.05$, $\phi = 6$, $Q_s = 0.314$.

The variation of the CRT affects the breakup of the drop due to the difference in the electric field and electric stress on the drop surface. Figure 6a,b show the isopotential lines for liquids with different CRTs at the moment of $t = 0$. The meniscus and the primary drop are nearly equipotential, but the potential changes along the neck. Figure 6c shows the velocity distribution along the z-direction for liquids with different CRTs. The velocity of the liquid in the meniscus and the main drop are low, therefore the surface charge densities can reach the electrostatic equilibrium state. But the velocity in the neck is high and changes significantly along the neck especially at both ends of the neck. During the breakup, the neck shrinks and stretches more rapidly than the charge density redistribution, so the electrostatic equilibrium state cannot be reached. Figure 6d shows the potential distributions along the z-direction. Because the migration speed of the charges in the liquid with large CRT is low, the potential drops faster along the neck.

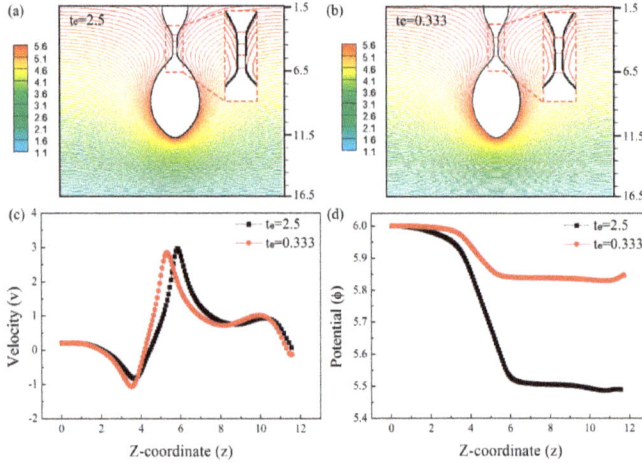

Figure 6. The isopotential lines at the instant when $t = 0$ for liquid with CRT of (a) $t_e = 2.5$ and (b) $t_e = 0.333$. (c) The velocity distribution for liquids with different CRTs at the instant when $t = 0$. (d) The potential distribution for liquids with different CRTs at the instant when $t = 0$. The other parameters of the liquids are $\varepsilon_r = 50$, $Oh = 0.05$, $\phi = 6$, $Q_s = 0.314$.

The decrease of the potential along the neck induces a large tangential electric field. The tangential electric field determines the electric stresses on the drop surface. The electric stress mainly exists at the surface of the liquid, and the stress balance at the surface of the liquid is [40]:

$$p + \tau_n = -2H \tag{10}$$

where τ_n is the normal electric stress, which is also called the electromechanical surface tension [26], $-2H$ represents the surface tension stress, which would be $2/r_0$ for a sphere drop of radius r_0. The surface tension stress points into the liquid and the normal electric stress points out of the liquid. The surface tension stress drives the shrinking of the neck and breakup, while the normal electric stress counters the surface tension stress. The larger the normal electric stress, the slower the shrinking of the neck. The tangential electric stress is negligible compared with the normal electric stress. The normal electric stresses consist of the electrostatic stress (τ_e) and polar stress (τ_p) [41], and the dimensionless normal electric stress can be expressed as:

$$\tau_n = \tau_e + \tau_p \tag{11}$$

$$\tau_e = q_s \cdot E_n \tag{12}$$

$$\tau_p = (\varepsilon_r - 1) \cdot E_s^2 \tag{13}$$

The dimensionless electrostatic stress (τ_e) is determined by the surface charge density (q_s) and normal electric field (E_n) on the drop surface and the dimensionless polar stress (τ_p) is proportional to the square of electric shear field (E_s). Figure 7a shows the profiles and the coordinates of the drops of different CRTs. Figure 7b shows E_s along the surface of the drops with different CRTs. E_s is high at both ends of the neck. This is caused by the large velocity change at both ends of the neck as shown in Figure 6c. E_s along the neck is stronger for the liquid with longer CRT. Figure 7c,d show the normal electric stresses, electrostatic stresses, and polar stresses along the drop surface for the liquid with

CRTs of $t_e = 2.5$ and $t_e = 0.333$. The stresses are large along the surface of the neck but small along the surface of the main drop and the meniscus. For the liquid with large CRT, the polar stress is much larger than the electrostatic stress and the normal electric stress is nearly equal to the polar stress. For the liquid with small CRT, both the electrostatic stress and the electric polar stress are small along the neck. So the normal electric stress of the liquid with large CRT along the neck is larger than that of the liquid with small CRT. The electric normal stress reaches its maximum at the upper end of the neck for the liquid with high CRT. Therefore, the shrinking of the neck and breakup of the upper end of the neck are delayed by the large electric normal stress as shown in Figure 5. The length of the neck is larger for liquid with large CRT and the volume of the resulting satellite drop is larger.

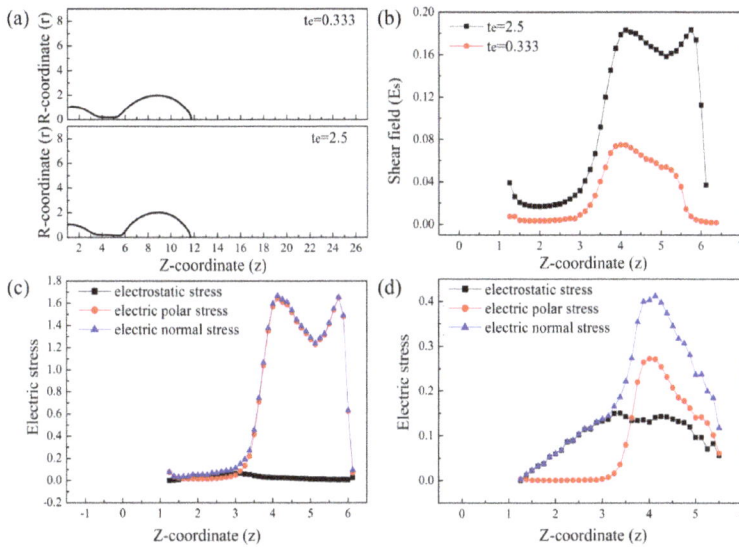

Figure 7. The comparisons of the electric stresses for liquids with different CRTs at the instant when $t = 0$. (**a**) The profiles of the drops of liquids with different CRTs. (**b**) The electric shear field along the surface of the drop for liquids with different CRTs. (**c**) The electric normal stress for liquid with CRT of $t_e = 2.5$. (**d**) The electric normal stress for liquid with CRT of $t_e = 0.333$. The other parameters of the liquids are $\varepsilon_r = 50$, $Oh = 0.05$, $\phi = 6$, $Q_s = 0.314$.

3.2. The Effect of Dielectric Constant on the Critical CRT

The critical CRT for the transition of the two different satellite drop flight behaviors is also affected by the dielectric constant of the liquid. Figure 8a shows that the critical CRT decreases with the dielectric constant of the liquid. When the CRT is larger than the critical value, the satellite drop merges with the meniscus. When the CRT is smaller than the critical value, the satellite drop separates from both the meniscus and the primary drop. Figure 8b shows the position of the satellite drop with time for different dielectric constants. The satellite drop merges with the meniscus for low dielectric constant but separates from the meniscus for high dielectric constants. The initial position of the satellite drop for the liquid with high dielectric constant is closer to the meniscus and the initial speed of the satellite drop is also higher.

Figure 9a shows the correlation between satellite drop volume and satellite drop charge with the relative dielectric constant of the liquid, where the CRTs of the liquids are the same. It can be seen that both the volume and charge decrease with the dielectric constant of the liquid. Figure 9b shows the ratio of charge to volume for the liquids with different dielectric constants. The ratio of the charge to volume does not change with the dielectric constant of the liquid. Thus, the ratio of the charge to volume will not influence the deceleration rate of the satellite drop according to Equation (9).

Figure 9c shows that the neck length decreases with the dielectric constant of the liquid at the moment the primary drop detaches from the neck. The satellite drop is farther away from the meniscus for the liquid with small dielectric constant, therefore the satellite drop is less likely to merge with the meniscus. Figure 9d shows the normal electric stress distribution along the drop surface. The normal electric stress along the neck is stronger for the liquid with small dielectric constant, thus the shrinking of the neck for liquid with small dielectric constant is slower, which leads to a longer neck.

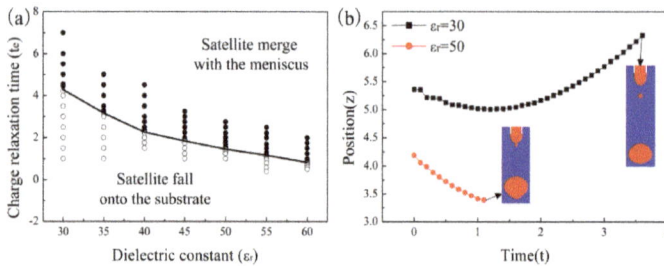

Figure 8. The satellite drop flight behavior for liquids with different dielectric constants. (**a**) The correlation between the critical CRT for the different satellite flight behaviors and dielectric constant. (**b**) The position of the satellite drop with time for liquids with CRT of $t_e = 2$. The other parameters are $Oh = 0.05$, $\phi = 6$, $Q_s = 0.314$.

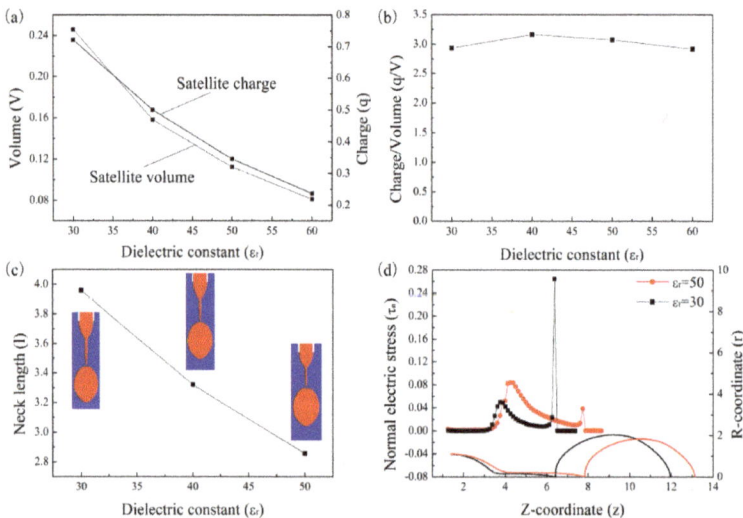

Figure 9. (**a**) The correlation of the satellite drop volume and satellite drop charge with dielectric constant. (**b**) The correlation between the charge-to-volume ratio of the satellite drop and dielectric constant. (**c**) The neck length for liquids with different dielectric constants. (**d**) The normal electric stress for liquids with different dielectric constants. The other parameters are $Oh = 0.05$, $\phi = 6$, $Q_s = 0.314$, $t_e = 2$.

3.3. The Effect of Supplied Flow Rate on the Critical CRT

Figure 10a,b shows the satellite drop formation process for different supplied flow rates. As the supplied flow rate increases, the upper end of the neck delays its breakup. However, when the CRT of the liquid is short enough, the satellite drop eventually separates from the primary drop and meniscus. Figure 10c shows the position of the satellite drop with time for different flow rates. When the flow rate is low, the satellite drop falls downward. When the supplied flow rate is high, the satellite drop moves upward and merges with the meniscus. For high flow rate, although the initial position of the satellite drop is low, the satellite drop is still close to the meniscus due to the large meniscus volume. So, the higher the supplied flow rate, the satellite drop is more prone to merge with the meniscus. Figure 11 shows that the critical CRT decreases with the supplied flow rate.

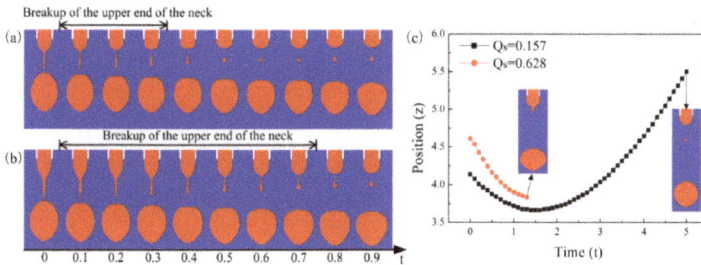

Figure 10. The satellite drop formation for supplied flow rate of (**a**) $Q_s = 0.157$ and (**b**) $Q_s = 0.628$. (**c**) The satellite drop position with time for different supplied flow rates. The other parameters are $\varepsilon_r = 50$, $Oh = 0.05$, $\phi = 6$, $t_e = 1.429$.

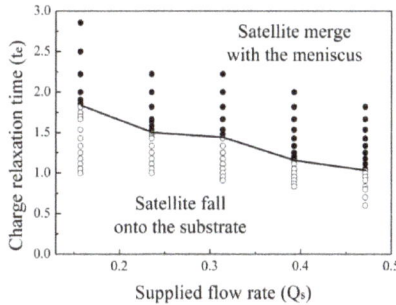

Figure 11. The critical CRT for the different satellite flight behaviors with different supplied flow rates. The other parameters are $\varepsilon_r = 50$, $Oh = 0.05$, $\phi = 6$.

4. Conclusions

The dripping mode of E-jet printing is numerically investigated to reveal the influence of the CRT on the generation and flight of satellite drops. The results show that the satellite drop merges with the meniscus when the CRT of the liquid is long. The charge-to-volume ratio of the satellite drop is small for liquid with large CRT, so the electric repulsion between the satellite drop and the meniscus is weak. This promotes the merging of the satellite drop with the meniscus for liquid with large CRT. The electric potential decreases along the neck due to rapid stretching prior to breakup, which induces a large tangential electric field as well as large normal electric stress along the neck surface. The normal electric stress delays the breakup of the upper neck, which makes the initial speed of the satellite drop larger and the initial position closer to the meniscus. This also promotes the merging of the satellite drop with the meniscus. The numerical results also show that the critical CRT is affected by the dielectric constant of the liquid and the supplied flow rate. The critical CRT distinguishing satellite merging or separation decreases with dielectric constant. As the dielectric constant of the liquid increases, the neck becomes shorter at the moment the primary drop detaches from the neck, while the charge-to-volume ratio of the satellite drop remains unchanged. For high dielectric constant, the initial position of the satellite drop is closer to the meniscus, therefore the satellite drop is more prone to merge with the meniscus. As the supplied flow rate increases, the breakup of the upper end of the neck delays, which promotes the satellite drop to merge with the meniscus. The findings in this work will help to choose appropriate parameters to absorb the satellite drop and improve the resolution of e-jet printing in the dripping mode.

Author Contributions: Conceptualization, L.G.; Funding acquisition, Y.D., Y.H., and Z.Y.; Investigation, L.G. and Y.H.; Writing—original draft, L.G.; Writing—review & editing, Y.D., W.D., Y.G, Y.H., and Z.Y.

Funding: This work is financially supported by the National Key Research and Development Program of China (2016YFB0401105), the National Natural Science Foundation of China (51635007, 51605180), Special Project of Technology Innovation of Hubei Province (2017AAA002), Program for HUST Academic Frontier Youth Team and Program for Guangdong Introducing Innovative and Entrepreneurial Teams (2017ZT07G331).

Acknowledgments: We thank the Flexible Electronics Manufacturing Laboratory in Comprehensive Experiment Center for Advanced Manufacturing Equipment and Technology at Huazhong University of Science and Technology for providing the e-jet printing equipment.

Conflicts of Interest: The authors declare no conflict of interest.

Appendix A

Dimensionless Numbers. The problem is non-dimensionalized using the nozzle radius $\tilde{R}$ as the length scale, capillary time $\tilde{t}_c = \left(\rho\tilde{R}^3/\gamma\right)^{1/2}$ as the time scale, capillary pressure $p_c = \gamma/\tilde{R}$ as the pressure scale, capillary velocity $\tilde{u}_c = \left(\gamma/\rho\tilde{R}\right)^{1/2}$ as the velocity scale, $\tilde{E}_0 = \left(\gamma/\left(\varepsilon_G\tilde{R}\right)\right)^{1/2}$ as the electric field scale, $\tilde{E}_0\tilde{R}$ as the electric potential scale, $\varepsilon_G\tilde{E}_0/\tilde{R}$ as volume charge density scale, and $\tilde{Q}_m = \gamma\varepsilon_G/\rho K$ as the supplied flow rate scale. There are seven dimensionless numbers in the governing equations: (1) Ohnesorge number $Oh = \mu/\left(\rho\gamma\tilde{R}\right)^{1/2}$; (2) dimensionless ion diffusivities $D^{\pm} = \omega^{\pm}k_BT/\left(\tilde{R}\tilde{u}_c\right)$, where k_B is the Boltzman constant, T is the temperature of the liquid; (3) dimensionless ion specific conductivities $\Lambda^{\pm} = \tilde{E}_0ew^{\pm}z^{\pm}/\tilde{u}_c$, where e is the elementary charge, $z^{\pm}$ are the valence of the ions, $z^+ = 1$ and $z^- = -1$ in the simulation; (4) permittivity ratio $\varepsilon_r = \varepsilon_L/\varepsilon_G$; (5) dimensionless charge relaxation time (CRT) $t_e = \tilde{t}_e/\tilde{t}_c$; (6) dimensionless nozzle-to-substrate distance $L = \tilde{L}/\tilde{R}$; (7) dimensionless applied voltage $\phi = \left(\varepsilon_G/\left(\gamma\tilde{R}\right)\right)^{1/2}\tilde{\phi}$.

References

1. Park, J.U.; Hardy, M.; Kang, S.J.; Barton, K.; Adair, K.; Mukhopadhyay, D.K.; Lee, C.Y.; Strano, M.S.; Alleyne, A.G.; Georgiadis, J.G.; et al. High-resolution electrohydrodynamic jet printing. *Nat. Commun.* **2007**, *6*, 782–789. [CrossRef] [PubMed]

2. Ye, D.; Ding, Y.J.; Duan, Y.Q.; Su, J.T.; Yin, Z.P.; Huang, Y.A. Large-Scale Direct-Writing of Aligned Nanofibers for Flexible Electronics. *Small* **2018**, *14*, 1703521. [CrossRef] [PubMed]

3. Huang, Y.A.; Duan, Y.Q.; Ding, Y.J.; Bu, N.B.; Pan, Y.Q.; Lu, N.S.; Yin, Z.P. Versatile, kinetically controlled, high precision electrohydrodynamic writing of micro/nano fibers. *Sci. Rep.* **2014**, *4*, 5949. [CrossRef] [PubMed]

4. Park, J.U.; Lee, S.; Unarunotai, S.; Sun, Y.G.; Dunham, S.; Song, T.; Ferreira, P.M.; Alleyene, A.G.; Paik, U.; Rogers, J.U. Nanoscale, electrified liquid jets for high-resolution printing of charge. *Nano Lett.* **2010**, *10*, 584–591. [CrossRef] [PubMed]

5. Basaran, O.A. Small-Scale Free Surface Flows with Breakup: Drop Formation and Emerging Applications. *AIChE J.* **2004**, *48*, 1842–1848. [CrossRef]

6. Juraschek, R.; Röllgen, F.W. Pulsation phenomena during electrospray ionization. *Int. J. Mass Spectrom.* **1998**, *177*, 1–15. [CrossRef]

7. An, S.; Lee, M.W.; Kim, N.Y.; Lee, C.; Al-Deyab, S.S.; James, S.C.; Yoon, S.S. Effect of viscosity, electrical conductivity, and surface tension on direct-current-pulsed drop-on-demand electrohydrodynamic printing frequency. *Appl. Phys. Lett.* **2014**, *105*, 214102. [CrossRef]

8. Lee, M.W.; Kang, D.K.; Kim, N.Y.; Kim, H.Y.; James, S.C.; Yoon, S.S. A study of ejection modes for pulsed-DC electrohydrodynamic inkjet printing. *J. Aerosol Sci.* **2012**, *46*, 1–6. [CrossRef]

9. Bu, N.B.; Huang, Y.A.; Wang, X.M.; Yin, Z.P. Continuously Tunable and Oriented Nanofiber Direct-Written by Mechano-Electrospinning. *Mater. Manuf. Process.* **2012**, *27*, 1318–1323. [CrossRef]

10. Deng, W.; Gomez, A. Full transient response of Taylor cones to a step change in electric field. *Microfluid. Nanofluid.* **2012**, *12*, 383–393. [CrossRef]

11. Wu, X.; Oleschuk, R.D.; Cann, N.M. Characterization of microstructured fibre emitters: In pursuit of improved nano electrospray ionization performance. *Analyst* **2012**, *137*, 4150–4161. [CrossRef] [PubMed]

12. Qin, H.T.; Dong, J.Y.; Lee, Y.S. AC-pulse modulated electrohydrodynamic jet printing and electroless copper deposition for conductive microscale patterning on flexible insulating substrates. *Robot. Comput.-Integr. Manuf.* **2017**, *43*, 179–187. [CrossRef]

13. Huang, Y.A.; Ding, Y.; Bian, J.; Su, Y.; Zhou, J.; Duan, Y.Q.; Yin, Z.P. Hyper-stretchable self-powered sensors based on electrohydrodynamically printed, self-similar piezoelectric nano/microfibers. *Nano Energy* **2017**, *40*, 432–439. [CrossRef]

14. Wang, X.; Xu, L.; Zheng, G.F.; Cheng, W.; Sun, D.H. Pulsed electrohydrodynamic printing of conductive silver patterns on demand. *Sci. China Technol. Sci.* **2012**, *55*, 1603–1607. [CrossRef]

15. Lim, S.; Park, S.H.; An, T.K.; Lee, H.S.; Kim, S.H. Electrohydrodynamic printing of poly(3,4-ethylenedioxythiophene):poly(4-styrenesulfonate) electrodes with ratio-optimized surfactant. *RSC Adv.* **2016**, *6*, 2004–2010. [CrossRef]

16. Park, J.U.; Lee, J.H.; Paik, U.; Lu, Y.; Rogers, J.A. Nanoscale patterns of oligonucleotides formed by electrohydrodynamic jet printing with applications in biosensing and nanomaterials assembly. *Nano Lett.* **2008**, *8*, 4210–4216. [CrossRef] [PubMed]

17. Shigeta, K.; He, Y.; Sutanto, E.; Kang, S.; Le, A.P.; Nuzzo, R.G.; Alleyne, A.G.; Ferreira, P.M.; Lu, Y.; Rogers, J.A. Functional protein microarrays by electrohydrodynamic jet printing. *Anal. Chem.* **2012**, *84*, 10012–10018. [CrossRef] [PubMed]

18. Sutanto, E.; Tan, Y.; Onses, M.S.; Cunningham, B.T.; Alleyne, A. Electrohydrodynamic jet printing of micro-optical devices. *Manuf. Lett.* **2014**, *2*, 4–7. [CrossRef]

19. Kim, B.H.; Onses, M.S.; Lim, J.B.; Nam, S.; Oh, N.; Kim, H.; Yu, K.J.; Lee, J.W.; Kim, J.H.; Kang, S.K.; et al. High-Resolution Patterns of Quantum Dots Formed by Electrohydrodynamic Jet Printing for Light-Emitting Diodes. *Nano Lett.* **2015**, *15*, 969–973. [CrossRef] [PubMed]

20. Galliker, P.; Schneider, J.; Eghlidi, H.; Kress, S.; Sandoghdar, V.; Poulikakos, D. Direct printing of nanostructures by electrostatic autofocussing of ink nanodroplets. *Nat. Commun.* **2012**, *3*, 1–9. [CrossRef] [PubMed]

21. Han, Y.; Wei, C.; Dong, J. Super-resolution electrohydrodynamic (EHD) 3D printing of micro-structures using phase-change inks. *Manuf. Lett.* **2014**, *2*, 96–99. [CrossRef]

22. Mao, M.; He, J.K.; Li, X.; Zhang, B.; Lei, Q.; Liu, Y.X.; Li, D.C. The emerging frontiers and applications of high-resolution 3D printing. *Micromachines* **2017**, *8*, 113. [CrossRef]

23. Cui, Z.; Han, Y.W.; Huang, Q.J.; Dong, J.Y.; Zhu, Y. Electrohydrodynamic printing of silver nanowires for flexible and stretchable electronics. *Nanoscale* **2018**, *10*, 6806–6811. [CrossRef] [PubMed]

24. Zhao, X.; Wang, X.; Lim, S.; Qi, D.; Wang, R.; Gao, Z.; Mi, B.; Chen, Z.; Huang, W.; Deng, W. Enhancement of the performance of organic solar cells by electrospray deposition with optimal solvent System. *Sol. Energy Mater. Sol. Cells* **2014**, *121*, 119–125. [CrossRef]

25. Duan, H.; Li, C.; Yang, W.; Lojewski, B.; An, L.; Deng, W. Near-Field Electrospray Microprinting of Polymer-Derived Ceramics. *J. Microelectromech. S.* **2013**, *22*, 1–3. [CrossRef]

26. Notz, P.K.; Basaran, O.A. Dynamics of Drop Formation in an Electric Field. *J. Colloid Interface Sci.* **1999**, *213*, 218–237. [CrossRef] [PubMed]

27. Collins, R.T.; Harris, M.T.; Basaran, O.A. Breakup of electrified jets. *J. Fluid Mech.* **2007**, *588*, 75–129. [CrossRef]

28. Lee, A.; Jin, H.; Dang, H.W.; Choi, K.H.; Ahn, K.H. Optimization of Experimental Parameters To Determine the Jetting Regimes in Electrohydrodynamic Printing. *Langmuir* **2013**, *29*, 13630–13639. [CrossRef] [PubMed]

29. Lee, M.W.; Kim, N.Y.; Yoon, S.S. On pinchoff behavior of electrified droplets. *J. Aerosol Sci.* **2013**, *57*, 114–124. [CrossRef]

30. Lopez-Herrera, J.M.; Ganan-Calvo, A.M.; Popinet, S.; Herrada, M.A. Electrokinetic effects in the breakup of electrified jets: A Volume-Of-Fluid numerical study. *Int. J. Multiph. Flow* **2015**, *71*, 14–22. [CrossRef]

31. Zhong, Y.; Fang, H.; Ma, Q.; Dong, X. Analysis of droplet stability after ejection from an inkjet nozzle. *J. Fluid Mech.* **2018**, *845*, 378–391. [CrossRef]
32. Ambravaneswaran, B.; Wilkes, E.D.; Basaran, O.A. Drop formation from a capillary tube: Comparison of one-dimensional and two-dimensional analyses and occurrence of satellite drops. *Phys. Fluids* **2002**, *14*, 2606–2621. [CrossRef]
33. Li, C.; Krueger, K.; Yang, W.; Duan, H.; Deng, W. Gas-focused liquid microjets from a slit. *Phys. Fluids* **2015**, *27*, 032101. [CrossRef]
34. Hoath, S.D.; Jung, S.; Hsiao, W.K.; Hutchings, I.M. How PEDOT: PSS solutions produce satellite-free inkjets. *Org. Electron.* **2012**, *13*, 3259–3262. [CrossRef]
35. Wijshoff, H. Drop dynamics in the inkjet printing process. *Curr. Opin. Colloid. Interface Sci.* **2018**, *36*, 20–27. [CrossRef]
36. Huo, Y.; Wang, J.; Zuo, Z.; Fan, Y. Visualization of the evolution of charged droplet formation and jet transition in electrostatic atomization. *Phys. Fluids* **2015**, *27*, 114105. [CrossRef]
37. López-Herrera, J.M.; Popinet, S.; Herrada, M.A. A charge-conservative approach for simulating electrohydrodynamic two-phase flows using volume-of-fluid. *J. Comput. Phys.* **2010**, *230*, 1939–1955. [CrossRef]
38. Saville, D.A. Electrohydrodynamics: The Taylor-Melcher Leaky Dielectric Model. *Annu. Rev. Fluid Mech.* **1997**, *29*, 27–64. [CrossRef]
39. Bober, D.B.; Chen, C.H. Pulsating electrohydrodynamic cone-jets: From choked jet to oscillating cone. *J. Fluid Mech.* **2011**, *689*, 552–563. [CrossRef]
40. Collins, R.T.; Jones, J.J.; Harris, M.T. Electrohydrodynamic tip streaming and emission of charged drops from liquid cones. *Nat. Phys.* **2008**, *4*, 149–154. [CrossRef]
41. Gañán-Calvo, A.M.; Rebollo-Muñoz, N.; Montanero, J.M. The minimum or natural rate of flow and droplet size ejected by Taylor cone–jets: Physical symmetries and scaling laws. *New J. Phys.* **2013**, *15*, 033035. [CrossRef]

micromachines

MDPI

Article

Design and Fabrication of an Artificial Compound Eye for Multi-Spectral Imaging

Axiu Cao [1,2], Hui Pang [1], Man Zhang [1], Lifang Shi [1,*], Qiling Deng [1] and Song Hu [1,*]

[1] Institute of Optics and Electronics, Chinese Academy of Sciences, Chengdu 610209, China; longazure@163.com (A.C.); ph@ioe.ac.cn (H.P.); zhangman881003@126.com (M.Z.); dengqiling@ioe.ac.cn (Q.D.)
[2] University of Chinese Academy of Sciences, Beijing 100049, China
*   Correspondence: shilifang@ioe.ac.cn (L.S.); husong@ioe.ac.cn (S.H.); Tel.: +86-028-8510-1178 (L.S.)

Received: 28 February 2019; Accepted: 22 March 2019; Published: 25 March 2019

Abstract: The artificial compound eye (ACE) structure is a new type of miniaturized, lightweight and intelligent imaging system. This paper has proposed to design a multi-spectral ACE structure to enable the structure to achieve multi-spectral information on the basis of imaging. The sub-eyes in the compound eye structure have been designed as diffractive beam splitting lenses with the same focal length of 20 mm, but with the different designed center wavelengths of 650 nm, 532 nm, and 445 nm, respectively. The proximity exposure lithography and reactive ion etching process were used to prepare the designed multi-spectral ACE structure, and the spectral splitting and multi-spectral imaging experiments were carried out to verify the multi-spectral imaging function of the structure without axial movement. Furthermore, the structure can be designed according to actual requirements, which can be applied to covert reconnaissance, camouflage identification, gas leakage or other fields.

Keywords: artificial compound eye; multi-spectral imaging; lithography; spectral splitting

---

1. Introduction

Compound eyes of arthropods such as ants, flies and bugs have attracted extensive research interests due to their unique features such as wide fields of view (FOV), high sensitivities to motions and infinite depths of field [1–8]. These compound eyes are integrated, lightweight and smart optical imaging systems composed of multiple individual lenses arranged on a hemispherical surface. Optical devices inspired by natural compound eyes exhibit great potentials in various applications such as surveillance cameras on micro aerial vehicles, high-speed motion detections, endoscopic medical tools, and image guided surgeries [9–11].

For years, several attempts were made to develop artificial compound eyes (ACEs) to realize imaging [12–15], which typically analyzes the spatial characteristics of the target by studying the shape information of the target. In this paper, the spectroscopy technology has been proposed to apply into the ACE structure combined with imaging. Based on the original two-dimensional spatial imaging, one-dimensional spectral information has been added in the structure. Compared with the traditional imaging technology, it can improve the accuracy and sensitivity of target detection, and expand the detection capability of traditional imaging systems for target detection, camouflage identification, complex background suppression, life state observation, environmental monitoring and other fields.

Multi-spectral imaging technology needs to divide the incident full-band or wide-band spectrum into several narrow-band spectrum so as to obtain images of different spectral bands by the corresponding detectors. Thereinto, spectral splitting technology is very important. Spectral splitting technology mainly includes interference spectroscopy [16,17] and traditional filter spectroscopy [18–20].

Interferometric imaging spectrometry detects the interferogram of target by combining two-beam interferometry with imaging technology and obtains spectral information by Fourier transform calculation. The structure of the system is complicated due to the addition of interferometric light paths.

The traditional spectral imaging technology is that the narrow and rectangular slits are imaged on the two-dimensional detector by the prism or grating of the dispersive element. The dispersion direction of the prism or grating is perpendicular to the long axis of the slit. The detector records the spatial image in one-dimensional direction, and the spectral data in the other one-dimensional direction is the spatial spectral image. Then, the spectral data of the whole target is obtained by the push-sweep method. It is inconvenient to operate. The other traditional narrow-band tuned filter imaging spectroscopy technology uses the scheme of camera and filter. Its principle is simple and there are many kinds, such as circular gradient filter, liquid crystal tunable filter and filter wheel. This spectral imaging technology obtains monochrome image of scene object through one snapshot, but it needs to change or modulate the filter to obtain the whole image data. In general, current multispectral cameras are larger than normal cameras because they must be equipped with a mechanism to the original imaging system for changing the spectral transmittance, such as a prism, a grating or a filter wheel or a liquid crystal tunable filter.

ACE is suitable to be used as the lens of a multi-spectral imaging system because optical information of different bands can be captured simultaneously by different units. A typical multi-spectral imaging system based on ACE put a multi-channel color filter in front of the microlens array to realize spectral separation [21]. However, it will increase the complexity of the system and also bring the alignment problem between the ACE and the filter. The position mismatch of ACE and filters will make the imaging quality decline. Therefore, it is desirable to integrate the ACE with multi-channel filter in order to improving the system performance. Then, a compact structure of integrating a microlens array with a multi-channel filter based on a Micro-Electro-Mechanical-System (MEMS) technique has been fabricated [22]. However, the fabrication of multi-channel filter is complex because of the repeated fabrication process of multi-channel filter based on pigment-dispersed method.

The diffractive beam splitting lenses can realize the function of imaging and spectroscopy at the same time. Diffraction optical spectrum imaging is a staring imaging technology based on the diffractive beam splitting lenses, which can simultaneously image the scene in the field of view and read it out by using a focal plane array detector at one time. Therefore, the structure is more compact and the optical path is much simpler because there is no optical-mechanical scanning component. At the same time, the whole input aperture can collect light, which is contrary to the narrow input slit of traditional spectral imaging. The luminous flux of the system is much higher than that of conventional dispersive spectral imaging technology. Meanwhile, diffraction optical spectrum imaging technology is very flexible. It can collect all or part of the band information by program control in the free band of the instrument. It can be used for spectrum detection from ultraviolet to far infrared band, which is limited only by detector array technology.

Generally, diffraction optical spectrum imaging has the advantages of compact structure, high throughput and staring imaging [23,24]. However, in the image space, the axial distance adjusting device should be used to adjust the detector to move along the optical axis; then, the corresponding two-dimensional spectral image sequence can be collected to achieve multi-spectral imaging.

In this paper, on the basis of making full use of the spectral imaging advantages of diffractive beam splitting lenses, different sub-eye imaging channels of the compound eye are designed to capture spectral information of different bands at the same distance avoiding the use of axial distance adjusting device, which can effectively reduce the dependence of the acquisition of different spectral information on the distance between the lens and the detector so that the ACEs can realize multi-spectral imaging function on the basis of imaging, which can be used in concealment reconnaissance, camouflage recognition, toxic gas leakage and other fields.

2. Principle and Structural Design

The ACE is composed of several sub-eye imaging channels, as shown in Figure 1. Different sub-eye imaging channels are designed as diffractive beam splitting lenses (D_i) with different designed parameters. The lenses have the same focal length (F) and different central wavelength (λ_{i_0}). In the imaging process, spectral information of different wavelengths can be captured by different sub-eyes at the same image plane.

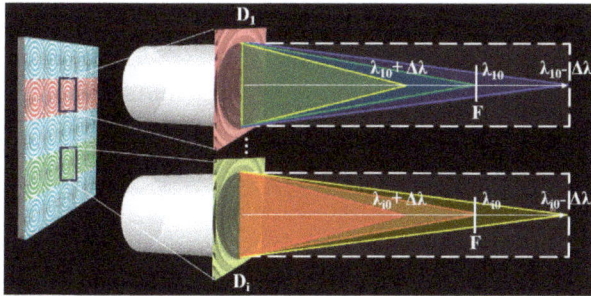

Figure 1. Multi-spectral imaging principle of artificial compound eye (ACE).

The arrangement of the sub-eye lenses in the multi-spectral ACE is shown in Figure 2. Along the Y direction, the sub-eye lenses are the same. In the X direction, the sub-eye lenses are designed for different wavelengths. The designed wavelengths (λ_{i_0}, $i = 1,2,3$) are 650 nm (R), 532 nm (G) and 445 nm (B), respectively. The aperture (D_i, $i = 1,2,3$) are the same as 2 mm, and the focal length (F) is also the same as 20 mm. Three sub-eye lenses with different central wavelengths are arranged periodically in the X direction with a period of 6 mm to form a complete ACE structure.

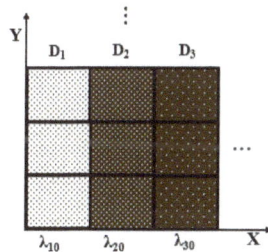

Figure 2. The arrangement of sub-eye lenses in multi-spectral ACE.

Different sub-eyes are composed of different annular structures. The maximum number of annular bands of the corresponding sub-eye lenses is determined by the aperture (D_i) of the sub-eye lens, the design wavelength (λ_{i_0}) and the focal length (F), which can be calculated by Formula (1). Then, according to Formula (2), the radius (r_{i_n}) corresponding to different number of annular bands (n_i) can be calculated to obtain the complete distribution of annular bands structure for different sub-eye lenses, and some radius values are shown in Table 1:

$$N_i = \frac{(D_i/2)^2}{F\lambda_{i0}}, i = 1, 2, 3; \tag{1}$$

$$r_{in} = \sqrt{n_i F \lambda_{i0}}, n_i = 1, 2 \cdots N_i; \tag{2}$$

Table 1. Radius values of multi-spectral structure in different annular bands.

n	$r_{1n}/\mu m$	$r_{2n}/\mu m$	$r_{3n}/\mu m$
1	114.0175	103.1504	94.33981
2	161.2452	145.8767	133.4166
3	197.4842	178.6617	163.4013
4	228.0351	206.3008	188.6796
5	254.9510	230.6513	210.9502
6	279.2848	252.6658	231.0844
7	301.6621	272.9102	249.5997
8	322.4903	291.7533	266.8333
9	342.0526	309.4511	283.0194
10	360.5551	326.1901	298.3287
11	378.1534	342.1111	312.8898
12	394.9684	357.3234	326.8027
13	411.0961	371.9140	340.1470
14	426.6146	385.9534	352.9873
15	441.588	399.4997	365.3765
16	456.0702	412.6015	377.3592
17	470.1064	425.2999	388.9730
18	483.7355	437.6300	400.2499
19	496.9909	449.6221	411.2177
20	509.902	461.3025	421.9005
...	...	...	...

3. Experiments and Discussion

3.1. Fabrication

The preparation technology of the proposed multi-spectral ACE was studied. The preparation process of diffractive beam splitting lens with a central wavelength of 532 nm is described in detail below. When the central wavelength is 532 nm, the etching depth (*h*) can be obtained by Formula (3). The refractive index (*n*) of silica material is 1.461 at 532 nm, so the etching depth is calculated as 577 nm:

$$h = \frac{\lambda}{2(n-1)}, n = 1.461@532m; \tag{3}$$

The fabrication process flow mainly includes photoresist spin coating, exposure, development and etching, as shown in Figure 3. Firstly, the silica material (JGS1) was selected as the substrate material, and the photoresist AZ1500 (20cp) (AZ 1500 Photoresist (20cp), Merck Electronic Materials, Shanghai, China) was spun on the surface of the substrate at a speed of 6000 rpm for 30 s as the initial structure (Figure 3a,b). Key parameters such as prebake temperature and prebake period were 90 °C and 10 min, respectively. Secondly, the annular bands structure was encoded as the mask data according to the design parameters, and the mask was prepared using the high-precision laser direct writing technology, as shown in Figure 4a. The UV lithography machine (URE-2000S/A) with a central wavelength of 365 nm was used. The exposure mode was proximity exposure with total exposure time of 10 s at an exposure power density of 2.5 mW/cm² (Figure 3c). Third, the development and post-exposure bake were carried out. The photoresist was developed in a solution of AZ300MIF (AZ300MIF Developer, Merck Electronic Materials, Shanghai, China) for 25 s, followed by 2 h post-exposure bakes at 120 °C, and the photoresist multi-spectral compound eye structure was obtained as shown in Figure 3d. Finally, etching of the silica was conducted, which transferred the patterns of the multi-spectral compound eye structure to the silica substrate (see Figure 4b). A reactive ion etching machine was used to transfer the structure into the substrate. The etching gases were SF_6 and CHF_3 in the proportion of 1:20 and the etching time was 15 min (see Figure 3e).

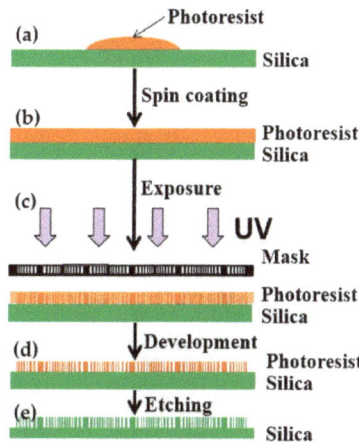

Figure 3. Preparation technology of multi-spectral ACE: (**a**) substrate preparation; (**b**) photoresist was spin-coated on the silica; (**c**) proximity exposure with designed mask; (**d**) photoresist-based structure was achieved after development; (**e**) the pattern was transferred to silica by etching.

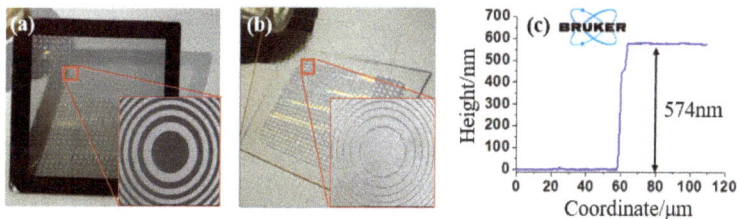

Figure 4. Key fabrication results, including: (**a**) mask; (**b**) prototype based on silica; (**c**) measurement of the fabricated structure by a step profilometer.

A step profilometer (Stylus Profiler System, Dektak XT, Bruker, Karlsruhe, Germany) was used for the measurement of the fabricated structure, and the cross-sectional profile was drawn by the scientific drawing and data analysis software of OriginLab (OriginPro 8.0, Northampton, MA, USA) with etching depth of 574 nm (see Figure 4c), which was consistent with the design result.

The structure of different diffractive beam splitting lenses corresponding to different wavelength was realized by multiple overlay exposure and etching technology. The lenses with different depths were obtained by changing etching times during the etching process.

The diffraction efficiency of the structure is related to the step number of the actual structure. The higher the step number is, the higher the diffraction efficiency is. When the number of steps is 2, the diffraction efficiency is about 40%; when the number of steps is 4, the diffraction efficiency is more than 80%; when the number of steps is more than 8, the diffraction efficiency can be more than 90%; if the number of steps is 32, the diffraction efficiency can be as high as 99%. Of course, with the increase of the number of processing steps, the difficulty of preparation process will also increase. In order to verify the principle of the proposed structure from the optical effect, the structure of the fabricated structure is only two steps, and the diffraction efficiency is slightly low at 40%. If the practical application of the structure requires higher diffraction efficiency, the structure with higher steps can be prepared.

3.2. Imaging Verification

The imaging light path was constructed to verify the spectral characteristics of the prepared multi-spectral ACE structure. The structure was irradiated by red, green and blue lasers of 650 nm,

532 nm and 445 nm, which were consistent with the design center wavelength, as shown in Figure 5. During the experiment, the expanding and collimation of the laser beam were firstly realized, so that the laser beam with a diameter of more than 6 mm can simultaneously illuminate the area with different designed center wavelengths on the multi-spectral compound eye structure. Secondly, the three-color beams were combined to make the three-channel beam coaxially illuminate the multi-spectral compound eye structure. Finally, by adjusting the distance Z between the photodetector and the structure, the spot distributions of different wavelengths at different distances were observed.

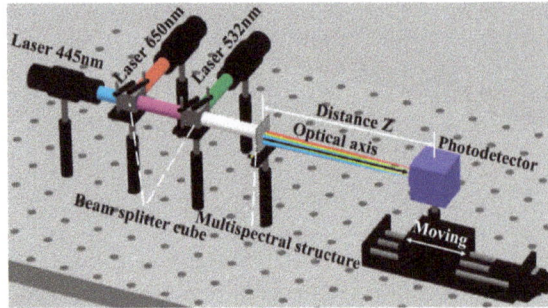

Figure 5. The light path diagram of red, green and blue laser beam combined coaxially illuminate the multi-spectral compound eye structure.

During the experiment, three laser beams were independently used to illuminate the multi-spectral compound eye structure. At the design focal length of 20 mm, laser beams of different wavelengths were converged by different sub-eye lenses, as shown in Figure 6. The sub-eye lens structure marked with λ_{10} corresponds to the designed wavelength of 650 nm, so when the red laser beam irradiated the structure, the detector detected the focus at 20 mm converged by the corresponding sub-eye structure, as shown in Figure 6c.

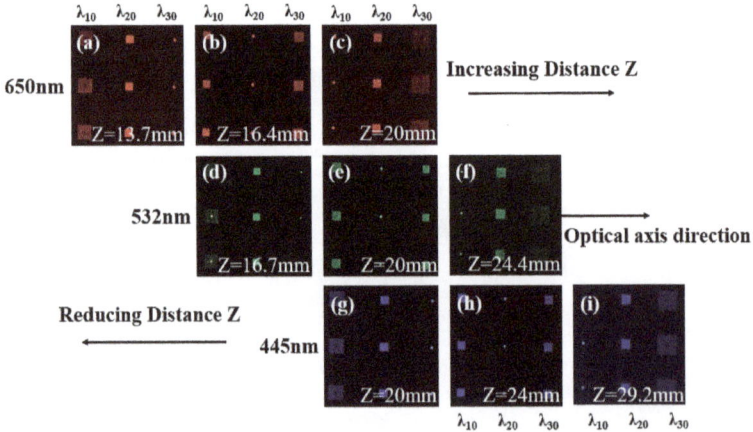

Figure 6. The spot distributions of red, green and blue laser after multi-spectral compound eye structure: (a–c) are the spots at the distance of 13.7 mm, 16.4 mm and 20 mm when the red laser beam with the wavelength of 650 nm irradiates through the structure, respectively; (d–f) are the spots at the distance of 16.7 mm, 20 mm and 24.4 mm when the green laser beam with the wavelength of 532 nm irradiates through the structure, respectively; (g–i) are the spots at the distance of 20 mm, 24 mm and 29.2 mm when the blue laser beam with the wavelength of 445 nm irradiates through the structure, respectively.

Accordingly, when the green laser beam irradiated the structure with λ_{20} corresponding to the designed wavelength of 532 nm, the detector detected the focus at 20 mm converged by the corresponding sub-eye structure, as shown in Figure 6e. When the blue laser beam irradiated the structure with λ_{30} corresponding to the designed wavelength of 445 nm, the detector detected the focus at 20 mm converged by the corresponding sub-eye structure, as shown in Figure 6g.

When tricolor laser beams of red, green and blue illuminate the multispectral structure at the same time, it can be seen that different laser beams of different wavelengths can be converged by the corresponding sub-eye structures of the same designed wavelengths, while the other laser beams disperse, as shown in Figure 7a. It is verified that the structure can realize the spectral splitting function for different wavelengths by different sub-eye imaging channels without axial movement. The intensity distributions of the focal spot sizes were analyzed and shown in Figure 7b–d. The ranges of the energy of the focal spots reduced from the highest to $1/e$ were calculated as the sizes of the focal spots. The focal spot sizes of the wavelengths of 650 nm, 532 nm and 445 nm were quantified as 88.8 μm, 66.6 μm and 59.2 μm, respectively.

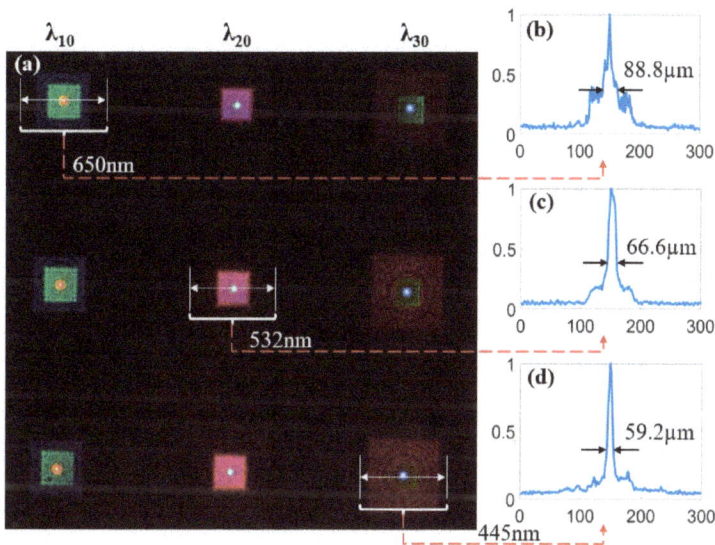

Figure 7. Distributions of the focal spots: (a) distributions of focal spots when red, green and blue lasers simultaneously irradiate the multi-spectral compound eye structure; (b) the intensity distribution of the focal spot at a wavelength of 650 nm; (c) the intensity distribution of the focal spot at a wavelength of 532 nm; (d) the intensity distribution of the focal spot at a wavelength of 445 nm.

Under the same focal length and aperture, the longer the wavelength is, the larger the focal spot size is. The experimental results are in agreement with the theory. Therefore, with the increase of focal spot size, the imaging quality will become worse. In addition, because the scattered spots of other wavelengths are superimposed on the focal spots, the sizes of the captured focal spots which affects the imaging quality are larger than that of the theoretical Airy spots. From the energy distribution curve, it can be seen that, for the focal spot of red light, the intensity of the diffuse spots produced by other wavelengths is larger than the other colors. This is also the reason for the poor effect of red image in the later imaging experiment.

At the same time, if the detector is moved axially, the spectral splitting function of each sub-eye lens can still be observed. For example, a sub-eye lens with a central wavelength of λ_{10} can converge red light of 650 nm at the distance of 20 mm, as shown in Figure 6c. With the increase of distance Z, the sub-eye lens converges green light at the distance of 24.4 mm, as shown in Figure 6f. As the distance

Z increases more, the blue light converges at the distance of 29.2 mm, as shown in Figure 6i. This is determined by the spectral splitting characteristics of the sub-eye lens. Its focal length is inversely proportional to the wavelength. With the increase of distance, the wavelength of the converged light decreases. The wavelength decreases from 650 nm to 532 nm, and then to 445 nm in the experiment, which is consistent with the spectral splitting theory.

Similarly, the sub-eye lens with a central wavelength of λ_{20} can converge green light of 532 nm at the distance of 20 mm, as shown in Figure 6e, With the increase of distance Z, the sub-eye lens converges shorter wavelength of blue light at the distance of 24 mm, as shown in Figure 6h. With the decrease of distance Z, the sub-eye lens converges longer wavelength of red light at the distance of 16.4 mm, as shown in Figure 6b. A sub-eye lens with a central wavelength of λ_{30} can converge blue light of 445 nm at a distance of 20 mm, as shown in Figure 6g. With the decrease of distance Z, sub-eye lens converges longer wavelength green light at the distance of 16.7 mm, as shown in Figure 6d. With the more decrease of distance Z, sub-eye lens converges the longest wavelength red light at the distance of 13.7 mm, as shown in Figure 6a.

Furthermore, in order to verify the imaging characteristics of the red, green and blue multi-spectral compound eye structure, the prepared structure was used to image the luminous object "E". The luminous object was white light in a wide band, and the image was captured by the photodetector. The overall experimental light path was shown in Figure 8. The object distance (*l*) between "E" and multi-spectral compound eye structure was set as 1850 mm. The distance between the detector and the multi-spectral compound eye structure was defined as the imaging distance (*l'*). According to the Gauss imaging formula, the imaging distance was calculated as 20.22 mm with the focal length of 20 mm for various wavelengths. The detector was moved along the axis to record the image, which was located on the corresponding imaging plane of the structure. The imaging effect of the prepared structure was effectively verified as shown in Figure 9. The detector obtained different wavelengths of red, green and blue spectral information at the same image plane, which verified the multi-spectral imaging performance of compound eye structure. It is worth mentioning that the spectral image received by the detector is the overlap image of the focus image formed by one wavelength on its focal plane and defocus images of other wavelengths at this position. Therefore, the light from other spectral channels will cause interference and blur to the image. A series of algorithms are needed to process the image in the following steps to obtain the target spectral image data cube. In order to eliminate spectral blurring and restore image quality, image space linearization based on a translation invariant model is needed, that is, the relative wavelength of spectral image should have an equal interval and equal frame. Then, the three-dimensional deconvolution and other reconstruction image processing techniques in three-dimensional optical slice microscopy technology need to be used to obtain clear spectral image information.

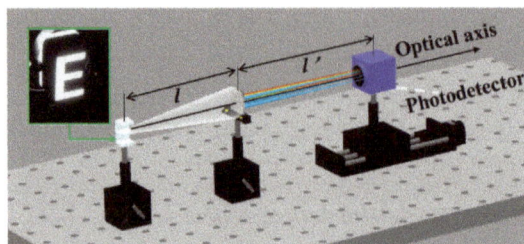

Figure 8. Multi-spectral imaging experimental verification.

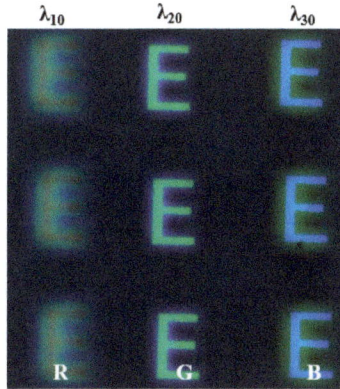

Figure 9. Red, Green and Blue multi-spectral imaging effects.

4. Conclusions

In this paper, a new multi-spectral ACE structure has been designed and fabricated by combining the diffractive beam splitting lenses with the ACE structure. Experiments of spectral splitting and imaging were carried out, and the red, green and blue spectral splitting and imaging were realized, which verified the multi-spectral imaging function of the structure. Because the ACE structure is a multi-aperture imaging structure composed of multiple sub-eye lenses, spectral imaging structures with different central wavelengths can be designed according to the needs.

Author Contributions: Conceptualization, A.C.; Data curation, A.C. and H.P.; Formal analysis, A.C.; Funding acquisition, L.S.; Investigation, A.C. and M.Z.; Supervision, Q.D. and S.H.; Validation, A.C.; Writing—original draft, A.C.; Writing—review and editing, L.S.

Acknowledgments: This research was supported by the National Natural Science Foundation of China (NSFC) (grant numbers 61605211, 51703227); The Instrument Development of Chinese Academy of Sciences; The National R&D Program of China (grant number 2017YFC0804900); The Applied Basic Research Programs of Department of Science and Technology of Sichuan Province (grant number 2019YJ0014); Youth Innovation Promotion Association, CAS; CAS "Light of West China" Program. The authors thank their colleagues for their discussions and suggestions for this research.

Conflicts of Interest: The authors declare no conflict of interest.

References

1. Jeong, K.H.; Kim, J.; Lee, L.P. Biologically inspired artificial compound eye. *Science* **2006**, *312*, 557–561. [CrossRef]
2. Luo, J.; Guo, Y.; Wang, X.; Fan, F. Design and fabrication of a multi-focusing artificial compound eyes with negative meniscus substrate. *J. Micromech. Microeng.* **2017**, *27*, 045011. [CrossRef]
3. Cao, A.; Shi, L.; Shi, R.; Deng, Q.; Du, C. Image process technique used in a large FOV compound eye imaging system. *Proc. SPIE* **2012**, *8558*, 85581K.
4. Song, Y.M.; Xie, Y.; Malyarchuk, V.; Xiao, J.; Jung, I.; Choi, K.J.; Lu, C.; Kim, R.-H.; Li, R.; Crozier, K.B.; et al. Digital cameras with designs inspired by the arthropod eye. *Nature* **2013**, *497*, 95–99. [CrossRef] [PubMed]
5. Jian, H.; He, J.; Jin, X.; Chen, X.; Wang, K. Automatic geometric calibration and three-dimensional detecting with an artificial compound eye. *Appl. Opt.* **2017**, *56*, 1296–1301. [CrossRef]
6. Li, L.; Hao, Y.; Xu, J.; Liu, F.; Lu, J. The Design and Positioning Method of a Flexible Zoom Artificial Compound Eye. *Micromachines* **2018**, *9*, 319. [CrossRef] [PubMed]
7. Kuo, W.K.; Lin, S.Y.; Hsu, S.W.; Yu, H.H. Fabrication and investigation of the bionic curved visual microlens array films. *Opt. Mater.* **2017**, *66*, 630–639. [CrossRef]
8. Moghimi, M.J.; Fernandes, J.; Kanhere, A.; Jiang, H. Micro-Fresnel-Zone-Plate Array on Flexible Substrate for Large Field-of-View and Focus Scanning. *Sci. Rep.* **2015**, *5*, 15861. [CrossRef] [PubMed]

9. Yoshimoto, K.; Yamada, K.; Watabe, K.; Kido, M.; Nagakura, T.; Takahashi, H.; Nishida, T.; Iijima, H.; Tsujii, M.; Takehara, T.; et al. Gastrointestinal tract volume measurement method using a compound eye type endoscope. *Proc. SPIE* **2015**, *9313*, 93131I.

10. Kagawa, K.; Horisaki, R.; Ogura, Y.; Tanida, J. A compact shape-measurement module based on a thin compound-eye camera with multiwavelength diffractive pattern projection for intraoral diagnosis. *Proc. SPIE* **2009**, *7442*, 74420U.

11. Kaadan, A.; Refai, H.; Lopresti, P. Wide-area and omnidirectional optical detector arrays using modular optical elements. *Appl. Opt.* **2016**, *55*, 4791–4800. [CrossRef] [PubMed]

12. Lin, J.; Kan, Y.; Jing, X.; Lu, M. Design and Fabrication of a Three-Dimensional Artificial Compound Eye Using Two-Photon Polymerization. *Micromachines* **2018**, *9*, 336. [CrossRef] [PubMed]

13. Liu, F.; Diao, X.; Li, L.; Hao, Y.; Jiao, Z. Fabrication and Characterization of Inhomogeneous Curved Artificial Compound Eye. *Micromachines* **2018**, *9*, 238. [CrossRef] [PubMed]

14. Cao, A.; Wang, J.; Pang, H.; Zhang, M.; Shi, L.; Deng, Q. Design and fabrication of a multifocal bionic compound eye for imaging. *Bioinspir. Biomim.* **2018**, *13*, 026012. [CrossRef]

15. Cao, A.; Shi, L.; Deng, Q.; Pang, H.; Man, Z.; Du, C. Structural design and image processing of a spherical artificial compound eye. *Optik* **2015**, *126*, 3099–3103. [CrossRef]

16. Sánchez, F.M.; Gál, C.; Eisenhauer, F.; Krabbe, A.; Haug, M.; Iserlohe, C.; Herbst, T.M. LIINUS/SERPIL: A design study for interferometric imaging spectroscopy at the LBT. *Proc. SPIE* **2008**, *7014*, 70147E.

17. Edelstein, J. Imaging interferometric spectroscopy for advanced missions. Spies International Symposium on Optical Science. *Proc. SPIE* **1996**, *2807*, 197–208.

18. Zhang, Y.H.; Yang, H.M.; Kong, C.H. Spectral imaging system on laser scanning confocal microscopy. *Opt. Precis. Eng.* **2014**, *22*, 1446–1453. [CrossRef]

19. Wu, W.-D. Spectral characteristics of displacement for parallel beam splitting prisms. *Laser Technol.* **2009**, *33*, 184–186.

20. Ura, S.; Sasaki, T.; Nishihara, H. Combination of grating lenses for color splitting and imaging. *Appl. Opt.* **2001**, *40*, 5819–5824. [CrossRef] [PubMed]

21. Shogenji, R.; Kitamura, Y.; Yamada, K.; Miyatake, S.; Tanida, J. Multispectral imaging using compact compound optics. *Opt. Express* **2004**, *12*, 1643–1655. [CrossRef] [PubMed]

22. Jin, J.; Di, S.; Yao, Y.; Du, R.; Du, R. Design and fabrication of filtering artificial-compound-eye and its application in multispectral imaging. *Proc. SPIE* **2013**, *8911*, 891106.

23. Hinnrichs, M.; Massie, M.A. New approach to imaging spectroscopy using diffractive optics. *Proc. SPIE* **1997**, 194–205. [CrossRef]

24. Hinnrichs, M. Simultaneous multispectral framing infrared camera using an embedded diffractive optical lenslet array. *Proc. SPIE* **2011**, *8012*, 150–154.

micromachines

MDPI

Article

2-Step Drop Impact Analysis of a Miniature Mobile Haptic Actuator Considering High Strain Rate and Damping Effects

Byungjoo Choi, Hyunjun Choi, You-sung Kang, Yongho Jeon and Moon Gu Lee *

Department of Mechanical Engineering, Ajou University, Su-won Si 16499, Korea; dasom@ajou.ac.kr (B.C.); reset0331@gmail.com (H.C.); gidalim89@ajou.ac.kr (Y.-s.K.); princaps@ajou.ac.kr (Y.J.)
* Correspondence: moongulee@ajou.ac.kr; Tel.: +82-31-219-2338

Received: 11 March 2019; Accepted: 7 April 2019; Published: 23 April 2019

Abstract: In recent times, the haptic actuators have been providing users with tactile feedback via vibration for a realistic experience. The vibration spring must be designed thin and small to use a haptic actuator in a smart device. Therefore, considerable interests have been exhibited with respect to the impact characteristics of these springs. However, these springs have been difficult to analyze due to their small size. In this study, drop impact experiments and analyses were performed to examine the damages of the mechanical spring in a miniature haptic actuator. Finally, an analytical model with high strain rate and damping effects was constructed to analyze the impact characteristics.

Keywords: haptic actuator; impact analysis; high strain rate effect; damping; 2-step analysis

---

1. Introduction

A haptic actuator is a component in mobile devices used to transmit tactile signals to the users. The main function of this component is to vibrate, when the mobile device receives messages, e-mails, social network service notifications, and so on. High definition haptic technology has been developed in recent years to transmit a host of dynamic expressions to the user. To realize this, a thin spring is required together with a strong electromagnet for achieving a high acceleration and wide frequency band. However, springs that deliver a realistic tactile feeling are less durable, as they are thin and structurally weak. In particular, drop impact-induced deformation is a typical type of damage observed in the springs, and several studies are being conducted in the industry to analyze this aspect. Due to the high velocity and small size, it is difficult to observe the impact behavior and identify the vulnerabilities of the springs. Therefore, a new analytical approach is required for the design that fulfils the reliability criterion of acceleration variation, which is only 10% when falling at 1.8 m.

Prior academic research has mainly focused on the drop impact tests and analysis of the finished product. The product level differs from the parts level, as it considers the key factors, such as the strain rate and damping modeling, in detail. However, the findings of some of these studies are be summarized to understand the research direction and analysis method.

Goyal et al. [1] designed a new impact tester and analyzed the multiple impacts of portable electronic products. They constructed an automated system for repeatability, which demonstrated that the impact direction could be regulated while fulfilling the free-drop conditions.

Lim et al. [2] used ABAQUS/Explicit to investigate the drop impact of an electronic pager. They analyzed the strain and impact force of the pager housing and verified the exceptional correlation between the results of the physical experiment and the computer-simulated analysis. Furthermore, they determined the impact direction and height that could result in critical damages.

Zhu [3] performed the finite element analysis to evaluate the reliability of a mobile phone. In this work, three impact modes (disengagement of the battery snap fit, preload spring failure, and ball grid array (BGA) solder crack) were numerically modeled and their correlation with the experiment was studied. Finally, the strain rate effect, impact contact force, and impact acceleration were individually confirmed in each mode.

Kim et al. [4] performed the drop impact simulation of a cell phone using the LS-DYNA explicit code. To verify the analytical model experimentally, both the global and local impact responses were confirmed with a high-speed camera. They also predicted the potential damage of the mobile phone using a statistical analysis in their experiment.

Karppinen et al. [5] compared the product level and board level drop tests, because the drop tests of handheld products demonstrated different results depending on the enclosure, impact orientation, strike surfaces, and mounting of the board. They also compared the mechanical impact responses, which were measured by laser vibrometry, and the acoustic excitation method was used for the analysis. Though the impact load delivered to the board was different in each test, the default failure mode was observed to be the same.

Mattila et al. [6] evaluated the drop impact responses of eight smartphones from various manufacturers. A small accelerometer and a strain gauge were attached inside the smartphones to measure their drop impact response. The maximum strain, average of the maximum strain rate, frequency of the mode shape, and the maximum deceleration were calculated for each product.

The haptic actuator spring is 9 mm in diameter and 0.3 mm in thickness and weighs only 0.1 g (Figure 1). The moving mass, which is approximately twice that of the spring mass, is attached to the spring resulting in serious damage after the primary impact due to an internal secondary impact. Choi et al. conducted a study related to the drop impact of haptic actuators [7]. They developed their finite element model using material properties, such as the microtensile strength and damping ratio, and compared the impact deformation and force using a drop test. However, research on the high strain rate effect and damping modeling, which can be used to improve the accuracy of an analytical model, are limited.

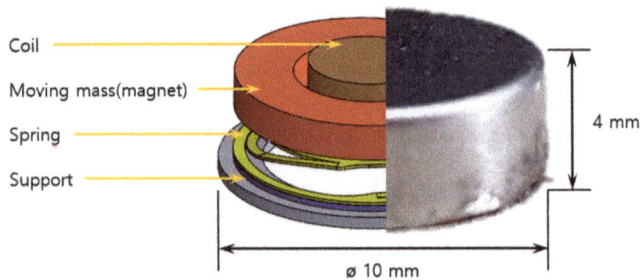

Figure 1. A haptic actuator for a smart device.

In this study, the Johnson–Cook model considering wave propagation was applied considering a high strain rate. A 2-step analysis was conducted for improving the calculation efficiency, and damping was modeled for each stage. Consequently, an analytical model was proposed after comparing the results from the analysis of a miniature haptic actuator with the respective experimental values.

2. Material Modeling

2.1. Haptic Spring

Regarding drop impact, a typical smartphone collides with the ground at a velocity of 3–6 m/s, depending on the initial height in the range of 0.5–1.8 m, and the inner elastic parts deform at higher velocities. In other words, parts that induce elastic deformation, such as a spring, must be modeled considering the strain rate effect. The ANSYS LS-DYNA utilizes the Johnson–Cook model [8] for modeling the strain rate effect, and the constitutive equation is presented as follows.

$$\sigma = (A + B\varepsilon^n)(1 + C\ln\dot{\varepsilon}) \qquad (1)$$

The first term represents the strain-hardening effect of the material, and the strain–stress curve is expressed by the Ludwick equation [9]. A is the initial yield strength, B is the strength coefficient, ε is the plastic strain, and n is the strain-hardening exponent. The second term represents the strain hardening caused by the strain rate effect. $\dot{\varepsilon}$ is the strain rate, and C represents the strength at which the strain rate varies.

The first term refers to the coefficient obtained by fitting a stress–strain curve plotted according to a quasi-static tensile test. The constant C from the second term can be calculated based on the stress–strain curve measured at various strain rates in the split Hopkinson pressure bar (SHPB) test proposed by Kolsky [10]. A round bar-type specimen is required in the test, as the SHPB test measures the deformation caused by an elastic wave between the input and output rods.

However, to enhance the durability of SUS301, the composition of chrome and nickel in SUS304 was reduced and cold-rolled as a plate. Therefore, we could not obtain round bar specimens for the SHPB tests, because only a thin sheet material resulted. To overcome this limitation, firstly, the Johnson–Cook model was constructed on the round bar specimens of SUS304 manufactured with similar materials and processes (Figure 2). The value of the constant C is 0.02 for SUS304, which was applied to SUS301, and the Johnson–Cook model was developed. The stress–strain curve of SUS301 is shown in Figure 3. We used the induced SUS301 model from the experimental SUS304 model. The stress–strain curve was derived from the research of Lee et al. [11]. A mismatch observed in the graph between the curves of Johnson Cook model ($\dot{\varepsilon} = 0.001$) and tensile test ($\dot{\varepsilon} = 0.001$) in the increase of the initial stress was compensated by using the modulus values in the analysis. The constants of the reference material SUS304 and spring material SUS301 used in the Johnson–Cook model are summarized in Table 1.

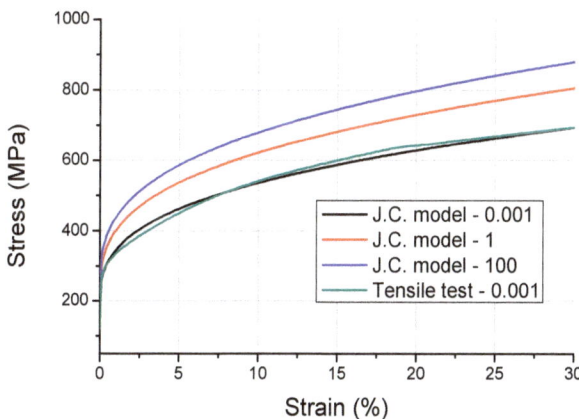

Figure 2. Stress–strain curve of SUS304 based on strain rate variations.

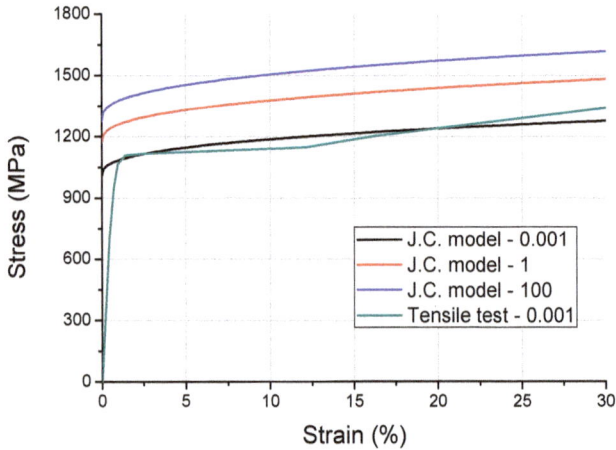

Figure 3. Stress–strain curve of SUS301 based on strain rate variations.

Table 1. Constants of Johnson–Cook model for the haptic spring.

Parameters	SUS304	SUS301
A (Initial yield strength)	215 MPa	1170 MPa
B (Hardening constant)	1945 MPa	1140MPa
n (Hardening exponent)	0.34	0.37
C (Strain constant)	0.02	0.02

2.2. Other Components

The other components determine the impact force delivered to the spring of the haptic actuator, and therefore, a detailed material modeling was required. The dummy mass comprised aluminum and polycarbonate, and the vibration motor comprised of a housing with a moving mass. The Johnson–Cook model was also applied to the parts absorbing the energy transmitted to the spring due to plastic deformation. The used material models of the dummy (Al6061-T6) and the sensor head (stainless steel) were from the ANSYS library and their constants are summarized in Table 2.

Table 2. Constants of Johnson–Cook model for the dummy and sensor.

Parameters	Al6061-T6	Stainless Steel
A (Initial yield strength)	270 MPa	792 MPa
B (Hardening constant)	138 MPa	510 MPa
n (Hardening exponent)	0.18	0.26
C (Strain constant)	0.002	0.014

The impact analysis produced dominant results on the velocity and mass of the drop object. The initial velocity was 5.88 m/s when impacting the ground from a height of 1.8 m. This was simply applied to the boundary condition. However, the mass required relatively detailed modeling. Firstly, the mass of polycarbonates attached to aluminum affected the occurrence of the primary impact in the dummy. Secondly, the moving mass attached to the spring affected the internal impact to the housing. The mass of the specimen, including the haptic actuator, was 114 g. The material properties of each of the specimens are summarized in Table 3.

The contact between the parts was modeled under two conditions, including "slightly expected slip" and "fixed with no slip".

Table 3. Elastic and physical properties of the specimens used in the drop impact model.

Part	Material	Density (kg/m³)	Bulk Modulus (GPa)	Shear Modulus (GPa)	Mass (g)
Spring	SUS301	7850	152	78	0.10
Moving mass	Stainless steel 4340	7830	159	81	0.19
Haptic housing	Stainless steel 4340	7830	159	81	0.41
Dummy body	Aluminum 6061-T6	2804	69	26	53.86
Dummy cover	Polycarbonate	1300	2	1	29.57
Sensor head	Stainless steel	7850	169	73	155.29

The former used the condition of "frictional contact". A static coefficient of 0.61 and a dynamic coefficient of 0.47 between aluminum and mild steel were applied at the interface between the dummy and sensor head for the external impact, while a static coefficient of 0.74 and a dynamic coefficient of 0.57 among the mild steel were applied at the interface between the haptic housing and moving mass for the internal impact. However, the analysis results were not affected to the friction coefficient, as there was no contact sliding during the impact behavior.

The latter used the condition of "bonded". The haptic actuator was fixed by spot welding the parts together. However, commonly used bonded conditions were applied to the interfaces, as the exact welding contact stiffness were not be known.

3. 2-Step Analysis Modeling

The impact analysis using the explicit solver required considerable amount of computation time and resources as it involved several iterations. When an electronic device is analyzed, the impact is first applied to the outside and then transmitted to the inside. Therefore, a full-scale analysis requires considerable amount of time to calculate these two impacts in the transmission process. The calculation of a single drop impact of the haptic actuator specimen required 6 h and used five CPU cores. However, dividing the impact into separate external and internal impacts and analyzing them as two steps reduced the calculation time to 0.5 h each. This method can be used for efficient and repeated analysis by changing the 3D model or material properties in the design process.

The drop impact model of the haptic actuator is shown in Figure 4. The specimen was modeled using a 20° tilt from the sensor surface to simulate an experiential worst case of the impact force. Under this condition, the center of gravity of the dummy phone was perpendicular to the ground. In addition, there was almost no drop impact under perpendicular to the ground. Therefore, the model was modified by titling the specimen sideways by 5°. This combination was applied, as these two angles (20° and 5°) were most frequently observed in a series of repeated impact tests.

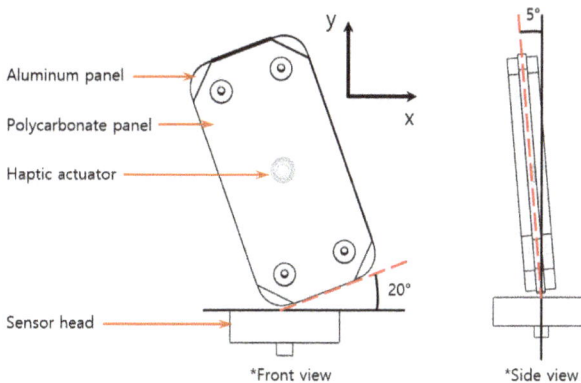

Figure 4. Drop impact analysis model.

3.1. First Step—External Impact

The purpose of this stage was to extract the velocity history of the attached surface of the haptic actuator, when the specimen collided with an impact force sensor (PCB, ICP®quartz force sensor, 200C50).

The impact force for the initial analytical model is shown in Figure 5. A peak force of 9.69 kN is observed in this figure; however, an undamped residual force vibration was confirmed. The abnormal attenuation was controlled by damping modeling, as energy dissipation could result in an abnormal rise in stress. The residual force vibration can be used to discover the dominant frequencies with high amplitude using fast Fourier transform (FFT), as shown in Figure 6. The modeling was performed to attenuate the main vibration frequency (14627 Hz) of the residual force vibration.

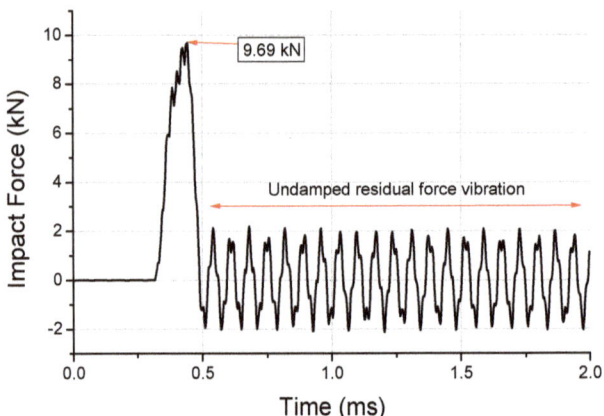

Figure 5. Analysis of the undamped impact force of first-step drop.

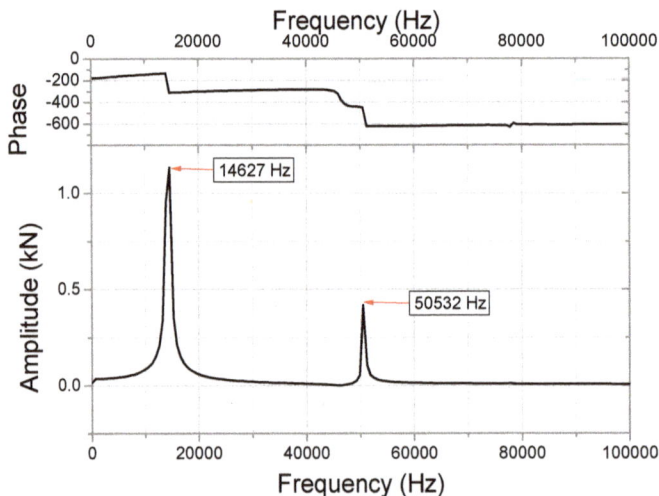

Figure 6. Fast Fourier transform (FFT) for residual force vibration (undamped area shown in Figure 5) in the frequency domain.

The impact force measured in the drop test is shown in Figure 7. The impact force is estimated to be 8.24 kN, when the specimen collides with the impact force sensor. The second and third peak forces in the residual force vibration are observed to be 1.58 and 1.19 kN, respectively. These values are used in the computation of the damping coefficient ($\xi = 0.045$) using the logarithmic decrement method.

The Rayleigh damping C used in the analysis is defined as follows [12]:

$$C = \alpha M + \beta K, \tag{2}$$

where M is the mass matrix, K is the stiffness matrix, and α and β are constants. As the dominant frequency of the residual force vibration and the damping ratio were measured during the test, the value of the constant α can be calculated as follows:

$$\alpha = 2\xi\omega, \tag{3}$$

where the value of α is 8301. The analyzed impact force obtained by applying this value is shown in Figure 7. The peak value of the force is observed to be 9.42 kN, which is 3% less than that before the damping modeling. An error of 14% resulted from the test, because the Johnson–Cook model of the aluminum dummy was not measured. Figure 8 compares the impact strain from the test and plastic strain from the analysis. It is evident from this figure that the type of deformation observed during the test agrees well with that observed during the analysis, and the maximum strain of 0.11 was confirmed by the analysis.

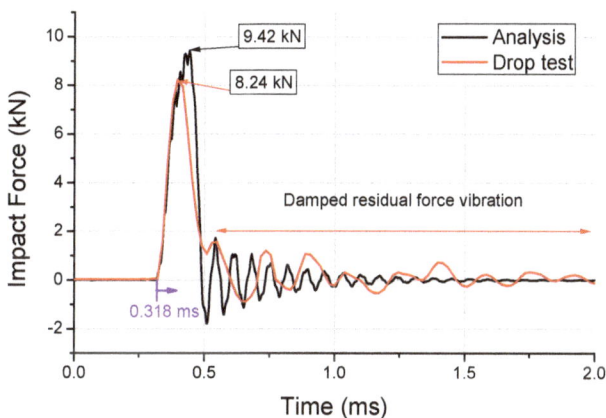

Figure 7. Damped impact force of the first-step drop-test and analysis.

3.2. Second Step—Internal Impact

The purpose of this stage was to identify the stress distribution and plastic strain in the haptic actuator based on the velocity history obtained from the first step.

The x-, y-, and z-axis velocity history of the haptic actuator attached to the surface of the dummy is shown in Figure 9. All the components initiate the drop at an initial velocity of −5.88 m/s in the y-axis. Therefore, no disturbance was experienced before the impact. The event is initiated at 0.340 ms, and the occurrence of elastic impact is observed over 0.115 ms. After a threshold of −2.63 m/s, the plastic impact is observed at the maximum repulsion rate of +4.27 m/s for a short duration of 0.01 ms. The y-axis velocity converges to an average of 2.57 m/s after the loss of contact, while the starting point of the velocity variation demonstrates a time delay of 0.022 ms from the impact contact of the dummy.

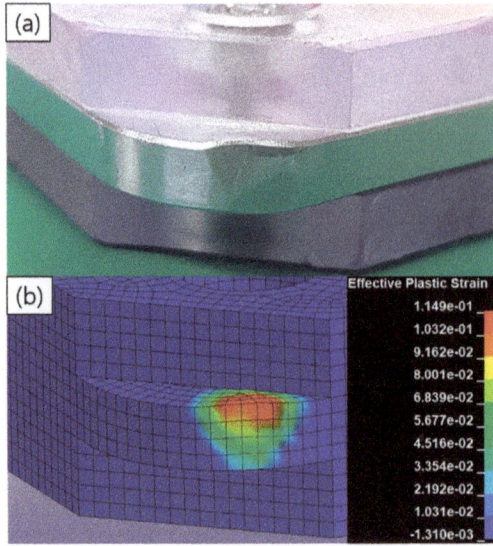

Figure 8. Plastic strain of the dummy after (**a**) the impact test and (**b**) analysis.

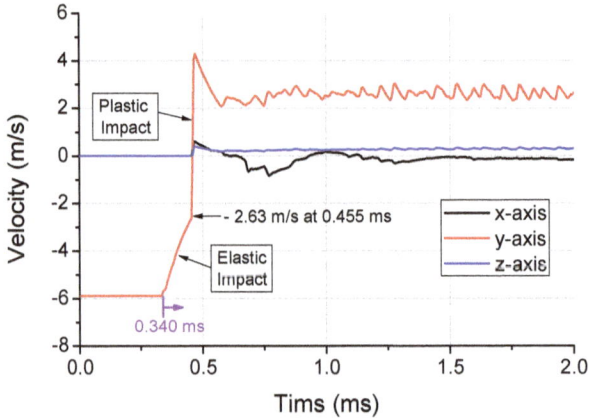

Figure 9. Velocity profile of haptic attachment surface to dummy in first-step analysis.

Generally, springs demonstrate a minimal amount of damping, as they induce elastic deformation for mechanical purposes. However, during the analysis of the input velocity history, the undamped spring (shown in Figure 10) exhibited no attenuation in residual vibration. Therefore, an FFT was performed in the frequency domain for modeling the damping. The results from this FFT are shown in Figure 11. At low frequencies, a giant wave is generated with high frequency vibrations. The damping was modeled using a high frequency (4410 Hz), which was dominant during the energy dissipation. The damping ratio (ξ) required for the modeling was 0.02, which was based on authors' previous measurements [7]. The constant α for the mass matrix is calculated as 1180 using Equation (3). The y-displacement of the damped spring is shown in Figure 10. The maximum equivalent stress of the spring is estimated to be 1803 MPa while undamped, which declines by 4% to 1724 MPa after the damping modeling. The moving mass attached to the spring collided with the housing in the −y direction, and this moment of stress distribution is shown in Figure 12. A high stress distribution of approximately 1700 MPa is observed in the upper notch, where tensile stress is applied to the spring.

In contrast, a relatively low stress distribution of approximately 1200 MPa is observed in the lower notch. The figure shows four sites, in which the damage is expected to exceed the yield stress of 1377 MPa. The yield stress is observed to be 18% higher than 1170 MPa at a strain rate of 0.001/s (shown in Figure 3) in the tensile tests.

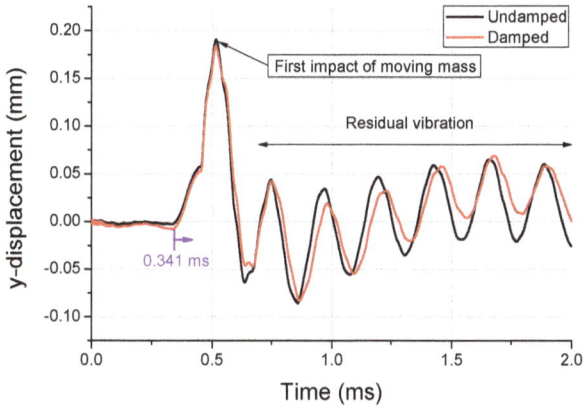

Figure 10. Y-displacement of the spring with and without damping in the second-step analysis.

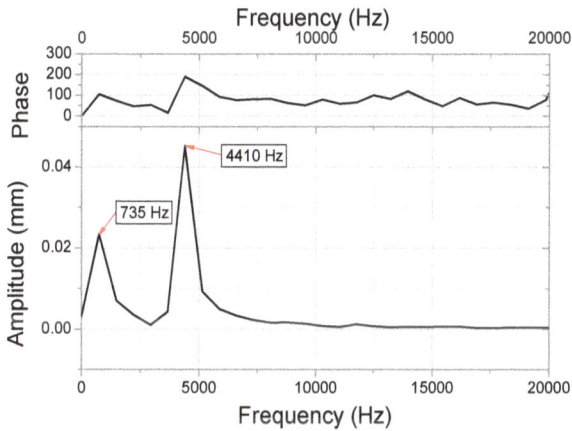

Figure 11. FFT for residual vibration (residual vibration area from Figure 9) in the frequency domain.

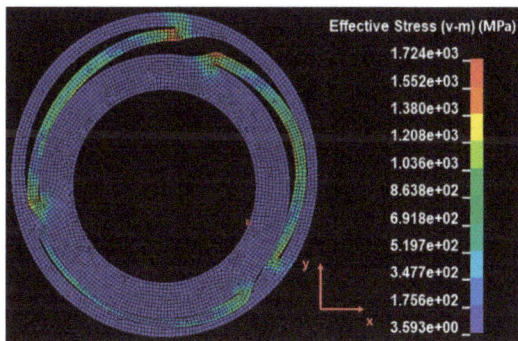

Figure 12. Effective stress of haptic spring at the moment of impact.

The effective plastic strain distribution of the spring is shown in Figure 13. At each position exceeding the yield, a maximum of 0.161 strain is observed for 3–6 elements, each with a size of 0.1 mm. The quantitative plastic strain of each element is shown in Figure 14. The occurrence of the plastic strain is observed mainly in the spiral notch region, where the upper and lower plates are connected. The upper site (+y direction in Figure 13) demonstrates a higher strain and wider stress distribution. In addition, it was confirmed that the notch connected to the upper plate demonstrated a higher plastic strain than the one in the lower plate, because the moving mass caused a secondary impact on the inner housing. The concentration of the plastic strain in the local region was confirmed by this analysis. If the impact load was applied repeatedly, the plastic deformation would increase possibly resulting in malfunction.

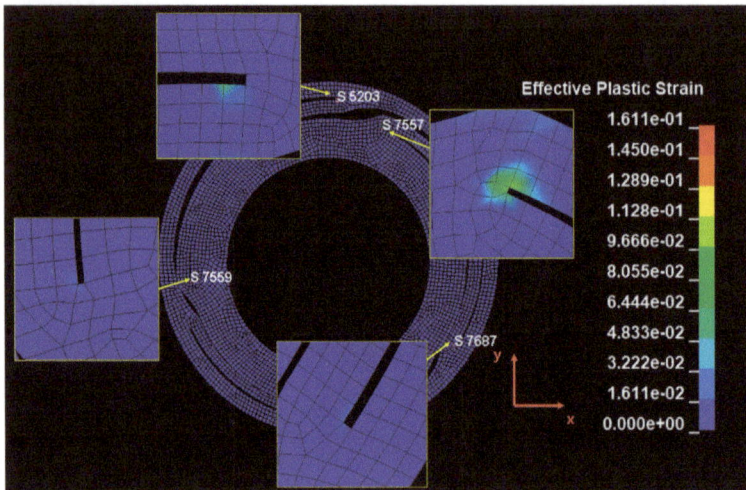

Figure 13. Effective plastic strain of haptic spring at steady state after impact (with number of elements).

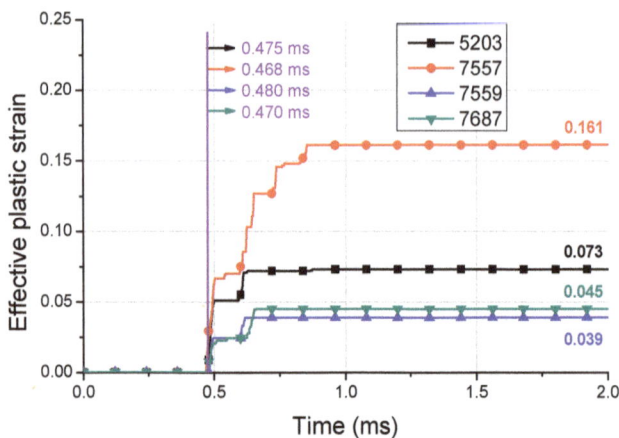

Figure 14. Effective plastic strain of the damage area (with the number of elements shown in Figure 10).

4. Drop Impact Tests

The haptic actuator must express a wide frequency at an acceleration of 0.5 g or more to deliver various tactile sensations. If the frequency history of the vibration signal changes, the operation of delivering a haptic expression to the user becomes impossible. More precisely, a vibration frequency and acceleration change of higher than 10% due to drop impact is regarded as a malfunction. Experiments were performed to determine the possibilities of malfunction resulting from a series of drop impacts, and the specimen was dropped from a height of 1.8 m to compare the results with those from the analysis. Two specimens were used in the drop tests, and the signals were measured with an acceleration sensor (PCB, Ceramic shear ICP®accel, 333B50) before and after the drop impact.

4.1. Driving Acceleration of the Haptic Actuator

A comparison of the vibration acceleration values of the haptic actuator before and after the drop impact tests is shown in Figure 15, and the quantitative data are summarized in Table 4. It is observed from the figure that the peak acceleration and frequency band of the No. 1 specimen decline by 5% and 29% after the first impact. However, no change is observed in the impact damage, as it is within the range of the measurement error. However, after the secondary drop impact, the acceleration decreases by 48%, and the frequency band decreases by 0.5 g owing to the damages in the spring. The acceleration of the No. 2 specimen reduces by 10% after the first impact and the frequency increases from 170.5–173 Hz. Moreover, the band width decreases by 22% after the first drop. The acceleration declined by 52% owing to the considerable amount of damage in the spring after the second drop, and the frequency band decreases to less than 0.5 g. The peak acceleration of the No. 2 specimen demonstrates a sharp decrease after the first impact due to the interference of the moving mass with adjacent components, which results from the spring deformation.

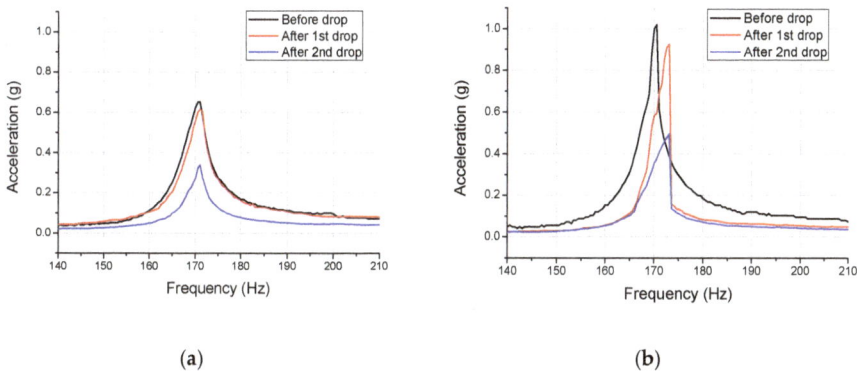

(a) (b)

Figure 15. Comparison of the vibration acceleration of the haptic actuator before and after the drop impact tests (**a**) No. 1 specimen (**b**) No. 2 specimen.

Table 4. Acceleration and frequency band variations due to the drop impact tests.

Specimen	Peak Acceleration (g)			Band Width at 0.5 g (Hz)		
	Before Drop	After 1st Drop	After 2nd Drop	Before Drop	After 1st Drop	After 2nd Drop
No. 1	0.65 at 171 Hz	0.62 at 171 Hz	0.34 at 171 Hz	3.5	2.5	0
No. 2	1.02 at 170.5 Hz	0.92 at 173 Hz	049 at 173 Hz	4.5	3.5	0

4.2. Correlation between the Analysis and Tests

The concentrated damage on the spring notch was confirmed by the single drop impact analysis (shown in Figure 13). The experimental results demonstrated that the vibration frequency was almost unchanged in the No. 1 specimen, while a variation was observed in case of No. 2 specimen. This difference can be explained using two factors. Firstly, the number of variables limited in the analysis but the impact tests were increased. In other words, it is necessary to increase the number of specimens to eliminate the uncertainty. It is also important to supplement the study with statistical analysis.

Secondly, the plastic deformation in the analysis was limited to a minor area of the notch, which implies the possibility of no variations in the vibration characteristics of the spring. However, if the stress and strain of the spring were repeated during the impact analysis irrespective of this fact, the vibration characteristics could change due to the accumulation of damages.

5. Conclusions

The drop impact experiments and analyses were performed to examine the spring damage of a miniature haptic actuator. The results obtained were as follows:

(1) The Johnson–Cook model was applied to the drop impact analysis of the haptic spring considering a high strain rate effect, and the computation time of the 2-step analysis was observed to be 86% less than that of the single-step analysis.

(2) The attenuation of the signal measured by the impact force sensor was applied to precisely model the impact force in the first-step analysis. The plastic strain of the impact contact area obtained from the analysis and experiment agreed well; however, the peak force was observed to be 14% higher in the analysis than in the experiment. The Johnson–Cook model of an aluminum dummy was required to improve the accuracy of the analysis.

(3) The damping coefficient (ξ) measured in the authors' previous study was used to model the damping of the spring in the second-step analysis. The occurrence of the maximum stress of the SUS301 spring (1724 MPa) and the greater-than-yield stresses of 1377 MPa at the four notches of the spring was confirmed using the strain constant C of SUS304. The maximum generated plastic strain of 0.161 was the same as the highest stress area.

(4) The acceleration changes after the first drop impact were different across the two specimens in the tests. The difference could be because of the increase in the number of variables in these experiments than analysis. Moreover, the plastic strain in the microscopic region may not possibly generate a variation in experimental acceleration. Due to the limited number of specimens used for the drop test, the statistical analysis of a greater number of experiments is needed for improved accuracy.

In conclusion, an analytical model with a high strain rate and damping effects was successfully constructed for analyzing the impact characteristics of the miniature haptic actuator.

Author Contributions: M.G.L., H.C., Y.-s.K., Y.J., and B.C. conceived and designed the experiments and simulations; B.C. modeled the analysis model and performed the simulations; H.C. performed the experiments; M.G.L., Y.J., and B.C. analyzed the data; B.C. and M.G.L. prepared the paper.

Funding: This work was supported by the National Research Foundation of Korea (NRF) grant funded by the Korea government (MSIT) (No. NRF-2018R1A2B2002683).

Conflicts of Interest: The authors declare no conflict of interest.

References

1. Goyal, S.; Buratynski, E.K. Method for Realistic Drop-Testing. *Int. J. Microcircuits Electron. Packag.* **2000**, *23*, 45–52.

2. Lim, C.T.; Teo, Y.M.; Shim, V.P.W. Numerical Simulation of the Drop Impact Response of a Portable Electronic Product. *IEEE Trans. Compon. Packag. Technol.* **2002**, *25*, 478–485.

3. Zhu, L. Modeling Technique for Reliability Assessment of Portable Electronic Product Subjected to Drop Impact Loads. In Proceedings of the 2003 Electronic Components and Technology Conference, New Orleans, LA, USA, 27–30 May 2003; pp. 100–104.

4. Kim, J.G.; Park, Y.K. Experimental Verification of Drop/Impact Simulation for a Cellular Phone. *Exp. Mech.* **2004**, *44*, 375–380. [CrossRef]

5. Karppinen, J.; Li, J.; Pakarinen, J.; Mattila, T.T.; Paulasto-Krockel, M. Shock impact reliability characterization of a handheld product in accelerated test and use environment. *Microelectron. Reliab.* **2012**, *52*, 190–198. [CrossRef]

6. Mattila, T.T.; Vajavaara, L.; Hokka, J.; Hussa, E.; Makela, M.; Halkola, V. Evaluation of the drop response of handheld electronic products. *Microelectron. Reliab.* **2014**, *54*, 601–609. [CrossRef]

7. Choi, B.; Kwon, J.; Jeon, Y.; Lee, M.G. Development of Novel Platform to Predict the Mechanical Damage of a Miniature Mobile Haptic Actuator. *Micromach.* **2017**, *8*, 156. [CrossRef]

8. Johnson, G.R.; Cook, W.H. Fracture characteristics of three metals subjected to various strains, strain rates, temperatures and pressures. *Eng. Fract. Mech.* **1985**, *20*, 31–48. [CrossRef]

9. Ludwik, P. *Element der Technologischen Mechanik*; Springer: New York City, NY, USA, 1909; pp. 1–57.

10. Kolsky, H. An Investigation of the Mechanical Properties of Materials at very High Rates of Loading. *Proc. Phys. Soc. Sect. B* **1949**, *62*, 676–700. [CrossRef]

11. Lee, W.; Zhang, C. Computational consistency of the material models and boundary conditions for finite element analyses on cantilever beams. *Adv. Mech. Eng.* **2018**, *10*. [CrossRef]

12. *LS-DYNA Keyword User's Manual*; Version 971; Livermore Software Technology Corporation (LSTC): Livermore, CA, USA; pp. 1–2206.

MDPI

St. Alban-Anlage 66

4052 Basel

Switzerland

Tel. +41 61 683 77 34

Fax +41 61 302 89 18

www.mdpi.com

Micromachines Editorial Office

E-mail: micromachines@mdpi.com

www.mdpi.com/journal/micromachines

www.ingramcontent.com/pod-product-compliance
Lightning Source LLC
Chambersburg PA
CBHW051723210326
41597CB00032B/5579